LONDON MATHEMATICAL SOCIETY STUDENT TEXTS

Managing Editor: Ian J. Leary,
Mathematical Sciences, University of Southampton, UK

51   Steps in commutative algebra: Second edition, R. Y. SHARP
52   Finite Markov chains and algorithmic applications, OLLE HÄGGSTRÖM
53   The prime number theorem, G. J. O. JAMESON
54   Topics in graph automorphisms and reconstruction, JOSEF LAURI & RAFFAELE SCAPELLATO
55   Elementary number theory, group theory and Ramanujan graphs, GIULIANA DAVIDOFF,
     PETER SARNAK & ALAIN VALETTE
56   Logic, induction and sets, THOMAS FORSTER
57   Introduction to Banach algebras, operators and harmonic analysis, GARTH DALES *et al*
58   Computational algebraic geometry, HAL SCHENCK
59   Frobenius algebras and 2D topological quantum field theories, JOACHIM KOCK
60   Linear operators and linear systems, JONATHAN R. PARTINGTON
61   An introduction to noncommutative Noetherian rings: Second edition, K. R. GOODEARL &
     R. B. WARFIELD, JR
62   Topics from one-dimensional dynamics, KAREN M. BRUCKS & HENK BRUIN
63   Singular points of plane curves, C. T. C. WALL
64   A short course on Banach space theory, N. L. CAROTHERS
65   Elements of the representation theory of associative algebras I, IBRAHIM ASSEM,
     DANIEL SIMSON & ANDRZEJ SKOWROŃSKI
66   An introduction to sieve methods and their applications, ALINA CARMEN COJOCARU &
     M. RAM MURTY
67   Elliptic functions, J. V. ARMITAGE & W. F. EBERLEIN
68   Hyperbolic geometry from a local viewpoint, LINDA KEEN & NIKOLA LAKIC
69   Lectures on Kähler geometry, ANDREI MOROIANU
70   Dependence logic, JOUKU VÄÄNÄNEN
71   Elements of the representation theory of associative algebras II, DANIEL SIMSON &
     ANDRZEJ SKOWROŃSKI
72   Elements of the representation theory of associative algebras III, DANIEL SIMSON &
     ANDRZEJ SKOWROŃSKI
73   Groups, graphs and trees, JOHN MEIER
74   Representation theorems in Hardy spaces, JAVAD MASHREGHI
75   An introduction to the theory of graph spectra, DRAGOŠ CVETKOVIĆ, PETER ROWLINSON &
     SLOBODAN SIMIĆ
76   Number theory in the spirit of Liouville, KENNETH S. WILLIAMS
77   Lectures on profinite topics in group theory, BENJAMIN KLOPSCH, NIKOLAY NIKOLOV &
     CHRISTOPHER VOLL
78   Clifford algebras: An introduction, D. J. H. GARLING
79   Introduction to compact Riemann surfaces and dessins d'enfants, ERNESTO GIRONDO &
     GABINO GONZÁLEZ-DIEZ
80   The Riemann hypothesis for function fields, MACHIEL VAN FRANKENHUIJSEN
81   Number theory, Fourier analysis and geometric discrepancy, GIANCARLO TRAVAGLINI
82   Finite geometry and combinatorial applications, SIMEON BALL
83   The geometry of celestial mechanics, HANSJÖRG GEIGES
84   Random graphs, geometry and asymptotic structure, MICHAEL KRIVELEVICH *et al*
85   Fourier analysis: Part I - Theory, ADRIAN CONSTANTIN
86   Dispersive partial differential equations, M. BURAK ERDOĞAN & NIKOLAOS TZIRAKIS
87   Riemann surfaces and algebraic curves, R. CAVALIERI & E. MILES
88   Groups, languages and automata, DEREK F. HOLT, SARAH REES & CLAAS E. RÖVER
89   Analysis on Polish spaces and an introduction to optimal transportation, D. J. H. GARLING
90   The homotopy theory of $(\infty, 1)$-categories, JULIA E. BERGNER

London Mathematical Society Student Texts 90

The Homotopy Theory of (∞, 1)-Categories

JULIA E. BERGNER
University of Virginia

CAMBRIDGE
UNIVERSITY PRESS

CAMBRIDGE
UNIVERSITY PRESS

University Printing House, Cambridge CB2 8BS, United Kingdom

One Liberty Plaza, 20th Floor, New York, NY 10006, USA

477 Williamstown Road, Port Melbourne, VIC 3207, Australia

314-321, 3rd Floor, Plot 3, Splendor Forum, Jasola District Centre, New Delhi - 110025, India

79 Anson Road, #06-04/06, Singapore 079906

Cambridge University Press is part of the University of Cambridge.

It furthers the University's mission by disseminating knowledge in the pursuit of education, learning and research at the highest international levels of excellence.

www.cambridge.org
Information on this title: www.cambridge.org/9781107101364
DOI: 10.1017/9781316181874

First published 2018

A catalogue record for this publication is available from the British Library

ISBN 978-1-107-10136-4 Hardback
ISBN 978-1-107-49902-7 Paperback

Ad Majorem Dei Gloriam

Contents

Preface   *page* xi
Acknowledgments   xiii

Introduction   1

1   Models for Homotopy Theories   4
   1.1   Some Basics in Category Theory   4
   1.2   Weak Equivalences and Localization   13
   1.3   Classical Homotopy Theory   16
   1.4   Model Categories   18
   1.5   Homotopy Categories   22
   1.6   Equivalences Between Model Categories   25
   1.7   Additional Structures on Model Categories   28

2   Simplicial Objects   34
   2.1   Simplicial Sets and Simplicial Objects   34
   2.2   Simplicial Sets as Models for Spaces   37
   2.3   Homotopy Limits and Homotopy Colimits   39
   2.4   Simplicial Model Categories   42
   2.5   Simplicial Spaces   45
   2.6   The Reedy Model Structure on Simplicial Spaces   47
   2.7   Combinatorial Model Categories   54
   2.8   Localized Model Categories   56
   2.9   Cartesian Model Categories   63

3   Topological and Categorical Motivation   66
   3.1   Nerves of Categories   66
   3.2   Kan Complexes and Generalizations   68
   3.3   Classifying Diagrams   71
   3.4   Higher Categories   74
   3.5   Homotopy Theories   78

4   Simplicial Categories   83
   4.1   The Category of Small Simplicial Categories   83
   4.2   Fixed-Object Simplicial Categories   84
   4.3   The Model Structure   86
   4.4   Proof of the Existence of the Model Structure   88
   4.5   Properties of the Model Structure   96
   4.6   Nerves of Simplicial Categories   99

5   Complete Segal Spaces   101
   5.1   Segal Spaces   102
   5.2   Segal Spaces as Categories Up to Homotopy   107
   5.3   Complete Segal Spaces   110
   5.4   Categorical Equivalences   113
   5.5   Dwyer–Kan Equivalences   116

6   Segal Categories   124
   6.1   Basic Definitions and Constructions   125
   6.2   Fixed-Object Segal Categories   130
   6.3   The First Model Structure   138
   6.4   The Equivalence With Complete Segal Spaces   145
   6.5   The Second Model Structure   148
   6.6   The Equivalence With Simplicial Categories   151

7   Quasi-Categories   157
   7.1   Basic Definitions   157
   7.2   Properties of Acyclic Cofibrations   160
   7.3   The Model Structure   166
   7.4   The Coherent Nerve and Rigidification Functors   171
   7.5   Necklaces and Their Rigidification   173
   7.6   Rigidification of Simplicial Sets   179
   7.7   Properties of the Rigidification Functor   187
   7.8   The Equivalence With Simplicial Categories   194
   7.9   The Equivalence With Complete Segal Spaces   207

8   Relative Categories   213
   8.1   Basic Definitions   213
   8.2   Subdivision Functors   218
   8.3   The Model Structure and Equivalence With Complete
        Segal Spaces   221

9   Comparing Functors to Complete Segal Spaces   233
   9.1   Classifying and Classification Diagrams   234
   9.2   Some Results for Simplicial Categories   236

9.3	Comparison of Functors	239
9.4	Complete Segal Spaces From Simplicial Categories	242
10	**Variants on $(\infty, 1)$-Categories**	**248**
10.1	Finite Approximations	248
10.2	Stable $(\infty, 1)$-Categories	251
10.3	Dendroidal Objects	254
10.4	Higher (∞, n)-Categories	256
	References	261
	Index	267

Preface

The starting point for this book was the mini-workshop on "The Homotopy Theory of Homotopy Theories", held in Caesarea, Israel, in May 2010, and the lecture notes from the talks given there. I was asked by a number of people if those notes might be turned into a more formal manuscript, and this book is the result. In the end, I have omitted some of the topics that were addressed at that workshop, simply because to have included them properly would have greatly increased the length. The topics included in the last chapter of this book were chosen to be in line with some of the applications that were discussed there.

Since 2010, our understanding of $(\infty, 1)$-categories has only increased, and they are being used in a wide range of applications. There are many directions I could have taken with this book, but I have chosen here to give a balanced treatment of the different models and the comparisons between them. I also look at these structures primarily from the viewpoint of homotopy theory, considering model structures and Quillen equivalences. In particular, I do not go into much detail on the treatment of $(\infty, 1)$-categories as generalizations of categories, nor to the development of standard categorical notions in this new context. Joyal and Lurie have treated this topic extensively, extending many categorical notions into the context of quasi-categories in particular [73, 88]. Much less has been done in other models with weak composition, but some work has been done in complete Segal spaces, for example [76].

A number of choices have been made in the presentation given here. While the goal is to give a thorough treatment of the different models, experts on the subject know that just about every model has some technicalities which are intuitively sensible but exceptionally messy to prove. In many cases, I have chosen to omit these technical points and simply refer the reader to the original reference. While this decision makes our book less comprehensive, the hope is that it allows the reader to get the big ideas and most of the details of the proofs without getting sidetracked into often unenlightening, if necessary, combinatorial arguments. I have also deliberately suppressed most set-theoretic points, but have tried to point out where they arise.

Finally, I should point out that our treatment does not treat all possible approaches to the subject. Most notably absent is Toën's axiomatic treatment [116], but recently there have also been more geometric models, for example by Ayala, Francis, and Rozenblyum [5, 6], as well as a formal categorical treatment by Riehl and Verity [105, 106, 107].

Acknowledgments

I would like to extend my gratitude to David Blanc, Emmanuel Farjoun, and David Kazhdan for organizing the workshop whose notes were the starting point for this book, and inviting me to be the primary speaker that year. I'd also like to thank Ilan Barnea, David Blanc, Boris Chorny, Emmanuel Farjoun, Yonatan Harpaz, Vladimir Hinich, Matan Prasma, and Tomer Schlanck for their talks during the workshop; while not everything that was discussed there ended up in this book, I am thankful for their contribution to the ideas that led to this work.

Converting 40 pages of lecture notes to a full-length book was a nontrivial task. I want to thank the people who have shared their enthusiasm for the project over the years, including John Greenlees, Mike Hill, Nick Kuhn, and Angélica Osorno. I would also like to thank Matthew Barber, Christina Osborne, Viktoriya Ozornova, Alex Sherbetjian, and Jacob West for their comments on various stages of this manuscript, and the many minor errors they helped me to identify. I'd also like to thank the staff at Cambridge University Press for their patience and help along the way.

Along the way, I was able to correct some mistakes and hopefully clarify some confusing points in some of my earlier work. Some of these difficulties were discovered in conversations with Matthew Barber, Clark Barwick, Christina Osborne, Luis Pereira, and Chris Schommer-Pries, and I thank each of them for bringing these issues to my attention. Discussions with Charles Rezk and Ieke Moerdijk, mostly in the context of ongoing collaborative work, have led to some improved versions of proofs, and I thank both of them for their helpful insights. I'd also like to thank Emily Riehl for bringing my attention to the description of cofibrant simplicial categories.

Finally, much of my own work that appears here had its origins in my PhD thesis. I'd like to thank Bill Dwyer for getting me started in this direction of research and for his continuing support and enthusiasm.

During the time I worked on this book, I was supported by NSF grants DMS-1105766 and DMS-1352298. Some of this work was done while I was a participant at the MSRI program on Algebraic Topology in Spring 2014, which was supported by NSF grant 0932078 000, and while I was a visitor at the Hausdorff Institute for Mathematics in Summer 2015.

Introduction

There are two ways to think about $(\infty, 1)$-categories. The first is that an $(\infty, 1)$-category, as its name suggests, should be some kind of higher categorical structure. The second is that an $(\infty, 1)$-category should encode the data of a homotopy theory. So we first need to know what a homotopy theory is, and what a higher category is.

We can begin with the classical homotopy theory of topological spaces. In this setting, we consider topological spaces up to homotopy equivalence, or up to weak homotopy equivalence. Techniques were developed for defining a nice homotopy category of spaces, in which we define morphisms between spaces to be homotopy classes of maps between CW complex replacements of the original spaces being considered. However, the general framework here is not unique to topology; an analogous situation can be found in homological algebra. We can take projective replacements of chain complexes, then chain homotopy classes of maps, to define the derived category, the algebraic analogue of the homotopy category of spaces.

The question of when we can make this kind of construction (replacing by some particularly nice kinds of objects and then taking homotopy classes of maps) led to the definition of a model category by Quillen in the 1960s [100]. The essential information consists of some category of mathematical objects, together with some choice of which maps are to be designated as weak equivalences; these are the maps we would like to think of as invertible but may not be. The additional data of a model structure, and the axioms this data must satisfy, guarantee the existence of a well-behaved homotopy category as we have in the above examples, with no set-theoretic problems arising.

A more general notion of homotopy theory was developed by Dwyer and Kan in the 1980s. Their simplicial localization [57] and hammock localization [56] constructions provided a method in which a category with weak equivalences can be assigned to a simplicial category, or category enriched in

1

simplicial sets. More remarkably, they showed that up to a natural notion of equivalence (now called Dwyer–Kan equivalence), every simplicial category arises in this way [55]. Thus, if a "homotopy theory" is just a category with weak equivalences, then we can think of simplicial categories as homotopy theories. In other words, simplicial categories provide a model for homotopy theories.

However, with Dwyer–Kan equivalences, the category of small simplicial categories itself forms a category with weak equivalences, and therefore has a homotopy theory. Hence, we have a "homotopy theory of homotopy theories". In fact, this category has a model structure, making it a homotopy theory in the more rigorous sense [27].

In practice, unfortunately, this model structure is not as nice as we might wish. It is not compatible with the monoidal structure on the category of simplicial categories, does not seem to have the structure of a simplicial model category in any natural way, and has weak equivalences which are difficult to identify for any given example. Therefore, a good homotopy theorist might seek an equivalent model structure with better properties.

An alternative model, that of complete Segal spaces, was proposed by Rezk [103]. Complete Segal spaces are simplicial diagrams of simplicial sets, satisfying some conditions which allow them to be thought of as something like simplicial categories but with weak composition. Their corresponding model category is cartesian, and is given by a localization of the Reedy model structure on simplicial spaces. Hence, the weak equivalences between fibrant objects are just levelwise weak equivalences of simplicial sets, and we have a good deal of extra structure that the model category of simplicial categories does not possess.

Meanwhile, in the world of category theory, simplicial categories were seen as models for $(\infty, 1)$-categories, or weak ∞-categories, with k-morphisms defined for all $k \geq 1$, that satisfy the property that, for $k > 1$, the k-morphisms are all weakly invertible. To see why simplicial categories provide a natural model, it is perhaps easier to consider instead topological categories, where we have a topological space of morphisms between any two objects. The 1-morphisms are just points in these mapping spaces. The 2-morphisms are paths between these points; at least up to homotopy, they are invertible. Then 3-morphisms are homotopies between paths, 4-morphisms are homotopies between homotopies, and we could continue indefinitely.

In the 1990s, Segal categories were developed as a weakened version of simplicial categories. They are simplicial spaces with discrete 0-space, and look like homotopy versions of the nerves of simplicial categories. They were first defined by Dwyer, Kan, and Smith [58], but developed from this categorical perspective by Hirschowitz and Simpson [70]. The model structure for Segal categories, begun in their work, was given explicitly by Pellissier [97].

Yet another model for $(\infty, 1)$-categories was given in the form of quasi-categories or weak Kan complexes, first defined by Boardman and Vogt [36]. They were developed extensively by Joyal, who defined many standard categorical notions, for example limits and colimits, within this more general setting. Although much of his work is still unpublished, the beginnings of these ideas can be found in [73]. The notion was adopted by Lurie, who established many of Joyal's results independently [88].

Finally, going back to the original motivation, Barwick and Kan proved that there is a model category on the category of small categories with weak equivalences; they instead use the term "relative categories" [11].

Comparisons between all these various models were conjectured by several people, including Toën [115] and Rezk [103]. In a slightly different direction, Toën proved that any model category satisfying a particular list of axioms must be Quillen equivalent to the complete Segal space model structure, hence axiomatizing what is meant to be a homotopy theory of homotopy theories, or homotopy theory of $(\infty, 1)$-categories [116].

Eventually, explicit comparisons were made, as shown in the following diagram:

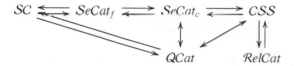

The single arrows indicate that Quillen equivalences were given in both directions, and these were established by Joyal and Tierney [74]. The Quillen equivalence between simplicial categories and quasi-categories was proved in different ways by Joyal, Lurie [88], and Dugger and Spivak [51, 52]. The Quillen equivalence between complete Segal spaces and relative categories was given by Barwick and Kan [11]. The zigzag across the top row was established by the present author [30]. The original model structure for Segal categories is denoted by $SeCat_c$; the additional one $SeCat_f$ was established for the purposes of this proof.

In short, the purpose of this book is to make sense of this diagram. What, explicitly, are simplicial categories, Segal categories, quasi-categories, complete Segal spaces, and relative categories? What is the model category corresponding to each, and how can they be compared to one another? The answers to these questions have all been known and are in the literature, but we bring them together here.

1

Models for Homotopy Theories

In this chapter, we introduce the main ideas of modeling homotopy theories. Since the main objective of this book is to understand the homotopy theory of $(\infty, 1)$-categories, this material allows us to put this idea into a rigorous framework. Most significantly, we explain the structure of a model category, as developed by Quillen. However, most of the material here is to be regarded as background, so very few proofs are given; we give numerous references, so that a reader unfamiliar with certain concepts can find more details elsewhere.

1.1 Some Basics in Category Theory

We begin with a brief review of some essential definitions in category theory.

Definition 1.1.1 A *category* C consists of:

- a collection of *objects*, $\mathrm{ob}(C)$, and
- for any $x, y \in \mathrm{ob}(C)$, a set of *morphisms*, denoted $\mathrm{Hom}_C(x, y)$, such that
- if $f \in \mathrm{Hom}_C(x, y)$ and $g \in \mathrm{Hom}_C(y, z)$, then there is a *composite* morphism $g \circ f \in \mathrm{Hom}_C(x, z)$, and
- given any object x in C, there is an identity morphism $\mathrm{id}_x \in \mathrm{Hom}_C(x, x)$.

If $f \in \mathrm{Hom}_C(x, y)$, we often write $f \colon x \to y$, and say that x is the *source* of f and that y is the *target* of f.

These data are required to satisfy the following two axioms.

- (Associativity) If $f \colon w \to x$, $g \colon x \to y$, and $h \colon y \to z$, then

$$h \circ (g \circ f) = (h \circ g) \circ f.$$

- (Unit) Given any $f: x \to y$, we have

$$f \circ \mathrm{id}_x = f = \mathrm{id}_y \circ f.$$

Example 1.1.2   The category of sets, denoted by *Sets*, has as objects all sets and as morphisms all functions between sets.

Example 1.1.3   The category of groups, denoted by *Gps*, has as objects all groups and as morphisms all group homomorphisms.

Definition 1.1.4   A category C is *small* if ob(C) is a set.

Example 1.1.5   Let $n \geq 0$ be a natural number. Consider the category $[n]$ with objects $0, 1, \ldots, n$ and morphisms defined by

$$\mathrm{Hom}_{[n]}(i, j) = \begin{cases} * & i \leq j \\ \varnothing & i > j. \end{cases}$$

Here, by $*$ we mean a one-element set. Then $[n]$ is an example of a small category. We can depict $[n]$ as

$$0 \to 1 \to 2 \to \cdots \to n.$$

Observe, in contrast, that neither the category of groups nor the category of sets is small.

In a category, we often distingush the morphisms which are invertible.

Definition 1.1.6   A morphism $f: x \to y$ in a category C is an *isomorphism* if there exists a morphism $g: y \to x$ such that $g \circ f = \mathrm{id}_x$ and $f \circ g = \mathrm{id}_y$. A category C is a *groupoid* if all its morphisms are isomorphisms.

We also note the following special kinds of objects that a category might possess.

Definition 1.1.7   An object $\varnothing$ of a category C is *initial* if, for any object c in C, there is a unique morphism $\varnothing \to c$ in C. Dually, an object $*$ is *terminal* if, for any object c of C, there is a unique morphism $c \to *$ in C. If an object is both initial and terminal, it is called a *zero object*.

Proposition 1.1.8   [4, 2.8] *Initial and terminal objects in a category are unique up to isomorphism.*

Let us look at a few ways to obtain new categories from ones we already have. One basic way is to reverse the direction of the morphisms.

Definition 1.1.9   Let C be a category. Its *opposite category* is the category C^{op} with the same objects as C and morphisms defined by

$$\text{Hom}_{C^{op}}(x, y) = \text{Hom}_C(y, x).$$

Definition 1.1.10   A *subcategory* $\mathcal{D}$ of a category C consists of a subclass of the objects of C and, for any objects x and y, a subset $\text{Hom}_{\mathcal{D}}(x, y) \subseteq \text{Hom}_C(x, y)$, such that $\mathcal{D}$ also satisfies the necessary conditions to be a category.

Definition 1.1.11   Let C be a category. A *full subcategory* of C is a category $\mathcal{D}$ whose objects form a subclass of the objects of C and for which $\text{Hom}_{\mathcal{D}}(c, c') = \text{Hom}_C(c, c')$.

Definition 1.1.12   Let C be a category and c an object of C. The category of *objects of C over c* has objects given by morphisms $d \to c$ in C and morphisms the maps $d \to d'$ in C making the diagram

commute. This category is denoted by $C \downarrow c$ or by C/c. Dually, the category of *objects of C under c* has objects given by morphisms $c \to d$ and morphisms the maps $d \to d'$ making the diagram

commute. This category is denoted by $c \downarrow C$.

We can also consider functions between categories.

Definition 1.1.13   Let C and $\mathcal{D}$ be categories. A *functor* $F\colon C \to \mathcal{D}$ assigns to any object x of C an object $F(x)$ of $\mathcal{D}$, and to any morphism $f\colon x \to y$ of C a morphism $F(f)$ of $\mathcal{D}$, such that

- $F(f)\colon F(x) \to F(y)$,
- $F(g \circ f) = F(g) \circ F(f)$, and
- $F(\text{id}_x) = \text{id}_{F(x)}$ for every object x of C.

Example 1.1.14   The collection of small categories and functors between them itself forms a category, which we denote by *Cat*.

We might ask whether two categories are essentially the same. While we could demand that categories only be considered equivalent if their objects and morphisms are in bijection with one another, we typically consider instead the following definition.

Definition 1.1.15 A functor $F: C \to \mathcal{D}$ is an *equivalence of categories* if:

1 for every object x and y of C, the map of sets $\mathrm{Hom}_C(x, y) \to \mathrm{Hom}_\mathcal{D}(Fx, Fy)$ is an isomorphism, and
2 F is essentially surjective, i.e., for any object d of $\mathcal{D}$, there exists an object c in C together with an isomorphism $F(c) \to d$ in $\mathcal{D}$.

In this case, we say the categories C and $\mathcal{D}$ are *equivalent*.

If a functor F satisfies the first condition for equivalence of categories, it is said to be *fully faithful*. It is *full* if each such map is surjective and *faithful* if each such map is injective.

We are often interested not just in a functor from one category to another, but in pairs of functors which go back and forth between two categories in a suitably compatible way.

Definition 1.1.16 Suppose that $F: C \to \mathcal{D}$ and $G: \mathcal{D} \to C$ are functors. The pair (F, G) is an *adjoint pair* of functors if, for any object x of C and object y of $\mathcal{D}$, there is a natural isomorphism

$$\mathrm{Hom}_\mathcal{D}(F(x), y) \cong \mathrm{Hom}_C(x, G(y)).$$

The functor F is called the *left adjoint* and the functor G is called the *right adjoint*. We often write an adjoint pair as

$$F: C \rightleftarrows \mathcal{D}: G$$

and employ the convention that the left adjoint always appears as the topmost arrow.

Example 1.1.17 There is a forgetful functor $\mathcal{G}ps \to \mathcal{S}ets$ which takes a group to its underlying set. This functor has a left adjoint, taking a set to the free group on that set. Such an adjoint pair is called a *forgetful-free adjunction*.

A functor from a small category $\mathcal{D}$ to an arbitrary category C can be thought of as picking out a configuration of objects and morphisms in C which have the shape of $\mathcal{D}$.

Definition 1.1.18 Let C be a category and $\mathcal{D}$ a small category. A $\mathcal{D}$-*diagram* in C is a functor $\mathcal{D} \to C$.

Within a category, we are often interested in objects which satisfy certain universal properties with respect to diagrams in that category. Hence, we turn to limits and colimits.

Definition 1.1.19   Let $D\colon \mathcal{D} \to C$ be a diagram. A *limit* for D is an object $\lim_{\mathcal{D}} D$ of C such that there are maps $\lim_{\mathcal{D}} D \to D(d)$ for every object d of $\mathcal{D}$, compatible in the sense that, if $d \to e$ is a morphism in $\mathcal{D}$, there is a commutative triangle

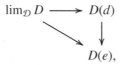

and these triangles are all compatible with one another. Furthermore, the object $\lim_{\mathcal{D}} D$ is universal in the sense that, if there exists any other object c of C together with such maps, then each map $c \to D(d)$ factors through $\lim_{\mathcal{D}} D$.

In particular, if a limit of a diagram exists, it is unique up to unique isomorphism.

Definition 1.1.20   A category C *has all small limits* if, for every diagram $D\colon \mathcal{D} \to C$, with $\mathcal{D}$ small, the limit $\lim_{\mathcal{D}} D$ exists.

We can similarly define what it means for a category to have all finite limits. We now give a few of the most common examples of limits.

Definition 1.1.21   A *product* is a limit of a diagram consisting of objects but no nonidentity morphisms. A *pullback* is a limit of a diagram of the form

$$(\bullet \to \bullet \leftarrow \bullet).$$

An *equalizer* is a limit of a diagram of the form

$$(\bullet \rightrightarrows \bullet).$$

We give two criteria for determining whether certain kinds of limits exist in a category.

Proposition 1.1.22   [4, 5.23] *If a category has pullbacks and a terminal object, then it has all finite limits.*

Proposition 1.1.23   [4, 5.24] *If a category has all small products and all equalizers, then it has all small limits.*

We similarly have the dual notion of colimit of a diagram.

Definition 1.1.24 Let $D: \mathcal{D} \to C$ be a diagram. A *colimit* for D is an object $\text{colim}_{\mathcal{D}} D$ of C such that there are maps $D(d) \to \text{colim}_{\mathcal{D}} D$ for every object d of $\mathcal{D}$, compatible in the sense that, if $d \to e$ is a morphism in $\mathcal{D}$, there is a commutative triangle

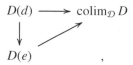

and these triangles are all compatible with one another. Furthermore, the object $\text{colim}_{\mathcal{D}} D$ is universal in the sense that, if there exists any other object c of C together with such maps, then each map $D(d) \to c$ factors through $\text{colim}_{\mathcal{D}} D$.

Again, if colimits exist, then they are unique up to unique isomorphism.

Definition 1.1.25 A category C *has all small colimits* if, for every diagram $D: \mathcal{D} \to C$, with $\mathcal{D}$ small, the colimit $\text{colim}_{\mathcal{D}} D$ exists.

Definition 1.1.26 A *coproduct* is a colimit of a diagram consisting of objects but no nonidentity morphisms. A *pushout* is a colimit of a diagram of the form

$$(\bullet \leftarrow \bullet \to \bullet).$$

A *coequalizer* is a colimit of a diagram of the form

$$(\bullet \rightrightarrows \bullet).$$

We will have need of the following kinds of colimits as well.

Definition 1.1.27 [90, IX.1] A nonempty category $\mathcal{D}$ is *filtered* if

1 for any two objects d and d' of $\mathcal{D}$, there exists an object e together with morphisms $d \to e$ and $d' \to e$, and
2 given two different morphisms $u, v: c \to d$, there exist an object e and morphism $w: d \to e$ such that $wu = wv$.

If $F: \mathcal{D} \to C$ is a functor with $\mathcal{D}$ a filtered category, then the colimit of F is called a *filtered colimit*. If $\mathcal{D}$ is a partially ordered set (so that there is only one possible morphism $i \to j$ in $\mathcal{D}$) which satisfies condition (1), then $\mathcal{D}$ is a *directed* poset and a colimit of a functor $F: \mathcal{D} \to C$ is a *directed colimit*.

Similarly to the case for limits, we have criteria for when a category has certain kinds of colimits.

Proposition 1.1.28 [4, 5.25] *If a category has all small coproducts and all coequalizers, then it has all small colimits.*

Proposition 1.1.29   [4, 9.14] *Left adjoint functors preserve colimits and right adjoint functors preserve limits.*

Remark 1.1.30   We can define initial and terminal objects in terms of limits and colimits. Consider the empty category $\varnothing$ with no objects. For any category C, the limit of the functor $\varnothing \to C$ (if it exists) is a terminal object of C; and the colimit of such a functor (if it exists) is an initial object of C.

Not only do we have functors between categories, but we can also consider interactions between different functors.

Definition 1.1.31   Let $F, G\colon C \to \mathcal{D}$ be functors. A *natural transformation* $\eta\colon F \to G$ is a family of morphisms $\eta_c\colon F(c) \to G(c)$ in $\mathcal{D}$, indexed over all objects c of C, such that, for any morphism $f\colon c \to c'$ in C, the diagram

$$
\begin{array}{ccc}
F(c) & \xrightarrow{\;\eta_c\;} & G(c) \\
{\scriptstyle F(f)}\big\downarrow & & \big\downarrow{\scriptstyle G(f)} \\
F(c') & \xrightarrow{\;\eta_{c'}\;} & G(c')
\end{array}
$$

commutes. A *natural isomorphism* is a natural transformation η such that each morphism η_c is an isomorphism in $\mathcal{D}$.

Given this definition, we can give an equivalent formulation of what it means to be an equivalence of categories.

Proposition 1.1.32   [4, 7.25] *A functor $F\colon C \to \mathcal{D}$ is an equivalence of categories if and only if there exists a functor $G\colon \mathcal{D} \to C$ together with natural isomorphisms $GF \cong \mathrm{id}_C$ and $FG \cong \mathrm{id}_{\mathcal{D}}$ to the respective identity functors.*

We can also use natural transformations to assemble the functors between two fixed categories into a category.

Example 1.1.33   Let C be a category and $\mathcal{D}$ a small category. There is a category of diagrams $C^{\mathcal{D}}$ whose objects are functors $\mathcal{D} \to C$ and whose morphisms are natural transformations.

If $\mathcal{D} = [1]$, the category depicted by

$$\bullet \to \bullet,$$

then $C^{[1]}$ is the category whose objects are morphisms of C.

Of particular interest are functors to the category of sets which are represented by an object, in the following sense.

Definition 1.1.34 Let C be a category. A functor $C \to Sets$ is *representable* if it is of the form $\mathrm{Hom}_C(-, c)$ for some object c of C.

Definition 1.1.35 Let $\mathcal{D}$ be a small category. The *Yoneda embedding* is the functor $y \colon \mathcal{D} \to Sets^{\mathcal{D}^{op}}$ which sends an object d of $\mathcal{D}$ to the representable functor $\mathrm{Hom}_{\mathcal{D}}(-, d)$ and a morphism $f \colon d \to d'$ to the natural transformation $\mathrm{Hom}_{\mathcal{D}}(-, d) \to \mathrm{Hom}_{\mathcal{D}}(-, d')$.

The following result gives some indication of why we like representable functors.

Lemma 1.1.36 (Yoneda lemma) [4, 8.2] *Let C be a small category and $F \colon C^{op} \to Sets$ a functor. Given any object c of C, there is an isomorphism $\mathrm{Hom}_{Sets^{C^{op}}}(y(c), F) \cong F(c)$ of sets which is natural in both F and c.*

Now we turn our attention to additional structures on categories. The first such structure has the additional data of a binary operation on the objects of a category.

Definition 1.1.37 A *monoidal category* is a category C which is equipped with

1. a tensor product functor $\otimes \colon C \times C \to C$, where the image of a pair of objects (x, y) is denoted by $x \otimes y$,
2. a *unit object I*,
3. for every $x, y, z \in \mathrm{ob}(C)$, an associativity isomorphism

$$a_{x,y,z} \colon (x \otimes y) \otimes z \to x \otimes (y \otimes z),$$

  natural in the objects x, y, and z, and
4. for every $x \in \mathrm{ob}(C)$, a left unit isomorphism $\ell_x \colon I \otimes x \to x$ and a right unit isomorphism $r_x \colon x \otimes I \to x$, both natural in x.

We further assume that the diagrams

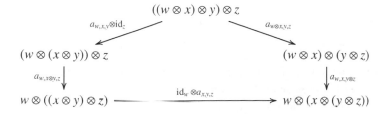

and

$$(x \otimes I) \otimes z \xrightarrow{\quad a_{x,I,y} \quad} x \otimes (I \otimes y)$$

with $r_x \otimes \mathrm{id}_y$ and $\mathrm{id}_x \otimes \ell_y$ mapping to $x \otimes y$

commute for any objects w, x, y, and z of C.

We denote such a monoidal category by $(C, \otimes, I)$ when we want to emphasize the tensor product and unit.

Definition 1.1.38   A monoidal category $(C, \otimes, I)$ is *symmetric* if, additionally, it is equipped with isomorphisms $s_{x,y} \colon x \otimes y \to y \otimes x$ for any objects x and y of C, natural in x and y, such that the diagrams

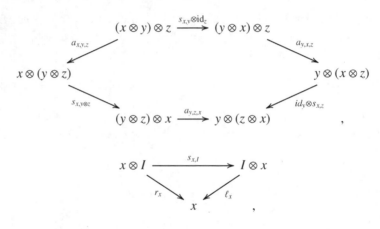

and

$$x \otimes y \xrightarrow{\quad s_{x,y} \quad} y \otimes x$$

with $\mathrm{id}_{x \otimes y}$ and $s_{y,x}$ mapping to $x \otimes y$

commute for all objects x, y, and z of C.

Definition 1.1.39   A symmetric monoidal category $(C, \otimes, I)$ is *closed* if, for any object y of C, the functor $- \otimes y \colon C \to C$ has a right adjoint.

The right adjoint is usually denoted by $(-)^y$; an object of C in the image of this functor is called an *internal hom object* of C. One can also define what it means for a nonsymmetric model category to be closed.

Now we turn to the case where the hom sets in a category have extra structure.

Definition 1.1.40 Let $(C, \otimes, I)$ be a monoidal category. A *category $\mathcal{D}$ enriched in C* consists of

1 a collection of objects, denoted by $\mathrm{ob}(\mathcal{D})$,
2 for any pair $x, y \in \mathrm{ob}(\mathcal{D})$, an object $\mathrm{Map}_{\mathcal{D}}(x, y)$ of C,
3 for every x, y, z in $\mathrm{ob}(\mathcal{D})$, a composition morphism

$$c_{x,y,z} : \mathrm{Map}_{\mathcal{D}}(x, y) \otimes \mathrm{Map}_{\mathcal{D}}(y, z) \to \mathrm{Map}_{\mathcal{D}}(x, z)$$

in C, and
4 for every $x \in \mathrm{ob}(\mathcal{D})$, a unit map $\eta_x : I \to \mathrm{Map}_{\mathcal{D}}(x, x)$ such that the diagrams

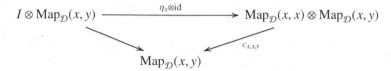

and

$$
\begin{array}{ccc}
\mathrm{Map}_{\mathcal{D}}(w, x) \otimes I & \xrightarrow{\ \mathrm{id} \otimes \eta_x\ } & \mathrm{Map}_{\mathcal{D}}(w, x) \otimes \mathrm{Map}_{\mathcal{D}}(x, x) \\
& \searrow \quad \qquad \swarrow{\scriptstyle c_{w,x,x}} & \\
& \mathrm{Map}_{\mathcal{D}}(w, x) &
\end{array}
$$

are commutative.

We assume that composition is associative, in that, for any $w, x, y, z \in \mathrm{ob}(\mathcal{D})$, the diagram

$$
\begin{array}{ccc}
\mathrm{Map}_{\mathcal{D}}(w, x) \otimes \mathrm{Map}_{\mathcal{D}}(x, y) \otimes \mathrm{Map}_{\mathcal{D}}(y, z) & \xrightarrow{\ c_{w,x,y} \otimes \mathrm{id}\ } & \mathrm{Map}_{\mathcal{D}}(w, y) \otimes \mathrm{Map}_{\mathcal{D}}(y, z) \\
{\scriptstyle \mathrm{id} \otimes c_{x,y,z}} \downarrow & & \downarrow {\scriptstyle c_{w,y,z}} \\
\mathrm{Map}_{\mathcal{D}}(w, x) \otimes \mathrm{Map}_{\mathcal{D}}(x, z) & \xrightarrow{\ c_{w,x,z}\ } & \mathrm{Map}_{\mathcal{D}}(w, z)
\end{array}
$$

commutes.

1.2 Weak Equivalences and Localization

The main idea of homotopy theory is that a category C may have morphisms which are not isomorphisms, but which we would like to regard as equivalences. We begin with a few classical examples.

Example 1.2.1   Let $\mathcal{T}op$ be the category of compactly generated Hausdorff spaces. In this category, the isomorphisms are the homeomorphisms. Recall that a map $f: X \to Y$ is a *homotopy equivalence* if there exists a map $g: Y \to X$ such that $f \circ g \simeq \mathrm{id}_Y$ and $g \circ f \simeq \mathrm{id}_X$. Thus, a homotopy equivalence is not necessarily a homeomorphism, since it may only admit an inverse up to homotopy.

For the remainder of this book, by "topological space" we always mean a compactly generated Hausdorff space.

Example 1.2.2   We can alternatively consider topological spaces and weak homotopy equivalences, or maps $f: X \to Y$ which induce isomorphisms

$$f_*: \pi_n(X) \to \pi_n(Y)$$

for all $n \geq 0$, where $\pi_n(X)$ denotes the nth homotopy group of X.

Example 1.2.3   Let R be a ring. Consider the category $Ch(R)$ of chain complexes of R-modules and chain maps. In an analogy with weak homotopy equivalences of spaces, we can consider the quasi-isomorphisms, maps which induce isomorphisms on all homology groups.

Let us now return to general theory. Let C be a category and W a collection of morphisms of C which we want to think of as isomorphisms but may not actually be so. We call these morphisms *weak equivalences*. Our goal is to define a category from C in which the weak equivalences are now isomorphisms, but which is as similar to C as possible.

As a first approach, we can look at Gabriel–Zisman localization [61].

Definition 1.2.4   Let C be a category and W a collection of morphisms in C. A *localization* of C with respect to W is a category $C[W^{-1}]$ together with a map

$$\gamma: C \to C[W^{-1}]$$

which takes the maps in W to isomorphisms and which is universal with respect to this property. In other words, if $\delta: C \to \mathcal{D}$ is a functor such that $\delta(w)$ is an isomorphism in $\mathcal{D}$ for every map w of W, then there is a unique map $C[W^{-1}] \to \mathcal{D}$ such that the diagram

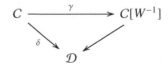

commutes.

The category $C[W^{-1}]$ has more morphisms than the original C. For example, zigzags with backward maps in W, such as

become morphisms in $C[W^{-1}]$. The localization of a category C with respect to weak equivalences is called the *homotopy category* of that category with respect to that choice of weak equivalences. We often denote it by $\mathrm{Ho}(C)$ when W is understood.

Example 1.2.5   If $C = \mathcal{G}p$ is the category of groups, we can define

$$S = \{G_1 \to G_2 \mid G_1/[G_1, G_1] \xrightarrow{\cong} G_2/[G_2, G_2]\},$$

where $[G_i, G_i]$ denotes the commutator subgroup of G_i. Then the localization $\mathcal{G}p[S^{-1}]$ is equivalent to the category $\mathcal{A}b$ of abelian groups.

Example 1.2.6   If $\mathcal{T}op$ is the category of topological spaces, we can consider the class H of homotopy equivalences, or the class W of weak homotopy equivalences. There are thus two homotopy categories to consider, $\mathcal{T}op[H^{-1}]$ and $\mathcal{T}op[W^{-1}]$. The latter is the more common one to consider, and thus it is what we refer to as the homotopy category of spaces and denote by $\mathrm{Ho}(\mathcal{T}op)$.

Example 1.2.7   If $C = Ch_{\geq 0}(\mathbb{Z})$ is the category of nonnegatively graded chain complexes of abelian groups, let S be the class of quasi-isomorphisms. Then $Ch_{\geq 0}(\mathbb{Z})[S^{-1}]$ is equivalent to $\mathcal{D}_{\geq 0}(\mathbb{Z})$, the bounded below derived category, formed by taking chain homotopy classes of maps and inverting the classes coming from quasi-isomorphisms [118, 10.3].

In general, the hom sets in $C[W^{-1}]$ may be too big, in that they may form proper classes rather than sets. There are different possible solutions to this difficulty, of which we mention three.

- Use the universe axiom, so that everything "large" becomes small in the next universe. While this approach can be used, it is not always helpful in practice.

- The approach used to define the derived category (mentioned in Example 1.2.7) is to show that W is a *multiplicative system*, so that:

  - the class W is closed under composition and contains all identity morphisms;

– (Ore's condition) given any morphism $a \to b$ in W and any other morphism $a \to c$ in C, there is a commutative square

with $c \to d$ in W, and, dually, given any morphism $f : c \to d$ in W and any other morphism $b \to d$ in C, then there is a commutative square as above with $a \to b$ in W; and

– if $f, g : a \to b$ are morphisms in C, then there exists a morphism s with source b such that $sf = sg$ if and only if there is a morphism t with target a such that $ft = gt$.

- Consider a model category structure on C. This approach is the one we consider here.

1.3 Classical Homotopy Theory

In the next section, we give the definition of model category. However, much of the terminology and even the spirit of the setup comes from classical homotopy theory. Therefore, in this section we look at some relevant facts from the homotopy theory of topological spaces.

We have already considered topological spaces together with weak homotopy equivalences, or maps which induce isomorphisms on all homotopy groups. We really want to consider the localization with respect to these maps, but we cannot ignore that homotopy equivalences have some helpful properties. In particular, they allow one to look at homotopy classes of maps between two topological spaces. Indeed, we propose to fix our set-theoretic problems with the localization by taking homotopy classes of maps, rather than all maps, as the morphisms in the homotopy category.

Unfortunately, simply taking homotopy classes of maps is insufficient, as this process is not homotopy invariant. We could have a weak homotopy equivalence $X \to X'$ which induces a map $[X', Y] \to [X, Y]$ which is not an isomorphism; we look at an example momentarily. If we really do not want to tell the difference between weakly homotopy equivalent spaces, then we have a problem.

We first look more closely at the similarities and differences between homotopy equivalences and weak homotopy equivalences. It is not a difficult exercise to check that every homotopy equivalence is a weak homotopy equivalence. However, the converse does not hold.

Example 1.3.1 Consider the Warsaw circle W, which is formed by replacing an arc of a circle by an infinitely oscillating curve, such as the graph of $\sin(1/x)$ near 0, and the collapse map $W \to *$ to a point. This map is a weak homotopy equivalence, since there are no nontrivial maps from any sphere to W. However, W is not contractible, so this map is not a homotopy equivalence. It is from such examples that we have trouble with homotopy classes of maps. For example, there are nontrivial homotopy classes of maps $W \to S^1$, whereas all maps from a point to S^1 are necessarily trivial.

Thus, we might ask when the two kinds of maps might coincide.

Theorem 1.3.2 (Whitehead's theorem) [66, 4.5] *If X and Y are CW complexes, then any weak homotopy equivalence $X \to Y$ is a homotopy equivalence.*

This theorem suggests that CW complexes are sufficiently well-behaved when we try to pass to homotopy classes of maps. But what do we do about other topological spaces? The following theorem helps.

Theorem 1.3.3 (CW approximation theorem) [66, 4.13] *Given any topological space X, there exists a CW complex X^c together with a weak homotopy equivalence $X^c \to X$.*

In other words, if we are only interested in topological spaces up to weak homotopy equivalence, then we can always replace any space with a CW complex.

Definition 1.3.4 The *homotopy category of topological spaces* $\mathrm{Ho}(\mathcal{T}op)$ has objects all topological spaces and morphisms defined by

$$\mathrm{Hom}_{\mathrm{Ho}(\mathcal{T}op)}(X, Y) = [X^c, Y^c].$$

The main idea here is that the category $\mathcal{T}op$ has enough structure so that we can define a well-behaved homotopy category with no set-theoretic difficulties. But what exactly is this structure, and how can we identify it in other categories with weak equivalences? Certainly we need to have a notion of homotopy between maps, at least between sufficiently nice objects. Furthermore, we need every object to be weakly equivalent to one of these nice objects, so they cannot be too restrictively defined.

Much of this structure can be identified via two additional classes of maps, which we identify here for the category $\mathcal{T}op$.

Definition 1.3.5　　A map $X \to Y$ of topological spaces is a *Serre fibration* if, for any CW complex A, a dotted arrow lift exists in any diagram of the form

where the left-hand vertical arrow is the inclusion.

If we allow A to be any topological space, not just a CW complex, then we get the related, but not identical, structure of a *Hurewicz fibration*.

Dually, we define a map $A \to B$ of topological spaces to be a *cofibration* if, for any Serre fibration $X \to Y$ which is also a weak homotopy equivalence, and any square commutative diagram

$$
\begin{array}{ccc}
A & \longrightarrow & X \\
\downarrow & \nearrow & \downarrow \\
B & \longrightarrow & Y
\end{array}
$$

the dotted arrow lift exists, making the diagram commute.

While it is less clear why we need Serre fibrations, we get some hint that cofibrations are important for obtaining the homotopy category from the following result.

Proposition 1.3.6　　[59, 8.9] *Let $\varnothing$ denote the empty space. Given a topological space X, the unique map $\varnothing \to X$ is a cofibration if and only if X is a retract of a CW complex.*

1.4　Model Categories

Model categories were first defined by Quillen [100]. The idea behind them is to give an axiomatization of the structure that a category with weak equivalences has to possess in order to be able to define a homotopy category in a similar way that we do for the category of topological spaces with weak homotopy equivalences.

Definition 1.4.1　　[69, 7.1.3], [100, I.1]　　A *model category* is a category $\mathcal{M}$ together with a choice of three classes of morphisms, *weak equivalences*, *fibrations*, and *cofibrations*. A map which is a (co)fibration and weak equivalence is called an *acyclic (co)fibration*. The category $\mathcal{M}$, together with these three classes, must satisfy the following five axioms.

(MC1) The category $\mathcal{M}$ has all small limits and colimits.

(MC2) If f and g are maps in $\mathcal{M}$ such that gf is defined, and if two of the maps f, g, and gf are weak equivalences, then so is the third.

(MC3) If f and g are maps in $\mathcal{M}$ such that f is a retract of g, and g is a weak equivalence, fibration, or cofibration, then so is f.

(MC4) Given a commutative solid arrow diagram

the dotted arrow lift exists if either

 (i) the map i is a cofibration and p is an acyclic fibration, or

 (ii) the map i is an acyclic cofibration and p is a fibration.

(MC5) Any map g in $\mathcal{M}$ can be factored in two ways:

 (i) as $g = qi$, where i is a cofibration and q is an acyclic fibration, and

 (ii) as $g = pj$, where j is an acyclic cofibration and p is a fibration.

Axiom (MC2) is often referred to as the *two-out-of-three property*. The factorizations in (MC5) are often assumed to be functorial, and in practice most model categories have functorial factorizations, but this assumption is not part of the original definition.

These axioms actually guarantee that any choice of two classes determines the third class. In any diagram of the form given in (MC4), if the dotted arrow lift exists, we say that i has the *left lifting property* with respect to p, and p has the *right lifting property* with respect to i. If we choose weak equivalences and one of the other classes, then the third is determined by the lifting properties given by axiom (MC4): the acyclic cofibrations have the left lifting property with respect to the fibrations, and the cofibrations have the left lifting property with respect to the acyclic fibrations. If the fibrations and cofibrations are specified, then the weak equivalences are determined by this axiom together with the fact that any weak equivalence must factor as an acyclic cofibration followed by an acyclic fibration.

In fact, if we choose weak equivalences and one other class, then we still have an overdetermined structure. Suppose we specify the weak equivalences and the fibrations. Although the cofibrations can be defined via the left lifting property with respect to the acyclic fibrations, there are still two different ways to define the acyclic cofibrations: either as the cofibrations which are also weak equivalences, or as the maps with the left lifting property with respect to the fibrations. Part of the process of determining that a model structure exists is to prove that these two notions coincide.

Example 1.4.2   Our two choices of weak equivalences in $\mathcal{T}op$ give two possible model structures. If the weak equivalences are homotopy equivalences, we can define the fibrations to be the Hurewicz fibrations and the cofibrations to be the Borsuk pairs. The result is Strøm's model structure [112].

If we take the weak equivalences to be the weak homotopy equivalences, the cofibrations to be the class of CW inclusions and their retracts, and the fibrations to be Serre fibrations, then we get the standard model structure described in the previous section and described in more detail by Dwyer and Spalinski [59, §8].

Example 1.4.3   For $Ch_{\geq 0}(\mathbb{Z})$ with weak equivalences the quasi-isomorphisms, then we can choose the cofibrations to be the monomorphisms and the fibrations to be the epimorphisms with projective kernels. We denote this model structure by $Ch_{\geq 0}(\mathbb{Z})_c$ [59, §7].

Example 1.4.4   More generally, for a ring R, let $Ch(R)$ denote the category of (unbounded) chain complexes of R-modules and chain maps. Then take the weak equivalences to be the quasi-isomorphisms and the fibrations to be the chain maps $C \to D$ such that each $C_n \to D_n$ is a surjective map of R-modules, resulting in a model structure which we denote by $Ch(R)_f$ and call the *projective model structure*. Dually, we could take the same weak equivalences but the cofibrations to be the chain maps $C \to D$ such that each $C_n \to D_n$ is a monomorphism of R-modules, and get a model structure called the *injective model structure*, denoted by $Ch(R)_c$ [71, §2.3].

Observe that, by axiom (MC1), any model category must have an initial object $\emptyset$ and a terminal object $*$.

Definition 1.4.5   An object X of a model category is *cofibrant* if the unique map $\emptyset \to X$ is a cofibration. It is *fibrant* if the unique map $X \to *$ is a fibration.

Definition 1.4.6   Let X be an object of a model category. A *cofibrant replacement* for X is a cofibrant object X^c together with a weak equivalence $X^c \to X$. A *fibrant replacement* for X is a fibrant object X^f together with a weak equivalence $X \to X^f$.

Proposition 1.4.7   *Every object X in a model category has a cofibrant replacement and a fibrant replacement.*

Proof   To show that X has a cofibrant replacement, consider the unique map $\emptyset \to X$. By (MC5)(i) we have a factorization $\emptyset \hookrightarrow X^c \to X$ where the second map is a weak equivalence. Dually, we can obtain a fibrant replacement for X

by factoring the unique map $X \to *$ as an acyclic cofibration followed by a fibration $X \hookrightarrow X^f \to *$. □

Iterating the process, we can always replace any object X with an object X^{cf} which is both fibrant and cofibrant.

Example 1.4.8 In the model category $\mathcal{T}op$ of topological spaces, all objects are fibrant. The cofibrant objects are exactly the retracts of CW complexes. (Observe that our notation for CW approximations in the previous section was chosen to coincide with that for cofibrant replacement.)

Example 1.4.9 [69, 7.6.5] If $\mathcal{M}$ is a model category and X is an object of $\mathcal{M}$, then the category $X \downarrow \mathcal{M}$ of objects under X as described in Example 1.1.12 has a model structure in which a map is a weak equivalence, fibration, or cofibration if it is one in $\mathcal{M}$. Dually, the category $\mathcal{M} \downarrow X$ of objects over X has a model structure defined analogously.

We have already discussed that a model structure is determined (if it exists) by its weak equivalences and either the fibrations or the cofibrations. Less immediate is the following result, which Riehl [104] attributes to unpublished work of Joyal.

Proposition 1.4.10 [104, 15.3.1] *If it exists, a model structure is determined by its cofibrations and its fibrant objects.*

We now give some basic closure properties of (acyclic) (co)fibrations.

Proposition 1.4.11 [59, 3.14] *Let $\mathcal{M}$ be a model category.*

1 *The class of cofibrations is closed under pushouts. In other words, if P is a pushout in a diagram*

and i is a cofibration, then so is j.
2 *The class of acyclic cofibrations is closed under pushouts.*
3 *The class of fibrations is closed under pullbacks. In other words, if Q is a pullback in a diagram*

$$
\begin{array}{ccc}
Q & \longrightarrow & X \\
{\scriptstyle q}\downarrow & & \downarrow{\scriptstyle p} \\
B & \longrightarrow & Y
\end{array}
$$

pullback in a diagram

and p is a fibration, then so is q.

4 The class of acyclic fibrations is closed under pullbacks.

We conclude this section with a lifting lemma.

Lemma 1.4.12   *Suppose that $A \to B$ is a cofibration, $C \to D$ is a fibration, and $B' \to B$ is a weak equivalence in a model category M. Then in the commutative diagram*

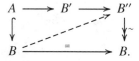

a lift $B \to C$ exists if and only if a lift $B' \to C$ exists.

Proof   If a lift $B \to C$ exists, then a lift $B' \to C$ is given by composition with the map $B' \to B$.

To prove the converse, first observe that the weak equivalence $B' \to B$ can be factored as a composite

$$B' \overset{\simeq}{\hookrightarrow} B'' \overset{\simeq}{\twoheadrightarrow} B$$

of an acyclic cofibration and an acyclic fibration. Therefore by model category axiom (MC4), a dotted arrow lift exists in the diagram

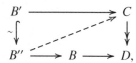

Applying axiom (MC4) again to the diagram

$$
\begin{array}{ccc}
B' & \longrightarrow & C \\
{\scriptstyle \sim}\downarrow & \nearrow & \downarrow \\
B'' & \longrightarrow B \longrightarrow & D,
\end{array}
$$

we obtain a dotted arrow lift. Then the composite $B \to B'' \to C$ is the desired lift in the original diagram. $\square$

1.5 Homotopy Categories

Let us look a little more closely at fibrant and cofibrant objects, and why they allow us to make sense of a homotopy category defined via homotopy classes of maps. We use the notation of [59]; the reader is referred there for a more detailed treatment, including proofs.

In the category of topological spaces, a *homotopy* between two maps $f, g :$ $X \to Y$ is defined to be a map $H: X \times I \to Y$ such that the diagram

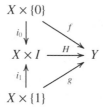

commutes. Here I denotes the closed interval $[0, 1]$, and the two vertical maps are inclusions.

In a more general categorical setting, we cannot assume the existence of an object that plays the role of an interval, but it is sufficient to have objects that behave sufficiently like the products $X \times I$, as given by the following definition.

Definition 1.5.1   Let $\mathcal{M}$ be a model category and A an object of $\mathcal{M}$. A *cylinder object* for A is an object $A \wedge I$ of $\mathcal{M}$ together with a factorization

$$A \amalg A \xrightarrow{\ i\ } A \wedge I \xrightarrow{\ \simeq\ } A$$

of the fold map $A \amalg A \to A$.

Observe that cylinder objects always exist, using axiom (MC5). One should not assume that the symbol I has any meaning on its own, other than to be suggestive of the role of intervals in defining cylinders in topology.

Definition 1.5.2   Two maps $f, g: A \to X$ in $\mathcal{M}$ are *left homotopic* if there exists a cylinder object $A \wedge I$ for A such that the sum map $f + g: A \amalg A \to X$ extends to a map $H: A \wedge I \to X$ called a *left homotopy*.

Proposition 1.5.3   [59, 4.7] *If A is cofibrant, then left homotopy defines an equivalence relation on* $\mathrm{Hom}_{\mathcal{M}}(A, X)$.

However, there is an equivalent way to define homotopy in topological spaces, using the adjunction between products and mapping spaces. A homotopy between maps $f, g: X \to Y$ can be described instead as a map $H: X \to Y^I$

such that the diagram

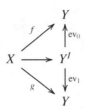

commutes. Here Y^I denotes the space of maps $I \to Y$, or paths in Y, and the two vertical maps are given by evaluation at the endpoints of I. Similarly to above, we can give a more general formulation for objects that play the role of Y^I.

Definition 1.5.4 Let X be an object in a model category $\mathcal{M}$. A *path object* for X is an object X^I of $\mathcal{M}$ together with a factorization

$$X \xrightarrow{\simeq} X^I \xrightarrow{p} X \times X$$

of the diagonal map $X \to X \times X$.

Again, one should be wary of reading too much into the notation X^I, suggesting the path space in the category of topological spaces. Nonetheless, we know such path objects also exist using (MC5).

Definition 1.5.5 Two maps $f, g \colon A \to X$ are *right homotopic* if there exists a path object X^I for X such that the product map $(f, g) \colon A \to X \times X$ lifts to a map $H \colon A \to X^I$ called a *right homotopy*.

Proposition 1.5.6 [59, 4.16] *If X is fibrant, then right homotopy defines an equivalence relation on* $\mathrm{Hom}_{\mathcal{M}}(A, X)$.

Now we bring together the two different notions of homotopy in a model category.

Lemma 1.5.7 [59, 4.21] *Let $f, g \colon A \to X$ be maps in $\mathcal{M}$.*

1 If A is cofibrant and f is left homotopic to g, then f is right homotopic to g.
2 If X is fibrant and f is right homotopic to g, then f is left homotopic to g.

The consequence of this lemma is that we can say that two maps $A \to X$ are simply *homotopic* if A is cofibrant and X is fibrant. We then write $[A, X]$ for the set of homotopy classes of maps $A \to X$. In order to incorporate composition, so that we get a well-defined category, we restrict to objects which are both fibrant and cofibrant.

Proposition 1.5.8 [59, 4.24] *Let A and X be objects of M which are both fibrant and cofibrant. Then a map f : A → X is a weak equivalence if and only if f has a homotopy inverse, i.e., there exists a map g : X → A such that fg and gf are homotopic to identity maps.*

Now we are ready to define the homotopy category associated to a model category.

Definition 1.5.9 Given a model category M, its *homotopy category* Ho(M) is the category with the same objects as M and morphisms defined by, for any objects X and Y,

$$\text{Hom}_{\text{Ho}(M)}(X, Y) = [X^{cf}, Y^{cf}].$$

Indeed, this construction of the homotopy category associated to a model category has the desired localization property.

Proposition 1.5.10 [59, 6.2] *The homotopy category* Ho(M) *is a localization of the model category M with respect to the weak equivalences. In other words, there is a functor $\gamma : M \to$ Ho(M) which is universal with respect to all functors taking the weak equivalences of M to isomorphisms.*

In this section, we have seen that a model structure on a category gives a means to obtain a well-defined homotopy category. However, it is important to note that the structure of a model category contains much more information which is lost when restricting to the homotopy category, so having this extra information is significant in its own right.

1.6 Equivalences Between Model Categories

Now we want to develop appropriate notions of functors between model categories and what it means for two model categories to be equivalent to one another. To begin, we want to consider adjoint pairs of functors

$$F : M_1 \rightleftarrows M_2 : G$$

between model categories, so that $\text{Hom}_{M_2}(FX, Y) \cong \text{Hom}_{M_1}(X, GY)$. However, we would like these functors to preserve the essential data of the model structures on M_1 and M_2.

Definition 1.6.1 An adjoint pair of functors (F, G) between model categories is a *Quillen pair* if F preserves cofibrations and G preserves fibrations.

Proposition 1.6.2 [69, 8.5.3] *The following are equivalent:*

1 The adjoint pair (F, G) is a Quillen pair.
2 The left adjoint F preserves cofibrations and acyclic cofibrations.
3 The right adjoint G preserves fibrations and acyclic fibrations.

Definition 1.6.3   A Quillen pair $F\colon M \rightleftarrows N\colon G$ is a *Quillen equivalence* if, for every cofibrant object X of M and fibrant object Y of N, a map $FX \to Y$ is a weak equivalence in N if and only if its adjoint map $X \to GY$ is a weak equivalence in M.

We give two more equivalent formulations of the notion of Quillen equivalence. A functor $F\colon M \to N$ *reflects weak equivalences* if, for any morphism $f\colon x \to y$ in M, if $F(f)$ is a weak equivalence in N, then f is a weak equivalence in M.

Proposition 1.6.4   [71, 1.3.16] *Let $F\colon M \rightleftarrows N\colon G$ be a Quillen pair. The following are equivalent:*

1 The Quillen pair (F, G) is a Quillen equivalence.
2 The left adjoint F reflects weak equivalences between cofibrant objects and, for every fibrant object Y of N, the map $F(GY)^c \to Y$ is a weak equivalence in N.
3 The right adjoint G reflects weak equivalences between fibrant objects and, for every cofibrant object X of M, the map $X \to G(FX)^f$ is a weak equivalence in M.

We say that two model categories are *Quillen equivalent* if there exists a chain of Quillen equivalences between them. Observe that, since adjoint pairs are not entirely symmetric, Quillen equivalences do not necessarily compose to a Quillen equivalence; we could have

$$M \rightleftarrows N \rightleftarrows P,$$

with the left and right adjoints going in opposite directions. Thus, model categories can be Quillen equivalent without having a single Quillen equivalence between them.

In general, a Quillen equivalence tells us a good deal about the similarity of structure between two model categories. In particular, we expect a Quillen equivalence to induce an equivalence between homotopy categories. We now develop the necessary language to make a precise statement. Recall the localization functor $\gamma\colon M \to \mathrm{Ho}(M)$.

Definition 1.6.5   [59, 9.1] Let M be a model category and N a category. Suppose that $F\colon M \to N$ is a functor. A *left derived functor* for F is a pair (LF, t), where LF: $\text{Ho}(\mathcal{M}) \to \mathcal{N}$ is a functor and t: $LF \circ \gamma \to F$ a natural transformation, which is universal from the left. In other words, for any other such pair (G, s), there is a factorization $G\gamma \to LF\gamma \to F$ with compatible natural transformations. Similarly, a *right derived functor* for F is a pair (RF, t), where RF: $\text{Ho}(\mathcal{M}) \to N$ is a functor and t: $F \to (RF) \circ \gamma$ is a natural transformation, which is universal from the right.

Definition 1.6.6 [59, 9.5] Let $\mathcal{M}$ and $\mathcal{N}$ be model categories and F: $\mathcal{M} \to \mathcal{N}$ be a functor. A *total left derived functor* for F is a functor $\mathbb{L}F$: $\text{Ho}(C) \to \text{Ho}(\mathcal{D})$ which is a left derived functor for the composite $\gamma \circ F$: $\mathcal{M} \to \text{Ho}(\mathcal{N})$. Similarly, a *total right derived functor* for F is a functor $\mathbb{R}F$: $\text{Ho}(\mathcal{M}) \to \text{Ho}(\mathcal{N})$ which is a right derived functor for the composite $\gamma \circ F$.

Proposition 1.6.7 [69, 8.5.8] *A Quillen pair F: $\mathcal{M} \leftrightarrows \mathcal{N}$: G induces*

1 a total left derived functor $\mathbb{L}F$: $\text{Ho}(\mathcal{M}) \to \text{Ho}(\mathcal{N})$, and
2 a total right derived functor $\mathbb{R}G$: $\text{Ho}(\mathcal{N}) \to \text{Ho}(\mathcal{M})$.

Proposition 1.6.8 [69, 8.4.23] *If F: $\mathcal{M} \rightleftarrows \mathcal{N}$: G is a Quillen equivalence, then the derived functors $\mathbb{L}F$: $\text{Ho}(\mathcal{M}) \rightleftarrows \text{Ho}(\mathcal{N})$: $\mathbb{R}G$ define an equivalence of categories.*

If two model categories are Quillen equivalent, then we say that they *model the same homotopy theory*. As we shall see throughout this book, there can be many models for a given homotopy theory, and in fact it can be advantageous to have multiple models.

Sometimes having different models can be as simple as having two model structures on the same category with the same class of weak equivalences. In Example 1.4.4, we have seen two different model structures on the category of chain complexes. Since the most fundamental information is given by the weak equivalences, we expect that these two model structures are equivalent. Indeed, using the identity functor in both directions gives a Quillen equivalence

$$Ch_{\geq 0}(R)_c \rightleftarrows Ch_{\geq 0}(R)_f.$$

But, we can also have different categories with equivalent homotopy categories. In the next chapter, we will define simplicial sets, which are combinatorial models for topological spaces. The fact that they do indeed model spaces is made precise via model categories: there is a model structure on the category $\mathcal{S}Sets$ of simplicial sets which is Quillen equivalent to the model structure on $\mathcal{T}op$.

1.7 Additional Structures on Model Categories

Although having a model structure on a category is very helpful, it is often desirable to have additional nice properties. Here we give a few such properties; more will be given in the next chapter.

Definition 1.7.1　[69, 13.1.1]　A model category $\mathcal{M}$ is:

1 *left proper* if every pushout of a weak equivalence along a cofibration is a weak equivalence; i.e., if P is a pushout in the diagram

$$
\begin{array}{ccc}
A & \xrightarrow{\ f\ } & X \\
{\scriptstyle i}\downarrow & & \downarrow \\
B & \xrightarrow{\ g\ } & P
\end{array}
$$

where i is a cofibration and f is a weak equivalence, then g is a weak equivalence;

2 *right proper* if every pullback of a weak equivalence along a fibration is a weak equivalence; i.e., if Q is a pullback in the diagram

$$
\begin{array}{ccc}
Q & \xrightarrow{\ f\ } & X \\
\downarrow & & \downarrow{\scriptstyle p} \\
B & \xrightarrow{\ g\ } & Y
\end{array}
$$

where p is a fibration and g is a weak equivalence, then f is a weak equivalence; and

3 *proper* if it is both left and right proper.

Example 1.7.2　[69, 13.1.11] The model category $\mathcal{T}op$ is proper.

Proposition 1.7.3　[69, 13.1.3] *Let $\mathcal{M}$ be a model category.*

1 *If every object of $\mathcal{M}$ is cofibrant, then $\mathcal{M}$ is left proper.*
2 *If every object of $\mathcal{M}$ is fibrant, then $\mathcal{M}$ is right proper.*

Proposition 1.7.4　[69, 13.5.4] *Let $\mathcal{M}$ be a left proper model category. Suppose that in the diagram*

$$
\begin{array}{ccccc}
C & \longleftarrow & A & \longrightarrow & B \\
{\scriptstyle \simeq}\downarrow & & \downarrow{\scriptstyle \simeq} & & \downarrow{\scriptstyle \simeq} \\
C' & \longleftarrow & A' & \longrightarrow & B'
\end{array}
$$

the vertical maps are weak equivalences and at least one map in each row is a cofibration. Then the induced map of pushouts

$$B \amalg_A C \to B' \amalg_{A'} C'$$

is a weak equivalence.

When a class of maps in a model category is defined via a lifting property with respect to another class, it can be difficult to verify whether a particular map really has the desired lifting property with respect to all such maps. For many examples of model categories, we can reduce to checking the lifting property with respect to a set of maps that, in some sense, generates the whole class.

Definition 1.7.5 A *transfinite composition* of a sequence

$$X_0 \to X_1 \to \cdots$$

in a model category $\mathcal{M}$ is given by the map $X_0 \to \operatorname{colim} X_i$.

Proposition 1.7.6 [69, 10.3.1] *Let $\mathcal{M}$ be a model category and S a set of maps in $\mathcal{M}$. The class of maps with the left lifting property with respect to S is closed under transfinite composition.*

Definition 1.7.7 Let $\mathcal{M}$ be a category and I a set of maps in $\mathcal{M}$. A *relative I-cell complex* is a (possibly transfinite) composition of pushouts along elements of I. Let I-*cell* denote the class of I-cell complexes in $\mathcal{M}$.

Definition 1.7.8 Let $\mathcal{M}$ be a category and I a class of morphisms in $\mathcal{M}$. An object A of $\mathcal{M}$ is *small relative to I* if, for any sequence

$$X_0 \to X_1 \to \cdots$$

with each $X_i \to X_{i+1}$ in I, the map

$$\operatorname{colim}_i \operatorname{Hom}_{\mathcal{M}}(A, X_i) \to \operatorname{Hom}_{\mathcal{M}}(A, \operatorname{colim}_i X_i)$$

is an isomorphism of sets. If I is the class of all morphisms of $\mathcal{M}$, then we simply say that the object A is *small*.

Strictly speaking, we should be careful with size issues here, taking small-ness relative to some cardinal. However, our goal is to give a basic treatment and so we bypass these technical details. We refer the reader to Hirschhorn [69, §10.4] and Hovey [71, §2.1.1] for a more precise treatment.

Definition 1.7.9 [69, 10.5.2] Let I be a set of maps in a category $\mathcal{M}$.

1 A map is an *I-fibration* if it has the right lifting property with respect to all maps in I.
2 A map is an *I-cofibration* if it has the left lifting property with respect to all the I-fibrations.

The I-fibrations are sometimes called *I-injectives*.

Definition 1.7.10   [69, 10.5.15] Let $\mathcal{M}$ be a category. A set of maps I in $\mathcal{M}$ *permits the small object argument* if the domains of the elements of I are small relative to I.

The following result is generally referred to as the "small object argument".

Proposition 1.7.11   [69, 10.5.16] *Let $\mathcal{M}$ be a category with all small limits and colimits and I a set of maps in $\mathcal{M}$ which permits the small object argument. Then there is a functorial factorization of any map in $\mathcal{M}$ into a relative I-cell complex followed by an I-fibration.*

The purpose of these definitions is to lead up to the following kind of model category.

Definition 1.7.12   [69, 11.1.2] A model category $\mathcal{M}$ is *cofibrantly generated* if

1 there exists a set I of cofibrations in $\mathcal{M}$ which permits the small object argument such that a map is an acyclic fibration if and only if it has the right lifting property with respect to every map in I, and
2 there exists a set J of acyclic cofibrations in $\mathcal{M}$ which permits the small object argument such that a map is a fibration if and only if it has the right lifting property with respect to every map in J.

The maps in I are called *generating cofibrations* and the maps in J are called *generating acyclic cofibrations*.

Example 1.7.13   The model category $\mathcal{T}op$ is cofibrantly generated, with generating cofibrations the boundary inclusions of disks

$$I = \{S^{n-1} \to D^n \mid n \geq 0\}$$

and generating acyclic cofibrations

$$J = \{D^n \times \{0\} \to D^n \times I \mid n \geq 0\}.$$

In particular, in a cofibrantly generated model category, the I-fibrations are precisely the acyclic fibrations, and the I-cofibrations are precisely the cofibrations; the J-fibrations are precisely the fibrations and the J-cofibrations are precisely the acyclic cofibrations.

Observe that a model category could have different possible choices for generating (acyclic) cofibrations. Knowing that a model category is cofibrantly generated only guarantees the existence of one choice of each kind.

Proposition 1.7.14 [71, 2.1.18] *Let C be a cofibrantly generated model category with I a set of generating cofibrations. Then any cofibration in C is a retract of a relative I-cell complex. In other words, any cofibration can be obtained as a retract of a transfinite composition of generating cofibrations.*

The analogous statement for acyclic cofibrations also holds.

The following two theorems, generally attributed to Kan, are useful to verify the existence of cofibrantly generated model categories.

Theorem 1.7.15 [69, 11.3.1] *Let M be a category which has all small limits and colimits, and let W be a class of maps in M that is closed under retracts and satisfies the two-out-of-three property. Suppose that I and J are sets of maps in M such that*

1 *both I and J permit the small object argument,*
2 *every J-cofibration is both an I-cofibration and an element of W,*
3 *every I-fibration is both a J-fibration and an element of W, and*
4 *one of the following conditions holds:*

  (i) *a map that is both an I-cofibration and an element of W is a J-cofibration, or*
  (ii) *a map that is both a J-fibration and an element of W is an I-fibration.*

Then there is a cofibrantly generated model structure on M in which W is the class of weak equivalences, I is a set of generating cofibrations, and J is a set of generating acyclic cofibrations.

Theorem 1.7.16 [69, 11.3.2] *Let M be a cofibrantly generated model category with set I of generating cofibrations and set J of generating acyclic cofibrations. Let N be a category with all small limits and colimits, and suppose $F: M \rightleftarrows N: G$ is a pair of adjoint functors. Let $FI = \{Fu \mid u \in I\}$ and $FJ = \{Fv \mid v \in J\}$. If*

1 *both FI and FJ permit the small object argument, and*
2 *the functor G takes relative FJ-cell complexes to weak equivalences,*

then there is a cofibrantly generated model structure on N in which FI is a set of generating cofibrations, FJ is a set of generating acyclic cofibrations, and the weak equivalences are morphisms that G takes to weak equivalences in M. Furthermore, the adjoint pair (F, G) is a Quillen pair.

We can strengthen the notion of cofibrantly generated model category in two different but useful ways. The first is that of a cellular model category, which we introduce here; the second is that of a combinatorial model category, which we define in the next chapter.

In the following definition, we again suppress certain set-theoretic details; we refer to [69, 10.8.1] for a more rigorous definition.

Definition 1.7.17   Let M be a cofibrantly generated model category with a set I of generating cofibrations. An object Z of M is *compact* if, for every relative I-cell complex $f: X \to Y$, every map from Z to Y factors through a small subcomplex.

Definition 1.7.18   A map $f: X \to Y$ in M is an *effective monomorphism* if it is the equalizer of the pair of natural inclusions $Y \rightrightarrows Y \amalg_X Y$.

Definition 1.7.19   Let M be a cofibrantly generated model category with a set I of generating cofibrations and a set J of generating acyclic cofibrations. This model structure is *cellular* if:

1 the domains and the codomains of the maps in I are compact,
2 the domains of the elements of J are small relative to I, and
3 the cofibrations are effective monomorphisms.

Example 1.7.20   [69, 12.1.4] The category $\mathcal{T}op$ with its usual model structure is a cellular model category.

We can also consider model structures on categories which themselves have more structure, but in these cases we want the extra data to be compatible with the model structure in an appropriate way. Here we look at monoidal model categories.

Definition 1.7.21   [71, 4.2.6] A *monoidal model category* is a closed monoidal category $(M, \otimes, I)$ with a model structure, such that the following conditions hold.

1 For any cofibrations $i: A \to B$ and $j: C \to D$ in M, the induced map

$$B \otimes C \amalg_{A \otimes C} A \otimes D \to B \otimes D$$

is a cofibration in M which is a weak equivalence if either i or j is.

2 For any cofibrant object A of $\mathcal{M}$, the maps $I^c \otimes A \to I \otimes A$ and $A \otimes I^c \to A \otimes I$ are weak equivalences.

If $\mathcal{M}$ is a closed symmetric monoidal category, and these conditions hold, then $\mathcal{M}$ is a *symmetric monoidal model category*, in which case we need only consider a single map in the second condition.

Example 1.7.22 The model category $\mathcal{T}op$ is a symmetric monoidal model category where the tensor product is the cartesian product and the unit object is a single point.

2

Simplicial Objects

Simplicial sets and other simplicial objects play a crucial role in modern homotopy theory, and in particular the structures that we consider later in this
book. The initial motivation for them arises in topology, where they are shown
to be combinatorial models for topological spaces. As such, they are natural
generalizations of simplicial complexes; the more flexible structure allows for
operations such as products and quotients. Simplicial sets have deep connec-
tions with category theory as well, which makes it unsurprising that they are
integral to most approaches to homotopical categories. In this chapter we also
look at several other constructions with model categories, continuing the ideas
from the previous chapter.

2.1 Simplicial Sets and Simplicial Objects

In this section, we begin with the basic definitions of simplicial sets and other
simplicial objects. For a conceptual introduction, the reader is referred to
Friedman [60]; more detailed treatments include Goerss and Jardine [62] and
May [91].

Definition 2.1.1 The category Δ has objects the finite ordered sets

$$[n] = \{0 \leq 1 \leq \cdots \leq n\}$$

and morphisms the weakly order-preserving functions.

The category Δ looks like

$$[0] \rightrightarrows [1] \rightrightarrows [2] \cdots.$$

The maps indicated are the generators for this category, and they satisfy simplicial relations. Specifically, Δ has *coface maps* $d^i : [n-1] \to [n]$, indexed by
$0 \leq i \leq n$, where d^i is the injective map which is not surjective precisely on the

object i. There are also *codegeneracy maps* $s^i \colon [n] \to [n-1]$ for $0 \le i \le n-1$, where each s_i is surjective and sends both i and $i+1$ in $[n]$ to i in $[n-1]$. One can check that these maps satisfy the following *cosimplicial identities*:

$$d^j d^i = d^i d^{j-1} \qquad i < j$$
$$s^j d^i = d^i s^{j-1} \qquad i < j$$
$$s^j d^j = \mathrm{id} = s^j d^{j+1}$$
$$s^j d^i = d^{i-1} s^j \quad i > j+1$$
$$s^j s^i = s^i s^{j+1} \qquad i \le j.$$

We can also take the opposite category Δ^{op}, where the arrows are formally reversed.

Definition 2.1.2 A *simplicial set* is a functor $K \colon \Delta^{op} \to Sets$.

Given a simplicial set K, for any object $[n]$ of Δ^{op}, we denote the set $K([n])$ by K_n and call its elements n-*simplices*. Since the arrows in a simplicial set K are reversed from those of Δ, there are $n+1$ *face maps* $d_i \colon K_n \to K_{n-1}$ and n *degeneracy maps* $s_i \colon K_{n-1} \to K_n$. The face and degeneracy maps satisfy the following *simplicial identities* which are dual to the cosimplicial identities given above:

$$d_i d_j = d_{j-1} d_i \qquad i < j$$
$$d_i s_j = s_{j-1} d_i \qquad i < j$$
$$d_j s_j = \mathrm{id} = d_{j+1} s_j$$
$$d_i s_j = s_j d_{i-1} \quad i > j+1$$
$$s_i s_j = s_{j+1} s_i \qquad i \le j.$$

A simplex is called *degenerate* if it is in the image of a degeneracy map.

Example 2.1.3 An important example of a simplicial set is the n-simplex $\Delta[n]$, defined for every $n \ge 0$. It is the representable functor $\mathrm{Hom}_\Delta(-, [n])$. In other words, for any $k \ge 0$, $\Delta[n]_k = \mathrm{Hom}_\Delta([k], [n])$.

There are two related simplicial sets that appear frequently. The first is the boundary of the n-simplex, denoted by $\partial\Delta[n]$. It is the simplicial set with all the nondegenerate simplices of $\Delta[n]$ except the n-simplex given by the identity map $[n] \to [n]$.

The other is the k-horn of the n-simplex, where $n \ge 1$ and $0 \le k \le n$, denoted by $V[n, k]$. It is obtained from $\partial\Delta[n]$ by further omitting the simplex defined by the injective map $[n-1] \to [n]$ in Δ where k is the only point not in the image.

We define a map $K \to L$ between simplicial sets to be a natural transformation of functors. Hence, we have a category $SSets$ of simplicial sets.

Using the Yoneda lemma, and the fact that the n-simplex $\Delta[n]$ is representable, for any simplicial set K and any $n \geq 0$, there is a natural isomorphism

$$K_n \cong \mathrm{Hom}_{SSets}(\Delta[n], K).$$

Example 2.1.4   Let C be a small category, and Cat the category of all small categories. Observe that the ordered set $[n]$ can also be regarded as a category with objects $0, 1, \ldots, n$ and a unique morphism $i \to j$ for $0 \leq i \leq j \leq n$. Define the *nerve* of C to be the simplicial set nerve(C) whose n-simplices consist of the set $\mathrm{Hom}_{Cat}([n], C)$.

Let $\alpha^i \colon [1] \to [n]$ be the map in Δ defined by $0 \mapsto i$ and $1 \mapsto i + 1$, for any $0 \leq i < n$. Then each α^i induces a map

$$\mathrm{Hom}_{Cat}([n], C) \to \mathrm{Hom}_{Cat}([1], C),$$

or, in other words, a function $\alpha_i \colon$ nerve(C)$_n \to$ nerve(C)$_1$. Since the images of the maps α^i overlap at single objects of $[n]$, we can assemble all the induced maps to obtain a *Segal map*

$$\varphi_n \colon \mathrm{nerve}(C)_n \to \underbrace{\mathrm{nerve}(C)_1 \times_{\mathrm{nerve}(C)_0} \cdots \times_{\mathrm{nerve}(C)_0} \mathrm{nerve}(C)_1}_{n}.$$

The fact that C is a category, and in particular has composition of morphisms, implies that this Segal map is an isomorphism of sets for any $n \geq 2$.

Segal maps can be defined more generally for any simplicial set K. The maps α^i induce inclusions of simplices $\Delta[1] \to \Delta[n]$ which in turn, using the natural isomorphism $K_n \cong \mathrm{Hom}_{SSets}(\Delta[n], K)$, induce set maps

$$K_n \to \underbrace{K_1 \times_{K_1} \cdots \times_{K_0} K_1}_{n}.$$

However, in general the Segal maps are not isomorphisms.

Although a simplicial set always has infinitely many degenerate simplices, the following result allows us to find, for any degenerate simplex, a unique nondegenerate simplex which gives rise to it.

Lemma 2.1.5   [69, 15.8.4]   *If K is a simplicial set and σ is a degenerate simplex of K, then there is a unique nondegenerate simplex τ of K and a unique iterated degeneracy map α such that $\alpha(\tau) = \sigma$.*

More generally, we can replace the category of sets with other categories to obtain more general simplicial objects.

Definition 2.1.6 Let C be a category. A *simplicial object* in C is a functor $\Delta^{op} \to C$.

Example 2.1.7 Let $\mathcal{G}p$ denote the category of groups and group homomorphisms. Then a *simplicial group* is a functor $\Delta^{op} \to \mathcal{G}p$.

Example 2.1.8 Let *SSets* denote the category of simplicial sets. Then a *bisimplicial set* is a functor $\Delta^{op} \to SSets$. Such an object can be regarded as a two-dimensional simplicial diagram of sets, with both horizontal and vertical face and degeneracy maps.

We also have the dual notion of cosimplicial objects.

Definition 2.1.9 Let C be a category. A *cosimplicial object* in C is a functor $\Delta \to C$.

A cosimplicial object has coface and codegeneracy operators which go in the reverse direction to the face and degeneracy operators in a simplicial object; namely, they go in the same direction as the original morphisms in Δ.

2.2 Simplicial Sets as Models for Spaces

The original motivation for defining simplicial sets was to obtain a combinatorial model for spaces. In this section, we make this idea more explicit by showing that simplicial sets and topological spaces have equivalent homotopy theories. We begin by defining functors relating the two categories.

Between simplicial sets and topological spaces, there is a geometric realization functor $| - | \colon SSets \to \mathcal{T}op$. We follow the definition given in [62, I.2].

First recall the definition of the *standard (topological) n-simplex*

$$\Delta^n = \left\{ (t_0, \ldots, t_n) \in \mathbb{R}^{n+1} \,\middle|\, t_i \geq 0, \sum_{i=0}^{n} t_i = 1 \right\}$$

regarded as a subspace of $\mathbb{R}^{n+1}$. We consider linear maps $\delta \colon \Delta^n \to \Delta^m$ which are induced by morphisms $\delta \colon [n] \to [m]$ in the category Δ, as follows. Define

$$(\delta(t_0, \ldots, t_n))_i = \begin{cases} 0 & \delta^{-1}(i) = \varnothing \\ \sum_{j \in \delta^{-1}(i)} t_j & \delta^{-1}(i) \neq \varnothing. \end{cases}$$

Definition 2.2.1 Given any simplicial set K, its *simplex category* $\Delta \downarrow K$ is the category of simplicial sets over the object K whose objects are the maps $\sigma \colon \Delta[n] \to K$.

Lemma 2.2.2   [62, I.2.1] *Given any simplicial set K, there is an isomorphism*

$$K \cong \operatorname{colim}_{\Delta \downarrow K} \Delta[n].$$

Definition 2.2.3   Given a simplicial set K, its *geometric realization* $|K|$ is defined as the colimit

$$|K| = \operatorname{colim}_{\Delta \downarrow K} \Delta^n.$$

Observe that $|\Delta[n]| \cong \Delta^n$. We also have the *singular functor* $S : \mathcal{T}op \to \mathcal{SS}ets$ defined by, for any topological space Y,

$$S(Y)_n = \operatorname{Hom}_{\mathcal{T}op}(\Delta^n, Y).$$

Proposition 2.2.4   *The singular functor S is right adjoint to the geometric realization functor $|-|$.*

Proof   Let K be a simplicial set and Y a topological space. Using the definitions of the two functors, we have isomorphisms

$$\begin{aligned}
\operatorname{Hom}_{\mathcal{T}op}(|K|, Y) &\cong \operatorname{colim}_{\Delta \downarrow Y} \operatorname{Hom}_{\mathcal{T}op}(|\Delta[n]|, Y) \\
&\cong \operatorname{colim}_{\Delta \downarrow Y} \operatorname{Hom}_{\mathcal{SS}ets}(\Delta[n], S(Y)) \\
&\cong \operatorname{Hom}_{\mathcal{SS}ets}(K, S(Y)). \qquad \square
\end{aligned}$$

We can use this adjoint pair to define the model structure on simplicial sets. Define a map $f : K \to L$ of simplicial sets to be a *weak equivalence* if its geometric realization $|f| : |K| \to |L|$ is a weak homotopy equivalence of spaces.

Theorem 2.2.5   [69, 13.1.3, 13.1.13], [71, 3.6.5], [100, II.3] *There is a proper, cofibrantly generated model structure on the category of simplicial sets in which*

1 *the weak equivalences are the maps $f : K \to L$ such that $|f| : |K| \to |L|$ is a weak homotopy equivalence of topological spaces,*
2 *the fibrations are the Kan fibrations, or maps with the right lifting property with respect to the horn inclusions $V[n, k] \to \Delta[n]$ for all $n \geq 1$ and $0 \leq k \leq n$, and*
3 *the cofibrations are the monomorphisms.*

In particular all objects are cofibrant, and the fibrant objects are called Kan complexes. *A set of generating cofibrations is*

$$I = \{\partial \Delta[n] \to \Delta[n] \mid n \geq 0\}$$

and a set of generating acyclic cofibrations is

$$J = \{V[n, k] \to \Delta[n] \mid n \geq 1, 0 \leq k \leq n\}.$$

Implicit in this statement is the fact that the simplicial sets $\partial\Delta[n]$ and $V[n,k]$ are small; indeed, all simplicial sets are small [71, 3.1.1].

Theorem 2.2.6   [71, 3.6.7], [100, I.4] *The adjoint pair*

$$| - |: SSets \leftrightarrows \mathcal{T}op: S$$

defines a Quillen equivalence of model categories.

In fact, we can say more here. The category of simplicial sets is closed symmetric monoidal, where the monoidal product is given by the cartesian product, and in fact the model structure *SSets* is symmetric monoidal [71, 4.2.8]. Furthermore, the geometric realization and singular functors both preserve products, so the Quillen equivalence between simplicial sets and topological spaces preserves products.

Because simplicial sets are so closely related to topological spaces, we can define, for example, the homotopy groups of a simplicial set to be the homotopy groups of their geometric realization. For sufficiently nice simplicial sets, homotopy groups can be defined directly, without passing to topological spaces, but we do not need this construction here. As a consequence, we sometimes apply topological terminology to simplicial sets. For example, a simplicial set is *weakly contractible* if its homotopy groups are all trivial.

We conclude this section with a result about the category of simplices of a given simplicial set.

Proposition 2.2.7   [69, 18.9.3] *Let K be a simplicial set and $\Delta \downarrow K$ its category of simplices. Then K is weakly equivalent to the nerve of $\Delta \downarrow K$.*

2.3 Homotopy Limits and Homotopy Colimits

We now need to consider a homotopy invariant version of the categorical construction of limits and colimits. Here, we need only consider the case of diagrams of simplicial spaces, and references include the original treatment of Bousfield and Kan [37] or the online notes of Dugger [50]. A more general treatment for arbitrary model categories can be found in Hirschhorn's book [69, §§18, 19].

We begin with a classical motivating example. Consider the diagram of topological spaces

$$* \leftarrow S^{n-1} \rightarrow *$$

for any $n \geq 1$. Since both maps collapse all of S^{n-1} to a point, the pushout of this diagram is also just a point. However, if we consider the diagram of boundary inclusions

$$D^n \leftarrow S^{n-1} \rightarrow D^n$$

the pushout is the n-sphere S^n. Although the two diagrams are levelwise homotopy equivalent, we obtain different spaces as the pushout. The idea is that the second diagram is better behaved, as the maps are cofibrations. Thus, we consider the homotopy pushout of either diagram to be S^n.

In general, homotopy pushouts of topological spaces or simplicial sets can be obtained by replacing the maps by cofibrations. Dually, homotopy pullbacks can be obtained by replacing maps in pullback diagrams by fibrations. However, taking homotopy limits or homotopy colimits over more complicated diagrams is not so simple.

Definition 2.3.1   Let $\mathcal{D}$ be a small category, and let $X \colon \mathcal{D} \to SSets$ be a diagram. Then its *homotopy colimit* hocolim$_{\mathcal{D}} X$ is defined to be the coequalizer of the diagram

$$\coprod_{c \to d} X_c \times \mathrm{nerve}(d \downarrow \mathcal{D})^{op} \rightrightarrows \coprod_{c} X_c \times \mathrm{nerve}(c \downarrow \mathcal{D})^{op}.$$

The coproducts range over the morphisms and objects of the indexing category $\mathcal{D}$.

Remark 2.3.2   As presented by Bousfield and Kan [37] one can alternatively define the homotopy colimit via the diagonal of a simplicial resolution of the diagram X. More specifically, given a diagram $X \colon \mathcal{D} \to SSets$, its homotopy colimit hocolim$_{\mathcal{D}} X$ is the diagonal of the bisimplicial set

$$[k] \mapsto \coprod_{d_0 \to \cdots \to d_k} X(d_0).$$

Now we present the dual definition of homotopy limit.

Definition 2.3.3   Let $\mathcal{D}$ be a small category, and let $X \colon \mathcal{D} \to SSets$ be a diagram. Then its *homotopy limit* holim$_{\mathcal{D}} X$ is defined to be the equalizer of the diagram

$$\prod_{c} X_c \times \mathrm{nerve}(c \downarrow \mathcal{D}) \rightrightarrows \prod_{c \to d} X_d \times \mathrm{nerve}(c \downarrow \mathcal{D}).$$

The products range over the objects and morphisms of the indexing category $\mathcal{D}$.

Analogously to homotopy colimits, homotopy limits can also be defined by taking a total complex of a cosimplicial resolution of the diagram X. Here, we have given concrete definitions in the context of simplicial sets, but homotopy limits and colimits can be defined in any model category. In this general setting, we have the following relationship between limits and colimits and their homotopy variants.

Proposition 2.3.4 [69, 18.3.8] *Let $\mathcal{D}$ be a small category and $\mathcal{M}$ a model category, and let $X: \mathcal{D} \to \mathcal{M}$ be a diagram. There are natural maps*

$$\mathrm{hocolim}_{\mathcal{D}} X \to \mathrm{colim}_{\mathcal{D}} X$$

and

$$\lim_{\mathcal{D}} X \to \mathrm{holim}_{\mathcal{D}} X.$$

We state a few general results that we will need later on.

Proposition 2.3.5 [69, 18.1.6] *Suppose $\mathcal{D}$ is a small category and $P: \mathcal{D} \to SSets$ is the constant diagram taking every object of $\mathcal{D}$ to $\Delta[0]$. Then $\mathrm{hocolim}_{\mathcal{D}} P$ is naturally isomorphic to* $\mathrm{nerve}(\mathcal{D}^{op})$.

Proposition 2.3.6 [69, 19.4.2]

1 Let $\mathcal{M}$ be a model category and C a small category. If $f: X \to Y$ is a map of C-diagrams in $\mathcal{M}$ such that each $f_\alpha: X_\alpha \to Y_\alpha$ is a weak equivalence of fibrant objects for every object α of C, then the induced map

$$\mathrm{holim}_C X \to \mathrm{holim}_C Y$$

is a weak equivalence of fibrant objects of $\mathcal{M}$.
2 Let $\mathcal{M}$ be a model category and C a small category. If $f: X \to Y$ is a map of C-diagrams in $\mathcal{M}$ such that each $f_\alpha: X_\alpha \to Y_\alpha$ is a weak equivalence of cofibrant objects for every object α of C, then the induced map

$$\mathrm{hocolim}_C X \to \mathrm{hocolim}_C Y$$

is a weak equivalence of cofibrant objects of $\mathcal{M}$.

Proposition 2.3.7 [37, XI, 4.3] *Let $X: C \times \mathcal{D} \to SSets$ be a functor. Then the three simplicial sets*

$$\mathrm{holim}_C(c \mapsto \mathrm{holim}_{\mathcal{D}} X_{c,-}),$$

$$\mathrm{holim}_{\mathcal{D}}(d \mapsto \mathrm{holim}_C X_{-,d}),$$

and

$$\mathrm{holim}_{C \times \mathcal{D}} X_{c,d}$$

are all weakly equivalent.

Definition 2.3.8 [69, 19.6.1] A functor $F: C \to \mathcal{D}$ between small categories is *homotopy initial* if, for every object α of $\mathcal{D}$, the simplicial set nerve$(F \downarrow \alpha)$ is contractible.

The following result is known as Quillen's Theorem A.

Theorem 2.3.9 [99], [69, 19.6.4] *Suppose $F: C \to \mathcal{D}$ is a homotopy initial functor between small categories. Then* nerve(F): nerve$(C) \to$ nerve$(\mathcal{D})$ *is a weak equivalence of simplicial sets.*

2.4 Simplicial Model Categories

The closed monoidal structure on *SSets* gives a notion of mapping objects between simplicial sets. More generally, a model category can be equipped with simplicial sets of morphisms between objects. To have such a structure, the category must first have an enrichment in simplicial sets, together with compatibility between the morphism sets and the mapping spaces.

Definition 2.4.1 A *simplicial category* is a category enriched in simplicial sets. In particular we have

1 for any objects X and Y of C, a simplicial set Map(X, Y),
2 for any objects X, Y, and Z of C, a composition map

$$\text{Map}(X, Y) \times \text{Map}(Y, Z) \to \text{Map}(X, Z),$$

3 for any object X of C a map $\Delta[0] \to \text{Map}(X, X)$ specifying the identity map, and
4 for any two objects X and Y of C, an isomorphism $\text{Map}(X, Y)_0 \cong \text{Hom}_C(X, Y)$ which is compatible with composition.

The spaces Map(X, Y) are often referred to as *mapping spaces*.

Remark 2.4.2 Observe that this terminology is potentially in conflict with that of the previous section, where one would expect a simplicial category to be a simplicial object in the category *Cat* of small categories. Indeed, the two notions are not identical. In particular, a simplicial object in *Cat* could be regarded as having a simplicial set of objects, which a category enriched in simplicial sets does not have. The two notions agree when we impose the additional condition on a simplicial object in *Cat* that all face and degeneracy maps are the identity on object sets. Throughout this book, the term "simplicial category" should be taken to be a category enriched in simplicial sets, so there should be no confusion caused by the abuse of terminology.

Definition 2.4.3 A *simplicial model category* is a model category M which is also a simplicial category, satisfying the following two axioms:

(MC6) For any objects X and Y of M and simplicial set K, there are objects $X \otimes K$ and Y^K of M together with isomorphisms of simplicial sets

$$\mathrm{Map}(X \otimes K, Y) \cong \mathrm{Map}(K, \mathrm{Map}(X, Y)) \cong \mathrm{Map}(X, Y^K)$$

which are natural in X, Y, and K.

(MC7) If $i\colon A \to B$ is a cofibration and $p\colon X \to Y$ is a fibration in M, then the induced map of simplicial sets

$$\mathrm{Map}(B, X) \to \mathrm{Map}(A, X) \times_{\mathrm{Map}(A, Y)} \mathrm{Map}(B, Y)$$

is a fibration which is a weak equivalence if either i or p is.

The map in axiom (MC7) is sometimes called the *pullback-corner map* of i and p. We can also use the following condition, using a *pushout-corner map*.

Proposition 2.4.4 [69, 9.3.7] *Axiom (MC7) is equivalent to the following condition: If $i\colon A \to B$ is a cofibration in M and $j\colon K \to L$ is an inclusion of simplicial sets, then the induced map*

$$A \otimes L \amalg_{A \otimes K} B \otimes K \to B \otimes L$$

is a cofibration in M which is a weak equivalence if either i or j is.

Example 2.4.5 The model structure for simplicial sets naturally has the structure of a simplicial model category. For K and L simplicial sets, we have $K \otimes L = K \times L$, and $\mathrm{Map}(K, L) = L^K$ is given by, for any $n \geq 0$,

$$\mathrm{Map}(K, L)_n = \mathrm{Hom}_{SSets}(K \times \Delta[n], L).$$

Observe that here the simplicial structure and the monoidal structure coincide.

The following result is a consequence of axiom (MC7).

Proposition 2.4.6 [69, 9.3.1] *Let $A \to B$ be a cofibration and X a fibrant object in a simplicial model category. Then the induced map*

$$\mathrm{Map}_M(B, X) \to \mathrm{Map}_M(A, X)$$

is a fibration of simplicial sets.

Let M be a simplicial model category, and consider, for objects X and Y in M, the mapping space $\mathrm{Map}_M(X, Y)$. In general, we can have a weak equivalence $X \to X'$ with $\mathrm{Map}(X, Y)$ not weakly equivalent to $\mathrm{Map}(X', Y)$, and similarly if Y is replaced by a weakly equivalent Y'. The problem here is similar to the one in defining the homotopy category; to get a homotopy invariant mapping space, we need to take a cofibrant replacement of X and a fibrant replacement for Y.

Definition 2.4.7   Let X and Y be objects in a simplicial model category. Define the *homotopy mapping space* to be $\mathrm{Map}^h_{\mathcal{M}}(X, Y) = \mathrm{Map}_{\mathcal{M}}(X^c, Y^f)$.

Mapping spaces and homotopy mapping spaces can also be defined for model categories which are not simplicial. One can either use the theory of simplicial and cosimplicial resolutions, as in chapter 17 of [69], or the mapping objects from the simplicial localization $\mathcal{L}\mathcal{M}$ which we define in Chapter 4 herein.

We conclude with a couple of nice facts about homotopy mapping spaces in model categories, simplicial or not. The first result assumes that the homotopy mapping space has been constructed using resolutions.

Proposition 2.4.8   [69, 17.4.3] *If X and Y are objects in a model category $\mathcal{M}$, then the homotopy mapping space $\mathrm{Map}^h_{\mathcal{M}}(X, Y)$ is a fibrant simplicial set.*

The following result can be interpreted as saying that the definition of homotopy function complex is the correct one.

Proposition 2.4.9   [69, 17.7.2] *Let X and Y be objects in a model category $\mathcal{M}$. Then*

$$\pi_0 \, \mathrm{Map}^h_{\mathcal{M}}(X, Y) = \mathrm{Hom}_{\mathrm{Ho}(\mathcal{M})}(X, Y).$$

In particular, just as one shows that, up to equivalence of categories, the homotopy category of a model category does not depend on the fibrations and cofibrations, but only on the weak equivalences, the homotopy type of a homotopy function complex only depends on the weak equivalences of the model category. For more details, see [69, §17.7].

We conclude this section with a result about the preservation of homotopy limits under taking homotopy mapping spaces. Observe that we can take ordinary mapping spaces due to the conditions on the objects involved.

Proposition 2.4.10   [69, 19.4.4] *Let $\mathcal{M}$ be a model category and C a small category.*

1 If X is a cofibrant object of $\mathcal{M}$ and $Y \colon C \to \mathcal{M}$ is a diagram of fibrant objects of $\mathcal{M}$, then there is a natural weak equivalence

$$\mathrm{Map}_{\mathcal{M}}(X, \mathrm{holim}_C \, Y_\alpha) \simeq \mathrm{holim}_C \, \mathrm{Map}_{\mathcal{M}}(X, Y_\alpha).$$

2 *If Y is a fibrant object of M and X : C → M is a diagram of cofibrant objects of M, then there is a natural weak equivalence*

$$\mathrm{Map}_{\mathcal{M}}(\mathrm{hocolim}_C X_\alpha, Y) \simeq \mathrm{holim}_C \mathrm{Map}_{\mathcal{M}}(X_\alpha, Y).$$

2.5 Simplicial Spaces

Consider the category $SSets^{\Delta^{op}} = Sets^{\Delta^{op} \times \Delta^{op}}$ of bisimplicial sets. In light of the comparison between simplicial sets and topological spaces, we abuse terminology and simply call them *simplicial spaces*. We can regard $SSets$ as a subcategory of $SSets^{\Delta^{op}}$, via the functor that takes a simplicial set K to the constant simplicial space with K at each level and all face and degeneracy maps the identity. We also denote this constant simplicial space by K.

There is another functor $SSets \to SSets^{\Delta^{op}}$ which takes a simplicial set K to a simplicial space K^t, where K^t_n is given by the discrete simplicial set K_n. The notation is meant to suggest that this simplicial space is the "transpose" of the constant one, since it is constant in the other simplicial direction. In particular, for each $n \geq 0$ the simplicial space $\Delta[n]^t$ is the appropriate generalization of the representable functor $\Delta[n]$ to the context of simplicial spaces. For this reason, it is sometimes denoted as a "free" object $F(n)$, for example in [103]. A consequence of this fact is that $\mathrm{Map}(\Delta[n]^t, X) \cong X_n$ for every $n \geq 0$.

Note that, in the case of the simplicial set $\Delta[0]$, or any discrete simplicial set, the two ways of considering it as a simplicial space agree. Although it is perhaps unnecessary, we sometimes continue to use the notation $\Delta[0]^t$ in a situation where it occurs as a special case of $\Delta[n]^t$ for more general n.

The category of simplicial spaces has a simplicial structure, where for a simplicial set K and simplicial space X, we have $(K \times X)_n = K \times X_n$. For simplicial spaces X and Y the simplicial set $\mathrm{Map}(X, Y)$ is given by the adjoint relation

$$\mathrm{Map}(X, Y)_n = \mathrm{Hom}_{SSets}(\Delta[n], \mathrm{Map}(X, Y)) = \mathrm{Hom}_{SSets^{\Delta^{op}}}(X \times \Delta[n], Y).$$

To define a model structure on the category of simplicial spaces, a natural choice for weak equivalences is that they be levelwise, so that a map $f : X \to Y$ is a weak equivalence of simplicial spaces if each $f_n : X_n \to Y_n$ is a weak equivalence of simplicial sets for each $n \geq 0$. We can then ask that either the fibrations or the cofibrations be defined levelwise; it is not possible to define them both in this way, since we would not get the necessary lifting conditions. We begin with the case where cofibrations are defined levelwise.

Theorem 2.5.1 [62, VIII, 2.4] *There is a simplicial cofibrantly generated* *model structure on the category of simplicial spaces where weak equivalences and cofibrations are levelwise, and fibrations have the right lifting property with respect to the maps which are both cofibrations and weak equivalences.*

We call this model structure the *injective model structure* and denote it by $SSets_c^{\Delta^{op}}$. Although this model structure is cofibrantly generated, the proof that it is so does not provide explicit generating sets of cofibrations and generating acyclic cofibrations.

Theorem 2.5.2 [69, 11.6.1, 13.1.4] *There is a proper simplicial cofibrantly generated model structure on the category of simplicial spaces where the weak equivalences and fibrations are given levelwise, and the cofibrations have the left lifting property with respect to the maps which are both fibrations and weak equivalences.*

We call this model structure the *projective model structure* and denote it by $SSets_f^{\Delta^{op}}$. Let us use the cofibrantly generated structure on $SSets$ to describe sets of generating cofibrations and generating acyclic cofibrations for the projective structure.

A cofibration should be a map with the left lifting property with respect to the acyclic fibrations. Here, acyclic fibrations are defined levelwise, so a map of simplicial spaces $X \to Y$ is an acyclic fibration if and only if, for each $n \geq 0$, the map $X_n \to Y_n$ is an acyclic fibration of simplicial sets. We know that such a map is characterized by the existence of a lift in any diagram of the form

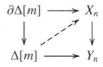

where $m \geq 0$. However, we can use the isomorphism $X_n \cong \mathrm{Map}(\Delta[n]^t, X)$ to ask instead for a lift in the diagram

$$
\begin{array}{ccc}
\partial\Delta[n] & \longrightarrow & \mathrm{Map}(\Delta[n]^t, X) \\
\downarrow & \nearrow & \downarrow \\
\Delta[m] & \longrightarrow & \mathrm{Map}(\Delta[n]^t, Y).
\end{array}
$$

Then we can apply the adjunction between mapping spaces and cartesian product to see that the existence of a lift in the above diagram of simplicial sets is equivalent to the existence of a lift in the diagram of simplicial spaces

$$
\begin{array}{ccc}
\partial\Delta[n] \times \Delta[n]^t & \longrightarrow & X \\
\downarrow & \nearrow & \downarrow \\
\Delta[m] \times \Delta[n]^t & \longrightarrow & Y
\end{array}
$$

for any $n, m \geq 0$. Thus, we can take

$$
\{\partial\Delta[m] \times \Delta[n]^t \to \Delta[m] \times \Delta[n]^t \mid m, n \geq 0\}
$$

as a set of generating cofibrations for $SSets_f^{\Delta^{op}}$. Similarly, we can take

$$
\{V[m, k] \times \Delta[n]^t \to \Delta[m] \times \Delta[n]^t \mid m \geq 1, 0 \leq k \leq m, n \geq 0\}
$$

as a set of generating acyclic cofibrations.

In either model structure on simplicial spaces, we have the following result, which is a special case of a more general fact on diagram categories.

Proposition 2.5.3   *Any simplicial space can be obtained as a homotopy colimit of simplicial spaces of the form $\Delta[n]^t$.*

The following corollary perhaps seems unnecessary but is useful in a later proof.

Corollary 2.5.4   *Any simplicial space can be obtained as a homotopy colimit of simplicial spaces of the form $\Delta[n]^t \times K$, where K is a simplicial set.*

2.6 The Reedy Model Structure on Simplicial Spaces

There is a third way to define a model structure with levelwise weak equivalences on the category of simplicial spaces, using the fact that Δ^{op} has the structure of a *Reedy category*. The theory of Reedy model structures was first developed by Reedy [101]. We give a summary here for the case of simplicial spaces. For more general simplicial objects, see Reedy's original paper [101]; for the more general theory of Reedy categories, see, for example, chapter 15 of [69].

Let Δ_n denote the full subcategory of Δ with objects $[0], [1], \ldots, [n]$. Let $SSets^{\Delta_n^{op}}$ be the category of all functors $\Delta_n^{op} \to SSets$. The inclusion functor $\Delta_n \to \Delta$ induces a truncation functor $tr_n \colon SSets^{\Delta^{op}} \to SSets^{\Delta_n^{op}}$ which has a left adjoint s_n and a right adjoint c_n.

Definition 2.6.1   Given a simplicial space X, define its *n-skeleton* to be the simplicial space $\mathrm{sk}_n(X) = s_n \circ tr_n(X)$. Define its *n-coskeleton* to be the simplicial space $\mathrm{cosk}_n(X) = c_n \circ tr_n(X)$.

Proposition 2.6.2   *For any $n \geq 0$, the pair $(\mathrm{sk}_n, \mathrm{cosk}_n)$ forms an adjoint pair of functors from the category $SSets^{\Delta^{op}}$ to itself.*

Proof   Let X and Y be simplicial spaces. Let $\mathrm{Hom}(-, -)$ denote a morphism set in the category of simplicial spaces and $\mathrm{Hom}_n(-, -)$ a morphism set in the category of *n*-truncated simplicial spaces. Then we have isomorphisms

$$\mathrm{Hom}(\mathrm{sk}_n X, Y) = \mathrm{Hom}(s_n \circ tr_n(X), Y)$$
$$\cong \mathrm{Hom}_n(tr_n(X), tr_n(Y))$$
$$\cong \mathrm{Hom}(X, c_n \circ tr_n(Y))$$
$$= \mathrm{Hom}(X, \mathrm{cosk}_n Y). \qquad \qquad \square$$

Example 2.6.3   Let X be an arbitrary simplicial space. Its 0-skeleton is the constant simplicial space given by X_0 in each simplicial degree. Its 0-coskeleton has $(n + 1)$ copies of X_0 in each degree.

Example 2.6.4   If we look at the even simpler case of simplicial sets, the $(n - 1)$-skeleton of the *n*-simplex $\Delta[n]$ is its boundary $\partial\Delta[n]$. Thus, the $(n - 1)$-skeleton of the simplicial space $\Delta[n]^t$ is $\partial\Delta[n]^t$. Using adjointness, we observe that, for any $n \geq 1$,

$$\mathrm{cosk}_{n-1}(X)_n = \mathrm{Map}(\Delta[n]^t, \mathrm{cosk}_{n-1}(X))$$
$$\cong \mathrm{Map}(\mathrm{sk}_{n-1} \Delta[n]^t, X)$$
$$= \mathrm{Map}(\partial\Delta[n]^t, X).$$

This description of the *n*-simplices of the $(n - 1)$-coskeleton of a simplicial space will be useful in what follows.

Remark 2.6.5   One can generalize this definition to the case where $n = -1$. Then, for any simplicial space X, its (-1)-skeleton $\mathrm{sk}_{-1}(X)$ is the initial simplicial space $\varnothing$, and its (-1)-coskeleton $\mathrm{cosk}_{-1}(X)$ is the terminal simplicial space $* = \Delta[0]$.

These skeleton and coskeleton functors can be used to define the cofibrations and fibrations for a model structure on simplicial spaces.

Theorem 2.6.6   [69, 15.3.4], [101, A]   *There is a proper, cofibrantly generated, simplicial model structure, called the* Reedy *model structure, on the category of simplicial spaces such that:*

1 the weak equivalences are levelwise weak equivalences of simplicial sets,

2 the cofibrations are the maps $X \to Y$ such that the induced maps

$$\mathrm{sk}_{n-1}(Y)_n \amalg_{\mathrm{sk}_{n-1}(X)_n} X_n \to Y_n$$

are cofibrations in SSets for all $n \geq 0$, and

3 the fibrations are the maps $X \to Y$ such that the induced maps

$$X_n \to \mathrm{cosk}_{n-1}(X)_n \times_{\mathrm{cosk}_{n-1}(Y)_n} Y_n$$

are fibrations of simplicial sets for all $n \geq 0$.

The simplicial structure here is the same as for the other two model structures.

While we refer the reader to Reedy's original paper [101] for the proof, we would like to understand the cofibrantly generated structure by producing generating sets of cofibrations and acyclic cofibrations. Let us begin with the generating acyclic cofibrations.

A generating acyclic cofibration should be a map with the left lifting property with respect to the fibrations. Let us use our description of the fibrations to produce such a set, using the cofibrantly generated structure for simplicial sets. First let $n = 0$. Since the (-1)-coskeleton of any simplicial space is the terminal object $*$, a fibration must first satisfy the condition that $X_0 \to Y_0$ must be a fibration of simplicial sets. In other words, we must have a lift in any diagram of the form

$$
\begin{array}{ccc}
V[m,k] & \longrightarrow & X_0 \\
\downarrow & \nearrow & \downarrow \\
\Delta[m] & \longrightarrow & Y_0,
\end{array}
$$

where $m \geq 1$ and $0 \leq k \leq m$. Using the fact that $X_0 = \mathrm{Map}_{SSets^{\Delta^{op}}}(\Delta[0]^t, X)$ and adjointness of the cartesian product and mapping spaces, the existence of such a lift is equivalent to the existence of a lift in the diagram

$$
\begin{array}{ccc}
V[m,k] \times \Delta[0]^t & \longrightarrow & X \\
\downarrow & \nearrow & \downarrow \\
\Delta[m] \times \Delta[0]^t & \longrightarrow & Y.
\end{array}
$$

Thus we conclude that our generating set must include maps of the form

$$V[m,k] \times \Delta[0]^t \to \Delta[m] \times \Delta[0]^t$$

for all $m \geq 1$ and $0 \leq k \leq m$.

Now we proceed to other values of n. Let

$$P = \text{cosk}_{n-1}(X)_n \times_{\text{cosk}_{n-1}(Y)_n} Y_n.$$

Then for $X \to Y$ to be a fibration we need a lift in any diagram of the form

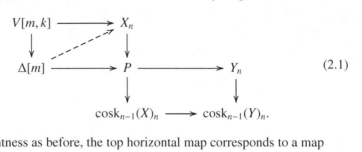

$$(2.1)$$

Using adjointness as before, the top horizontal map corresponds to a map

$$V[m,k] \times \Delta[n]^t \to X,$$

and similarly for the bottom horizontal map. The bottom horizontal map, however, is determined by two compatible maps, by the universal property for pullbacks. The first is $\Delta[m] \to Y_n$ which corresponds to the map

$$\Delta[m] \times \Delta[n]^t \to Y.$$

The second is $\Delta[m] \to \text{cosk}_{n-1}(X)_n = \text{Map}(\partial\Delta[n]^t, X)$, making use of Example 2.6.4, which corresponds to the map

$$\Delta[m] \times \partial\Delta[n]^t \to X.$$

Putting all this data together, we see that the existence of a lift in the previous diagram is equivalent to the existence of a lift in the diagram

$$
\begin{array}{ccc}
V[m,k] \times \Delta[n]^t \cup \Delta[m] \times \partial\Delta[n]^t & \longrightarrow & X \\
\downarrow & & \downarrow \\
\Delta[m] \times \Delta[n]^t & \longrightarrow & Y,
\end{array}
$$

where the union in the upper left-hand corner is taken over the overlap, which is $V[m,k] \times \partial\Delta[n]$.

Thus, a choice for the generating acyclic cofibrations is the set

$$J_R = \{V[m,k] \times \Delta[n]^t \cup \Delta[m] \times \partial\Delta[n]^t \to \Delta[m] \times \Delta[n]^t\}$$

where $n \geq 0$, $m \geq 1$, and $0 \leq k \leq m$. Observe that we have included $n = 0$ here; using the fact that $\partial\Delta[0] = \varnothing$, this description gives exactly the maps that we obtained above.

We claim that the generating cofibrations can be described analogously, so that we have

$$I_R = \{\partial\Delta[m] \times \Delta[n]^t \cup \Delta[m] \times \partial\Delta[n]^t \to \Delta[m] \times \Delta[n]^t\}$$

for all $m, n \geq 0$. However, to use the same procedure as above, we need to know the following result.

Proposition 2.6.7   [101, 1.4] *A map $f \colon X \to Y$ is an acyclic fibration of simplicial spaces if and only if, for every $n \geq 0$, the map*

$$X_n \to \mathrm{cosk}_{n-1}(X)_n \times_{\mathrm{cosk}_{n-1}(Y)_n} Y_n$$

is an acyclic fibration of simplicial sets.

Since Reedy proves the dual statement for acyclic cofibrations, we include here the proof, using the following two lemmas.

Lemma 2.6.8   [101, 1.1] *There is a pullback diagram*

$$
\begin{array}{ccc}
\mathrm{cosk}_n(X)_k & \longrightarrow & \mathrm{cosk}_{n-1}(X)_k \\
\downarrow & & \downarrow \\
\prod_{\alpha \colon [k]\to[n]} X_n & \longrightarrow & \prod_{\alpha \colon [k]\to[n]} \mathrm{cosk}_{n-1} X_n
\end{array}
$$

where the products are taken over all face maps $[k] \to [n]$ in Δ^{op} and the vertical maps are products of face maps.

The second lemma is dual to [101, 1.3].

Lemma 2.6.9   *Suppose, in any model category, that we have a commutative diagram*

$$
\begin{array}{ccc}
A_2 \longrightarrow A_1 \longleftarrow A_3 \\
\downarrow \qquad \downarrow \qquad \downarrow{\scriptstyle f_3} \\
B_2 \longrightarrow B_1 \longleftarrow B_3
\end{array}
$$

such that the maps f_3 and

$$A_2 \to B_2 \times_{B_1} A_1$$

are acyclic fibrations. Then the induced map on pullbacks

$$A_2 \times_{A_1} A_3 \to B_2 \times_{B_1} B_3$$

is an acyclic fibration.

Proof   The induced map on pullbacks can be written as the composite of acyclic fibrations

$$A_2 \times_{A_1} A_3 \to (B_2 \times_{B_1} A_1) \times_{A_1} A_3 \cong B_2 \times_{B_1} A_3 \to B_2 \times_{B_1} B_3. \qquad \square$$

Proof of Proposition 2.6.7   Suppose that $f\colon X \to Y$ is an acyclic fibration in the Reedy model structure. Then we know that each map $f_n\colon X_n \to Y_n$ is a weak equivalence of simplicial sets and that each map

$$X_n \to \operatorname{cosk}_{n-1}(X)_n \times_{\operatorname{cosk}_{n-1}(Y)_n} Y_n$$

is a fibration of simplicial sets. We need to show that the latter maps are also weak equivalences. We use induction.

Let $n = 0$. Then we know that $\operatorname{cosk}_{-1}(X) = * = \operatorname{cosk}_{-1}(Y)$. Therefore, we have that

$$X_0 \to \operatorname{cosk}_{-1}(X)_0 \times_{\operatorname{cosk}_{-1}(Y)_0} Y_0 = Y_0$$

is a weak equivalence by assumption. Observe also that the map $\operatorname{cosk}_{-1}(X)_k \to \operatorname{cosk}_{-1}(Y)_k$ is a weak equivalence (in fact, an isomorphism) for all $k \geq 0$.

Now suppose we know that the maps

$$X_n \to \operatorname{cosk}_{n-1}(X)_n \times_{\operatorname{cosk}_{n-1}(Y)_n} Y_n$$

and $\operatorname{cosk}_{n-1}(X)_k \to \operatorname{cosk}_{n-1}(Y)_k$ are weak equivalences, where $k \geq n$. Consider the commutative diagram

$$
\begin{array}{ccccc}
\prod_{\alpha\colon [k]\to[n]} X_n & \longrightarrow & \prod_{\alpha\colon [k]\to[n]} \operatorname{cosk}_{n-1}(X)_n & \longleftarrow & \operatorname{cosk}_{n-1}(X)_k \\
\downarrow & & \downarrow & & \downarrow \\
\prod_{\alpha\colon [k]\to[n]} Y_n & \longrightarrow & \prod_{\alpha\colon [k]\to[n]} \operatorname{cosk}_{n-1}(Y)_n & \longleftarrow & \operatorname{cosk}_{n-1}(Y)_k
\end{array}
$$

where, as in Lemma 2.6.8, the products are taken over face maps. Observe that, by the inductive hypothesis, the rightmost vertical map is an acyclic fibration, and that the map from the top left-hand corner to the pullback of the left-hand square is also an acyclic fibration. Therefore, by Lemma 2.6.9, the induced map on pullbacks, which is

$$\operatorname{cosk}_n(X)_k \to \operatorname{cosk}_n(Y)_k$$

by Lemma 2.6.8, is an acyclic fibration. In particular, we can take $k = n + 1$. Therefore, if we look at

$$X_{n+1} \to \operatorname{cosk}_n(X)_{n+1} \times_{\operatorname{cosk}_n(Y)_{n+1}} Y_{n+1} \to Y_{n+1},$$

the right-hand map is a weak equivalence. But the composite is a weak equivalence by assumption. Therefore, the left-hand map is also a weak equivalence, which is what we wanted to show.

Conversely, if each map

$$X_n \to \mathrm{cosk}_{n-1}(X)_n \times_{\mathrm{cosk}_{n-1}(Y)_n} Y_n$$

is an acyclic fibration of simplicial sets, then

$$X_n \to \mathrm{cosk}_{n-1}(X)_n \times_{\mathrm{cosk}_{n-1}(Y)_n} Y_n \to Y_n$$

is a composite of acyclic fibrations, establishing that $f \colon X \to Y$ is an acyclic fibration in the Reedy model structure. (Observe that the arguments above showing that the right-hand map is an acyclic fibration hold under our hypotheses for the converse.) $\square$

One of the most important properties of the Reedy model structure on the category of simplicial spaces is the following.

Theorem 2.6.10   [69, 15.8.7, 15.8.8] *The Reedy and injective model structures on the category of simplicial spaces are the same. In particular, Reedy cofibrations are precisely monomorphisms, and all objects are Reedy cofibrant.*

While the injective model structure is often helpful to consider due to the nice characterization of cofibrant objects, it can be problematic to use in practice because it is only known to be cofibrantly generated in an abstract sense, without explicit generating sets. While the Reedy model structure seems to have a more complicated structure, it has the benefit of easily described generating sets. The fact that the two model structures agree allows us to make use of both properties: having a nice cofibrantly generated structure and a simple description of cofibrations and cofibrant objects.

The following result and its corollaries will be used frequently in later chapters.

Proposition 2.6.11   *If X is a Reedy fibrant simplicial space, then the map*

$$(d_1, d_0) \colon X_1 \to X_0 \times X_0$$

is a fibration of simplicial sets.

Proof   Since X is Reedy fibrant, the map $X \to \Delta[0]$ is a fibration. Using the diagram (2.1) in this case, a lift exists in the diagram

$$
\begin{array}{ccc}
V[m,k] & \longrightarrow & X_1 \\
\downarrow & \nearrow & \downarrow \\
\Delta[m] & \longrightarrow & \mathrm{cosk}_0(X)_1.
\end{array}
$$

But $\mathrm{cosk}_0(X)_1 \cong X_0 \times X_0$, and the map to it from X_1 is precisely (d_1, d_0). $\qquad \square$

Corollary 2.6.12 *If X is Reedy fibrant, then the face maps $d_i \colon X_1 \to X_0$ for $i = 0, 1$ are fibrations.*

Proof Either face map $d_i \colon X_1 \to X_0$ can be written as the composite of (d_1, d_0) with the projection $\mathrm{pr}_i \colon X_0 \times X_0 \to X_0$, and both these maps are fibrations. $\qquad \square$

Taking iterations of the previous two results, we obtain the following.

Corollary 2.6.13 *For Reedy fibrant simplicial spaces, the iterated pullback of the diagram*

$$
X_1 \xrightarrow{d_0} X_0 \xleftarrow{d_1} \cdots \xrightarrow{d_0} X_0 \xleftarrow{d_1} X_1
$$

is a homotopy pullback. In particular, for the Segal map

$$
X_n \to \underbrace{X_1 \times_{X_0} \cdots \times_{X_0} X_1}_{n}
$$

the right-hand side is a homotopy pullback.

Although we do not go through the details here, we note that Δ is also a Reedy category, and the category of cosimplicial spaces has a model structure analogous to the model structure described above for simplicial spaces.

2.7 Combinatorial Model Categories

While we have already discussed cellular model categories, another convenient class of well-behaved model categories is given by those that are combinatorial. The following definition is originally due to J.H. Smith.

Definition 2.7.1 [49, 2.2] A category C is *locally presentable* if

1 it admits all small colimits, and
2 there is a set S of small objects of C such that any object of C can be obtained as the colimit of a small diagram with objects in S.

Definition 2.7.2 [49, 2.1] A model category $\mathcal{M}$ is *combinatorial* if it is cofibrantly generated and its underlying category is locally presentable.

Example 2.7.3 The model category $SSets$ is combinatorial.

We will have need of a recognition theorem for combinatorial model categories, for which we need some further terminology. In the following definition, we have again omitted some set-theoretic technicalities.

Definition 2.7.4 A category C is *accessible* if:

1 the category C has all filtered colimits, and
2 there is a set of compact objects that generate the category under directed colimits.

A functor between accessible categories is *accessible* if it preserves filtered colimits.

Given a category C, recall that we can take the category $C^{[1]}$ whose objects are the morphisms of C.

Definition 2.7.5 [21, 1.14] Let C be a locally presentable category and $\mathcal{W}$ a class of morphisms in C. Then $\mathcal{W}$ is an *accessible class of maps* if the full subcategory of $C^{[1]}$ whose objects are in $\mathcal{W}$ is an accessible category and is closed under filtered colimits.

Proposition 2.7.6 [21, 1.18] *Suppose C and $\mathcal{D}$ are locally presentable categories and W an accessible class in $\mathcal{D}$. Let $F: C^{[1]} \to \mathcal{D}^{[1]}$ be an accessible functor. Then the class of morphisms of C which are mapped to W by F also form an accessible class.*

We have the following specific example.

Proposition 2.7.7 [51, C.4] *The class of acyclic fibrations in $SSets$ is an accessible class of maps.*

Proof Consider the category $SSets^{[1]}$ of morphisms of simplicial sets and the functor $G: SSets^{[1]} \to Sets^{[1]}$ which sends a map $f: X \to Y$ to the pullback-corner map restricted to 0-simplices

$$\coprod_{n \geq 0} \left(X^{\Delta[n]} \right)_0 \to \left(X^{\partial\Delta[n]} \times_{Y^{\partial\Delta[n]}} Y^{\Delta[n]} \right)_0 .$$

Notice that G preserves filtered colimits because the functors $(-)^{\Delta[n]}$ and $(-)^{\partial\Delta[n]}$ do, and hence G is an accessible functor. The class of Kan acyclic fibrations is exactly the inverse image under G of the surjections. One can check that the category of surjections in *Sets* is an accessible class; hence by Proposition 2.7.6 the acyclic fibrations form an accessible class of maps. □

Theorem 2.7.8 [21, 1.7, 1.15] *Let C be a locally presentable category, $\mathcal{W}$ a class of morphisms, and I a set of morphisms of C. Suppose further that*

1 *the class $\mathcal{W}$ is closed under retracts and satisfies the two-out-of-three property,*
2 *the I-fibrations are in $\mathcal{W}$,*
3 *the class of I-cofibrations which are in $\mathcal{W}$ is closed under pushout and transfinite composition, and*
4 *$\mathcal{W}$ is an accessible class.*

Then C has a cofibrantly generated model structure in which $\mathcal{W}$ is the class of weak equivalences and I is a set of generating cofibrations.

2.8 Localized Model Categories

Sometimes we obtain a new model structure on a category by making more maps weak equivalences. In other words, we localize with respect to some set of maps which were not previously weak equivalences, but which we would like to be. Some assumptions need to be made so that the result is still a model category.

Recall that we have a notion of homotopy mapping spaces even in model categories which are not necessarily simplicial. We use these homotopy mapping spaces to give a definition of an object being local, in the sense that the source and target of a map do not distinguish that object differently. Then we consider all such maps which behave the same way with respect to local objects.

Definition 2.8.1 Let M be a model category and T a set of maps in M.

1 A *T-local object* in M is a fibrant object W in M such that the induced map

$$\mathrm{Map}^h(B, W) \to \mathrm{Map}^h(A, W)$$

is a weak equivalence for every $f\colon A \to B$ in T.
2 A *T-local equivalence* $g\colon X \to Y$ in M is a map such that

$$\mathrm{Map}^h(Y, W) \to \mathrm{Map}^h(X, W)$$

is a weak equivalence for every T-local object W.

If T consists of a single map f, then we call a T-local object f-*local*, and similarly consider f-*local equivalences*.

Theorem 2.8.2 [10, 4.7], [69, 4.1.1] *Let $\mathcal{M}$ be a left proper, cellular or combinatorial model category and T a set of maps in $\mathcal{M}$. There exists a model structure $\mathcal{L}_T \mathcal{M}$ on the same underlying category such that:*

1 the weak equivalences of $\mathcal{L}_T \mathcal{M}$ are the T-local equivalences in $\mathcal{M}$,
2 the cofibrations of $\mathcal{L}_T \mathcal{M}$ are precisely the cofibrations of $\mathcal{M}$, and
3 the fibrant objects of $\mathcal{L}_T \mathcal{M}$ are precisely the T-local objects of $\mathcal{M}$.

The model category $\mathcal{L}_T \mathcal{M}$ is left proper and cellular or combinatorial (agreeing with $\mathcal{M}$), and if $\mathcal{M}$ is a simplicial model category, then so is $\mathcal{L}_T \mathcal{M}$.

The model category $\mathcal{L}_T \mathcal{M}$ is often called the *left Bousfield localization* of $\mathcal{M}$ with respect to the set of maps T, or simply the *Bousfield localization*. There is a dual version, which is either called the right Bousfield localization, or the Bousfield colocalization, but we do not use it here.

Example 2.8.3 [69, §1.5] Consider the model category $\mathcal{T}op$ and, for some $n \geq 0$, the single boundary inclusion $f_n \colon S^{n-1} \to D^n$. Then a topological space Z is f_n-local precisely when $\pi_i(Z) = 0$ for all $i > n$ and every choice of basepoint. A map $g \colon X \to Y$ is an f_n-local equivalence if and only if g induces isomorphisms $\pi_i(X) \to \pi_i(Y)$ for all $i \leq n$. A fibrant replacement in this localized model structure is thus an nth Postnikov approximation.

Since a model category $\mathcal{L}_T \mathcal{M}$ obtained from localization in this way is still cofibrantly generated, we can assume that it has a functorial fibrant replacement functor L_T, for example constructed via the small object argument. Given any object X, the corresponding object $L_T X$ is considered to be a *localization* of X, since it is by definition T-local.

Proposition 2.8.4 [69, 4.3.6] *A morphism between T-local objects in $\mathcal{L}_T \mathcal{M}$ is a T-local equivalence if and only if it is a weak equivalence in $\mathcal{M}$.*

Proposition 2.8.5 *Let $\mathcal{M}$ be a model category and T a set of maps in $\mathcal{M}$. Suppose that $X \colon C \to \mathcal{M}$ is a diagram such that, for each object α of C, the object X_α of $\mathcal{M}$ is T-local. Then $\operatorname{holim}_\alpha X_\alpha$ is also T-local.*

Proof Let $A \to B$ be a map in T. Since each X_α is T-local, we know that we have isomorphisms $\operatorname{Map}^h(B, X_\alpha) \to \operatorname{Map}^h(A, X_\alpha)$. Since by Proposition 2.4.8 mapping spaces are fibrant, then we still have a weak equivalence

$$\operatorname{holim}_\alpha \operatorname{Map}^h(B, X_\alpha) \to \operatorname{holim}_\alpha \operatorname{Map}^h(A, X_\alpha)$$

by Proposition 2.3.6. But then we still have a weak equivalence when we commute the mapping space with the homotopy limit by Proposition 2.4.10, so we obtain a weak equivalence

$$\mathrm{Map}^h(B, \mathrm{holim}_\alpha\, X_\alpha) \to \mathrm{Map}^h(A, \mathrm{holim}_\alpha\, X_\alpha).$$

Varying over all maps in T, we conclude that $\mathrm{holim}_\alpha\, X_\alpha$ is T-local by definition. □

Proposition 2.8.6 *Suppose $\mathcal{M}_1$ and $\mathcal{M}_2$ are model structures on the same underlying category with the same class of weak equivalences but different fibrations and cofibrations. Let T be a set of maps in this underlying category. Then the weak equivalences in $\mathcal{L}_T \mathcal{M}_1$ and $\mathcal{L}_T \mathcal{M}_2$ agree.*

Proof Recall, as discussed in the paragraph after Proposition 2.4.9, that homotopy mapping spaces are defined in terms of the underlying category and the weak equivalences and do not depend on the fibrations and cofibrations. Using the definition of local equivalence, it follows that the weak equivalences of a localized model category $\mathcal{L}_T \mathcal{M}$ are determined only by the set T and by the weak equivalences of the original model category $\mathcal{M}$. □

The following result is helpful for characterizing fibrant objects in a localized model structure. Note, however, that the set of maps produced is only sufficient for identifying fibrant objects; it need not be a set of generating cofibrations for characterizing all fibrations.

Proposition 2.8.7 [69, §4.2] *Let $\mathcal{L}_T \mathcal{M}$ be a localization of a simplicial model category $\mathcal{M}$, where the maps in T are cofibrations between cofibrant objects. Then an object Z is T-local if and only if it has the right lifting property with respect to the set of maps*

$$\{A \otimes \Delta[m] \cup_{A \otimes \partial\Delta[m]} B \otimes \partial\Delta[m] \to B \otimes \Delta[m] \mid A \to B \text{ in } T, m \geq 0\}.$$

Proof An object Z is T-local if and only if, for every map $A \to B$ in T, the induced map

$$\mathrm{Map}(B, Z) \to \mathrm{Map}(A, Z)$$

is an acyclic fibration of simplicial sets. (Since Z is fibrant and A and B are assumed to be cofibrant, the homotopy mapping space is just the mapping space.) This condition is equivalent to the existence of a lift in any diagram of the form

$$
\begin{array}{ccc}
\partial\Delta[m] & \longrightarrow & \mathrm{Map}(B, Z) \\
\downarrow & \nearrow & \downarrow \\
\Delta[m] & \longrightarrow & \mathrm{Map}(A, Z)
\end{array}
$$

where $m \geq 0$. Using the simplicial structure on $\mathcal{M}$, the existence of that lift is equivalent to a lift in the diagram

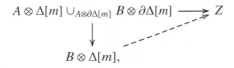

completing the proof. □

Under certain conditions, we can also think of applying localization to a Quillen pair of model categories. Here we apply the left derived functor of a left Quillen functor to the set of maps with respect to which we localize.

Theorem 2.8.8 [69, 3.3.20] *Suppose that $\mathcal{M}$ and $\mathcal{N}$ are model categories and $F: \mathcal{M} \rightleftarrows \mathcal{N}: G$ is a Quillen pair. Suppose that T is a class of maps in $\mathcal{M}$ such that the localized model structure $\mathcal{L}_T \mathcal{M}$ exists, and furthermore that the localized model structure $\mathcal{L}_{\mathbb{L}FT} \mathcal{N}$ exists.*

1 The adjoint pair (F, G) still defines a Quillen pair

$$F: \mathcal{L}_T \mathcal{M} \rightleftarrows L_{\mathbb{L}FT} \mathcal{N}: G.$$

2 If (F, G) defines a Quillen equivalence between $\mathcal{M}$ and $\mathcal{N}$, then it also defines a Quillen equivalence on the localized model structures.

We also need a version of local objects in which we consider isomorphisms, rather than weak equivalences, on mapping spaces.

Definition 2.8.9 Let $\mathcal{M}$ be a simplicial model category, and let T be a set of cofibrations in $\mathcal{M}$. A fibrant object Y of $\mathcal{M}$ is *strictly T-local* if, for every morphism $f: A \to B$ in T, the induced map on mapping spaces

$$f^*: \mathrm{Map}(B, Y) \to \mathrm{Map}(A, Y)$$

is an isomorphism of simplicial sets. A map $g: C \to D$ in $\mathcal{M}$ is a *strict T-local equivalence* if, for every strictly T-local object Y in $\mathcal{M}$, the induced map

$$g^*: \mathrm{Map}(D, Y) \to \mathrm{Map}(C, Y)$$

is an isomorphism of simplicial sets.

We use this definition specifically in the projective model structure on simplicial spaces, and the following lemma is useful.

Lemma 2.8.10 [28, 5.6] *Consider the category of all simplicial spaces* $SSets^{\Delta^{op}}$ *and the full subcategory of strictly local diagrams with respect to the set of maps* $T = \{f : A \to B\}$. *The forgetful functor from the category of strictly local diagrams to the category of all diagrams has a left adjoint.*

Proof Without loss of generality, assume that we have just one map f in T; otherwise replace f by $\coprod_\alpha f_\alpha$. Suppose we have a diagram X in $SSets^{\Delta^{op}}$ which is not strictly local, so that the map

$$f^* : \operatorname{Map}(B, X) \to \operatorname{Map}(A, X)$$

is not an isomorphism. We first modify X to an object for which f^* is surjective. Define X' to be the pushout in the diagram

$$
\begin{array}{ccc}
\coprod_{n \geq 0} \coprod_{A \times \Delta[n] \to X} A \times \Delta[n] & \longrightarrow & X \\
\downarrow & & \downarrow \\
\coprod_{n \geq 0} \coprod_{A \times \Delta[n] \to X} B \times \Delta[n] & \longrightarrow & X',
\end{array}
$$

where each coproduct is taken over all maps $A \times \Delta[n] \to X$ for each $n \geq 0$. Then we further modify X so that f^* is injective; define X'' to be the pushout

$$
\begin{array}{ccc}
\coprod_{n \geq 0} \coprod (B \coprod_A B) \times \Delta[n] & \longrightarrow & X' \\
\downarrow & & \downarrow \\
\coprod_{n \geq 0} \coprod B \times \Delta[n] & \longrightarrow & X'',
\end{array}
$$

again where the second coproduct is over all maps $(B \coprod_A B) \times \Delta[n] \to X'$, and where the map

$$B \coprod_A B \to B$$

is the fold map.

In the construction of X', for any strictly local object Y we obtain a pullback diagram

$$
\begin{array}{ccc}
\operatorname{Map}(X', Y) & \longrightarrow & \operatorname{Map}(\coprod B, Y) \\
\cong \downarrow & & \downarrow \cong \\
\operatorname{Map}(X, Y) & \longrightarrow & \operatorname{Map}(\coprod A, Y)
\end{array}
$$

showing that the map $X \to X'$ is a strict local equivalence since $f : A \to B$ is.

In the construction of X'', we obtain a similar diagram. To show that the map

$X' \to X''$ is a strict local equivalence, we first consider the pullback diagram

$$\begin{array}{ccc} \mathrm{Map}(X'', Y) & \longrightarrow & \mathrm{Map}(\coprod B, Y) \\ \downarrow & & \downarrow \\ \mathrm{Map}(X', Y) & \longrightarrow & \mathrm{Map}(\coprod(B \coprod_A B), Y). \end{array}$$

It suffices to show that the right-hand vertical arrow is an isomorphism.

Recall that the object $B \coprod_A B$ is defined as the pushout in the diagram

$$\begin{array}{ccc} A & \longrightarrow & B \\ \downarrow & & \downarrow \\ B & \longrightarrow & B \coprod_A B, \end{array}$$

which enables us to look at the pullback diagram

$$\begin{array}{ccc} \mathrm{Map}(B \coprod_A B, Y) & \longrightarrow & \mathrm{Map}(B, Y) \\ \downarrow & & \downarrow{\scriptstyle\cong} \\ \mathrm{Map}(B, Y) & \longrightarrow & \mathrm{Map}(A, Y). \end{array}$$

Hence the map

$$B \to B \coprod_A B$$

is a strict local equivalence. But, this map fits into a composite

$$B \underset{id}{\Longrightarrow} B \coprod_A B \longrightarrow B.$$

Since the identity map is a strict local equivalence, it follows that the map

$$B \coprod_A B \to B$$

is a strict local equivalence, since it can be shown that the strictly local equivalences satisfy the two-out-of-three property.

Therefore, we obtain a composite map $X \to X''$ which is a strict local equivalence. However, we still do not know that the map

$$\mathrm{Map}(B, X'') \to \mathrm{Map}(A, X'')$$

is an isomorphism. So, we repeat this process, taking a (possibly transfinite) colimit to obtain a strictly local object $\widetilde{X}$ such that there is a local equivalence $X \to \widetilde{X}$. If necessary, take a fibrant replacement of $\widetilde{X}$.

It remains to show that the functor which takes a diagram X to the local diagram $\widetilde{X}$ is left adjoint to the forgetful functor. So if J is the forgetful functor from the category of strictly local diagrams to the category of all diagrams and K is the functor we have just defined, we claim that

$$\text{Map}(X, JY) \cong \text{Map}(KX, Y)$$

for any diagram X and strictly local diagram Y. But, proving this statement is equivalent to showing that

$$\text{Map}(X, Y) \cong \text{Map}(\widetilde{X}, Y),$$

which was shown above for each step, and still holds for the colimit. In particular, the map $X \to \widetilde{X} = KX$ and the identity $Y = JY$ induce natural isomorphisms

$$\text{Map}(KX, Y) \to \text{Map}(X, Y) \to \text{Map}(X, JY),$$

and the restriction of this composite to the 0-simplices of each object,

$$\text{Hom}(KX, Y) \to \text{Hom}(X, JY),$$

is exactly the isomorphism we need to show that K is left adjoint to J.   □

Lemma 2.8.11   [29, 4.1] *Let L be a strict localization functor for a model category M. Given a small diagram of objects X_α of M,*

$$L(\text{hocolim } X_\alpha) \simeq L \text{ hocolim } L(X_\alpha).$$

Proof   It suffices to show that, for any strictly local object Y, there is a weak equivalence of simplicial sets

$$\text{Map}(L(\text{hocolim}_\alpha LX_\alpha), Y) \simeq \text{Map}(L \text{ hocolim}_\alpha X_\alpha, Y).$$

This fact follows from the following series of weak equivalences:

$$\begin{aligned}
\text{Map}(L \text{ hocolim}_\alpha LX_\alpha, Y) &\simeq \text{Map}(\text{hocolim}_\alpha LX_\alpha, Y) \\
&\simeq \text{holim}_\alpha \text{Map}(LX_\alpha, Y) \\
&\simeq \text{holim}_\alpha \text{Map}(X_\alpha, Y) \\
&\simeq \text{Map}(\text{hocolim}_\alpha X_\alpha, Y) \\
&\simeq \text{Map}(L \text{ hocolim}_\alpha X_\alpha, Y).
\end{aligned}$$

□

2.9 Cartesian Model Categories

For a model structure on a category which is monoidal for the cartesian product, we can consider the compatibility of the model structure with the monoidal structure. Here we will only be interested in the special case in which the monoidal product is the cartesian, or categorical, product. In this case, we have the following rewording of the definition of a closed monoidal category.

Definition 2.9.1 A category C is *cartesian closed* if it has finite products and, for any two objects X and Y of C, an internal function object Y^X, together with a natural isomorphism

$$\text{Hom}_C(Z, Y^X) \cong \text{Hom}_C(Z \times X, Y)$$

for any third object Z of C.

In other words, a category C with finite products is cartesian closed if, for any object X of C, the functor $X \times -: C \to C$ has a right adjoint, which we denote by $(-)^X$.

If a cartesian closed category additionally has a model structure, we can ask if these two structures are compatible, in the following sense.

Definition 2.9.2 [102, 2.2] A model category $\mathcal{M}$ is *cartesian* if its underlying category is cartesian closed, its terminal object is cofibrant, and the following equivalent conditions hold.

1 If $f: A \to A'$ and $g: B \to B'$ are cofibrations in $\mathcal{M}$, then the induced map

$$h: A \times B' \amalg_{A \times B} A' \times B \to A' \times B'$$

is a cofibration. If either f or g is a weak equivalence, then so is h.
2 If $f: A \to A'$ is a cofibration and $p: X' \to X$ is a fibration in $\mathcal{M}$, then the induced map

$$q: (X')^{A'} \to (X')^A \times_{X^A} X^{A'}$$

is a fibration. If either f or p is a weak equivalence, then so is q.

We refer to the first of these conditions as the *pushout-product condition*, just as in the analogous case of a simplicial model category.

Example 2.9.3 The model category of simplicial sets is cartesian, where for any simplicial sets X and Y, $Y^X = \text{Map}(X, Y)$ is defined by

$$(Y^X)_n = \text{Hom}(X \times \Delta[n], Y)$$

for any $k \geq 0$. In this case, the cartesian structure coincides with the simplicial structure.

Example 2.9.4   The Reedy and projective model structures on the category of simplicial spaces are cartesian, where the internal hom object Y^X is given by

$$(Y^X)_n = \mathrm{Map}(\Delta[n]^t, Y^X) \cong \mathrm{Map}(X \times \Delta[n]^t, Y).$$

Using the definition of cartesian, one can check that, if W is a fibrant simplicial space in either model structure and X is any simplicial space, then W^X is also fibrant in the same model structure.

Following Rezk [103] we use the following convenient criterion for when a localization of the Reedy model structure on simplicial spaces is still cartesian closed.

Proposition 2.9.5   [103, 9.2]   *Consider the category of simplicial spaces equipped with the Reedy model structure, and let T be a set of maps between simplicial spaces. Suppose that for each T-local object W, the simplicial space $W^{\Delta[1]^t}$ is also T-local. Then the T-local model structure on the category of simplicial spaces is compatible with the cartesian closure.*

Proof   Let W be a T-local simplicial space; by hypothesis, $W^{\Delta[1]^t}$ is also T-local. Iterating this application of the hypothesis, we know that $W^{(\Delta[1]^t)^k}$ is T-local for any $k \geq 1$. Since for any k, the simplicial set $\Delta[k]$ is a retract of $(\Delta[1])^k$, we can conclude that the simplicial space $W^{\Delta[k]^t}$ is a retract of $W^{(\Delta[1]^t)^k}$. Therefore, $W^{\Delta[k]}$ is T-local for any k.

Recall from Corollary 2.5.4 that any simplicial space X can be obtained as a homotopy colimit (in the Reedy model structure) of simplicial spaces of the form $\Delta[k]^t \times K$ for K an arbitrary simplicial set. Therefore, any simplicial space W^X can be written as a homotopy limit of simplicial spaces of the form $W^{\Delta[k]^t \times K} = (W^{\Delta[k]^t})^K$.

The Reedy model structure on simplicial spaces is a simplicial model category, and this simplicial structure is retained after localizing with respect to T. Therefore, each object $(W^{\Delta[k]^t})^K$ must be T-local. Since T-local objects are preserved under homotopy limits by Proposition 2.8.5, we can conclude that any simplicial space of the form W^X, with W T-local and X arbitrary, is T-local.

It remains to establish the pushout-product condition. Suppose that $i\colon X \to Y$ is a cofibration and that $j\colon U \to V$ is a T-local acyclic cofibration. We need to prove that the induced map

$$U \times Y \coprod_{U \times X} V \times X \to V \times Y$$

is a T-local equivalence. It suffices to prove that, for any T-local object W, the diagram of simplicial sets

$$\begin{array}{ccc} \mathrm{Map}(V \times Y, W) & \longrightarrow & \mathrm{Map}(V \times X, W) \\ \downarrow & & \downarrow \\ \mathrm{Map}(U \times Y, W) & \longrightarrow & \mathrm{Map}(U \times X, W) \end{array}$$

is a homotopy pullback square. However, this diagram is equivalent to

$$\begin{array}{ccc} \mathrm{Map}(V, W^Y) & \longrightarrow & \mathrm{Map}(V, W^X) \\ \downarrow & & \downarrow \\ \mathrm{Map}(U, W^Y) & \longrightarrow & \mathrm{Map}(U, W^X). \end{array}$$

Since we have established that both W^X and W^Y are T-local, the vertical maps are weak equivalences; we also know they are fibrations. Therefore the diagram is a homotopy pullback. $\qquad\qquad\square$

3

Topological and Categorical Motivation

The purpose of this chapter is to motivate the notion of $(\infty, 1)$-categories from two different perpectives: the first more topological and homotopy-theoretic in nature, and the second more categorical.

For the first perspective, we start with a well-known and well-understood construction, the nerve of a category, and use a weakness of this construction to motivate more refined approaches to thinking of a category as a simplicial object. In doing so, we obtain a way of thinking about what it means to be a "category up to homotopy". The two different ways in which we modify the nerve construction point toward two of the models for $(\infty, 1)$-categories and hint at the comparison between them.

We then instead consider the problem of understanding higher categories and give a definition of what an $(\infty, 1)$-category should be. In particular, this approach justifies the name of $(\infty, 1)$-categories, although they turn out to consist of the same information as categories up to homotopy.

3.1 Nerves of Categories

We have seen that simplicial sets are designed to model topological spaces. Simplicial sets are also closely related to categories, via the nerve, but the relationship is not quite so clean. However, understanding how to refine the relationship between simplicial sets and categories leads to interesting new ideas, in particular that of categories up to homotopy. In this section, we review the nerve functor and consider some key examples.

We recall from Example 2.1.4 the following definition.

Definition 3.1.1 The *nerve* of a small category C is the simplicial set nerve(C)

66

defined by

$$\text{nerve}(C)_n = \text{Hom}_{Cat}([n], C).$$

In other words, the n-simplices of the nerve are n-tuples of composable morphisms in C. In particular, the nerve of the category $[n]$ is just the n-simplex $\Delta[n]$.

Example 3.1.2 [69, 14.1.4] Let G be a group, regarded as a category with one object. Then $\text{nerve}(G)$ is a simplicial set whose geometric realization is the classifying space of G, denoted by BG.

Proposition 3.1.3 *If $F: C \to D$ is an equivalence of categories, then the induced map $\text{nerve}(F): \text{nerve}(C) \to \text{nerve}(D)$ is a weak equivalence of simplicial sets.*

Sketch of proof Using Proposition 1.1.32, there must be an inverse equivalence $G: D \to C$ together with natural isomorphisms $GF \cong \text{id}_C$ and $FG \cong \text{id}_D$. Then one can check that the induced map $\text{nerve}(G): \text{nerve}(D) \to \text{nerve}(C)$ is a homotopy inverse to $\text{nerve}(F)$. □

Let us look at the nerves of some simple categories.

Example 3.1.4 Consider $[0]$, the category with a single object and only the identity morphism, which looks like •, and let I be the category with two objects and a single isomorphism between them, depicted as • ↔ •. Then we can include $[0]$ into I (in two different but equivalent ways), and this functor defines an equivalence of categories. The nerves of the categories $[0]$ and I are both contractible.

However, the converse statement to Proposition 3.1.3 is false: a functor between categories which is not an equivalence may still induce a weak equivalence of nerves.

Example 3.1.5 Consider the category $[1]$ which is depicted as • → •. Then $\text{nerve}([1])$ is contractible, but $[1]$ is not equivalent to either $[0]$ or I from the previous example.

The problem here is that weak equivalences of simplicial sets are given by weak homotopy equivalences of spaces after geometric realization, so we do not remember in which direction a 1-simplex pointed. In particular, we do not remember whether or not a 1-simplex in the nerve came from an isomorphism in the original category. The isomorphism information is crucial, however, in the definition of equivalence of categories, as we saw in the above examples. If we assume that all morphisms are invertible, then the converse statement does follow.

Proposition 3.1.6   *A functor $F: C \to D$ between groupoids is an equivalence of categories if and only if* nerve(F): nerve$(C) \to$ nerve$(\mathcal{D})$ *is a weak equivalence of simplicial sets.*

In this case, one can prove the converse statement using the extra information that all morphisms in the groupoids are invertible. The lack of this information is precisely where the more general statement fails. This proposition suggests that when we think about simplicial sets as models for topological spaces, there is a closer connection to groupoids than to more general categories. We explore this idea from different perspectives throughout this chapter.

We consider two possible approaches for distinguishing between nerves of nonequivalent categories.

1 We can change our definition of weak equivalence between simplicial sets so that fewer morphisms are weak equivalences. We would like to do so in such a way that nerves of nonequivalent categories are not weakly equivalent in this new sense. We take this approach in Section 3.2.

2 We can take a more refined version of the nerve construction so that it does distinguish isomorphisms from other morphisms. The output of the new construction is a simplicial space rather than a simplicial set. We take this approach in Section 3.3.

3.2 Kan Complexes and Generalizations

Recall the following simplicial sets: the n-simplex $\Delta[n] = \mathrm{Hom}(-, [n])$; its boundary $\partial\Delta[n]$, obtained by leaving out the identity map $[n] \to [n]$; and, for any $0 \le k \le n$, the horn $V[n, k]$ obtained by removing the kth face from $\partial\Delta[n]$.

Example 3.2.1   When $n = 2$, we can think of $\partial\Delta[2]$, the boundary of the 2-simplex $\Delta[2]$, as

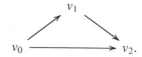

Then the horn $V[2, 0]$ can be depicted as

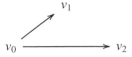

whereas $V[2, 1]$ looks like

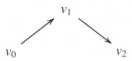

and $V[2, 2]$ looks like

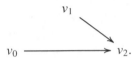

A basic question in the study of simplicial sets is when, given one of these horn diagrams in a simplicial set, it can be filled in to a full simplex.

Definition 3.2.2 A simplicial set X is a *Kan complex* if any map $V[n, k] \to X$ can be extended to a map $\Delta[n] \to X$. In other words, a lift exists in any diagram

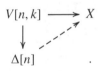

We have the following nice class of Kan complexes.

Proposition 3.2.3 *The nerve of a groupoid is a Kan complex.*

Idea of the proof Let G be a groupoid. When $n = 2$, having a lift in the diagram

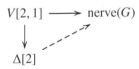

means that any pair of composable morphisms has a composite. In fact, such a lift exists for the nerve of any category, not just a groupoid.

However, the 1-simplices in the image of a map $V[2, 0] \to \text{nerve}(G)$ are not composable; they have a common source instead. Therefore, finding a lift to $\Delta[2] \rightarrow X$ is equivalent to the existence of a left inverse to the morphism giving rise to the 1-simplex $v_0 \rightarrow v_1$. (Here we use the vertex labels as given in Example 3.2.1.) Similarly, the existence of a lift in the $V[2, 2]$ case is equivalent to the existence of a right inverse to the morphism which is sent to the 1-simplex $v_1 \rightarrow v_2$. Since G is a groupoid, and hence all morphisms have (two-sided) inverses, all such lifts exist. □

However, we can actually say more about nerves of groupoids. Since composition in a category is unique, if a simplicial set K is the nerve of a groupoid, then the lifts along horns are actually unique.

Proposition 3.2.4 *A Kan complex K is the nerve of a groupoid if and only if the lift in each diagram*

is unique for every $0 \le k \le n$.

Thus we can think of Kan complexes as generalizations of groupoids, where lifts along horns are no longer unique. Thus, they can be thought of as "groupoids up to homotopy".

If we look at the model structure we have been considering on simplicial sets, the fibrant objects are exactly the Kan complexes. In particular, if we take the nerve of a category and then a fibrant replacement of it, we get something that behaves like the nerve of a groupoid, at least up to homotopy. Here we see a deeper explanation, then, of why our usual notion of weak equivalence of simplicial sets cannot recover enough information about categories from their nerves to know whether a 1-simplex came from an isomorphism or not.

How can we characterize simplicial sets of categories that are not necessarily groupoids? Let us return to the argument used in the sketch of the proof of Proposition 3.2.3. The fact that G was a groupoid was needed to get lifts along $V[2, k] \rightarrow \Delta[2]$ when $k = 0$ and $k = 2$, but not for $k = 1$. Considering higher values of n, we distinguish between *inner horns* $V[n, k]$, where $0 < k < n$, and *outer horns*, where $k = 0$ or $k = n$. Then a unique lift exists in any diagram

$$
\begin{array}{ccc}
V[n, k] & \longrightarrow & \text{nerve}(C) \\
\downarrow & \nearrow & \\
\Delta[n] & &
\end{array}
$$

for $0 \le k \le n$ if C is a groupoid but only for $0 < k < n$ if C is a category which is not necessarily a groupoid.

So, it seems that we want to consider simplicial sets which look like nerves of categories up to homotopy, in the same sense that Kan complexes can be thought of as groupoids up to homotopy. Thus, we make the following definition, first made by Boardman and Vogt [36].

Definition 3.2.5 A simplicial set K is an *inner Kan complex* or a *quasi-category* if a lift exists in any diagram

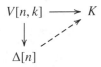

for $0 < k < n$.

Proposition 3.2.6 *A quasi-category is the nerve of a category if and only if the lifts in Definition 3.2.5 are all unique.*

What if we had another model structure on the category of simplicial sets where the fibrant objects were the quasi-categories? Then we would expect fewer morphisms to be weak equivalences, and nerves of categories would be fibrant. There is such a model structure which we discuss in Chapter 7. Since quasi-categories look like nerves of categories up to homotopy, it is sensible to think that they should model categories up to homotopy, or $(\infty, 1)$-categories.

Remark 3.2.7 We have chosen here to modify the weak equivalences in the category of simplicial sets, leading to the definition of quasi-categories. One could also change the notion of weak equivalence of categories, so that two categories are weakly equivalent if and only if their nerves are equivalent. Such a notion of weak equivalences was given by Thomason, and indeed he proved that there is a model structure on the category of small categories with these weak equivalences, and it is Quillen equivalent to the standard model structure on the category of simplicial sets, where the weak equivalences are the equivalences of categories [114].

3.3 Classifying Diagrams

We now turn to our second approach, namely, changing the nerve construction. The idea is to distinguish information about invertible morphisms from the rest of the morphisms of a category. The following definition is due to Rezk.

Definition 3.3.1 [103, 3.5] The *classifying diagram NC* of a category C is the simplicial space defined by

$$(NC)_n = \text{nerve}(\text{iso}(C^{[n]})),$$

where $C^{[n]}$ denotes the category of functors $[n] \to C$ whose objects are length-n chains of composable morphisms in C, and where iso denotes the maximal subgroupoid functor.

How does this definition help distinguish isomorphisms? When $n = 0$, $(NC)_0 = \text{nerve}(\text{iso}(C))$ is the nerve of the maximal subgroupoid of C. In particular, this simplicial set only detects information about isomorphisms in C. When $n = 1$, we have $(NC)_1 = \text{nerve}(\text{iso}(C^{[1]}))$. The objects of $\text{iso}(C^{[1]})$ are morphisms in C, and the morphisms of $\text{iso}(C^{[1]})$ are given by commutative squares

More generally, $(NC)_{n,m}$ is the set of diagrams of the form

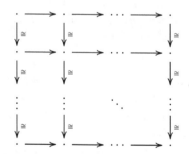

where there are n horizontal arrows in each row and m vertical arrows in each column.

If $C = G$ is a group, then applying this classifying diagram construction results in a simplicial space which is equivalent to the constant simplicial space consisting of the nerve of G at each level. In particular, since all morphisms are invertible, we obtain essentially no new information at level 1 that we did not have already at level 0. We illustrate with a basic example. For simplicity, in this section we denote the nerve of a group G by BG.

Example 3.3.2 Let $G = \mathbb{Z}/2$. Then $(NG)_0$ is just $B\mathbb{Z}/2$. The simplicial set $(NG)_1$ has two 0-simplices, given by the two elements of G. However, these two objects of $G^{[1]}$ are isomorphic, and the automorphism group of either one of them is $\mathbb{Z}/2$. Thus, $(NG)_1$ is also equivalent to $B\mathbb{Z}/2$.

If C is a groupoid with more than one object but only one isomorphism class of objects, then all objects of C have isomorphic automorphism groups. We denote the automorphism group of an object x in C by $\operatorname{Aut}(x)$. If we apply the classifying diagram construction, we obtain a simplicial space weakly equivalent to the constant simplicial space consisting of $B\operatorname{Aut}(x)$ at each level, for any representative object x.

If the groupoid C has more than one isomorphism class $\langle x \rangle$, then its classifying diagram is weakly equivalent to the constant simplicial space $\coprod_{\langle x \rangle} B\operatorname{Aut}(x)$.

We now turn to the case of a more general category C. Since in the classifying diagram NC, the simplicial set $(NC)_0$ picks out the isomorphisms of C only, we still obtain $\coprod_{\langle x \rangle} B\operatorname{Aut}(x)$ at level 0. However, if C is not a groupoid, then there is new information at level 1. It instead looks like

$$\coprod_{\langle x \rangle, \langle y \rangle} \coprod_{\langle \alpha \colon x \to y \rangle} B\operatorname{Aut}(\alpha)$$

where α ranges over the isomorphism classes of elements of $\operatorname{Hom}(x, y)$.

We give a more detailed treatment of these simplicial spaces in Theorem 9.4.1.

Let us return to the categories whose nerves we compared in Examples 3.1.4 and 3.1.5.

Example 3.3.3 Consider the category $[1] = (\bullet \to \bullet)$. If $\{e\}$ denotes the trivial group, then $(N[1])_0 \simeq B\{e\} \amalg B\{e\}$ and $(N[1])_1 \simeq B\{e\} \amalg B\{e\} \amalg B\{e\}$. In particular, $N[1]$ is not levelwise equivalent to the classifying diagram of the category $[0] = (\bullet)$, which would be the constant simplicial space $B\{e\}$. Thus, we can see that the classifying diagram is more refined than the nerve in distinguishing between these two categories.

Now recall the category I with two objects and a single isomorphism between them. Then $(NI)_0$ has the homotopy type of two points and $(NI)_1$ has the homotopy type of three points. In particular, the two spaces are not the same, and therefore the simplicial space NI is not levelwise equivalent to $N[1]$. Therefore we see that the classifying diagram distinguishes between $[1]$ and I, since it distinguishes the morphism of $[1]$ which is not an isomorphism.

In fact, we have the following more general result.

Proposition 3.3.4 *A functor $F\colon C \to \mathcal{D}$ is an equivalence of categories if and only if the induced map of classifying diagrams $N(F)\colon N(C) \to N(\mathcal{D})$ is a levelwise weak equivalence of simplicial sets.*

Proof Since each category iso($C^{[n]}$) is a groupoid, we can apply Proposition 3.1.6 to see that each $N(F)_n : N(C)_n \to N(\mathcal{D})_n$ is a weak equivalence if and only if each functor $F^{[n]} : C^{[n]} \to \mathcal{D}^{[n]}$ is an equivalence of categories. One can check that each $F^{[n]}$ is an equivalence of categories if and only if F is. □

We have several nice facts about classifying diagrams of categories.

1 The simplicial sets $(NC)_n$ are determined by $(NC)_0$ and $(NC)_1$, in that

$$(NC)_n \cong \underbrace{(NC)_1 \times_{(NC)_0} \cdots \times_{(NC)_0} (NC)_1}_{n}.$$

This property essentially arises from the composition of morphisms in C.

2 The subspace of $(NC)_1$ arising from isomorphisms in C is weakly equivalent to $(NC)_0$. We can see this fact more precisely by comparing the classifying diagrams of C and of iso(C). Consider the diagram of simplicial sets

$$
\begin{array}{ccc}
N \operatorname{iso}(C)_1 & \xrightarrow{\ i\ } & NC_1 \\
{\scriptstyle s_0}\big\uparrow{\scriptstyle \simeq} & & \big\uparrow{\scriptstyle s_0} \\
N \operatorname{iso}(C)_0 & \xrightarrow{\ =\ } & NC_0.
\end{array}
$$

The left vertical morphism is a weak equivalence, since iso(C) is a groupoid, and thus, since i is an inclusion, we have a weak equivalence $s_0(NC_0) \simeq i(N \operatorname{iso}(C)_1)$. In other words, the difference between NC_0 and NC_1 consists entirely of information about nonisomorphisms in C.

3 The simplicial set $(NC)_{*,0}$ is exactly nerve(C).

If we weaken the first property, then we get a homotopical version of a category. Allowing the isomorphism in (1) to be a weak equivalence leads to the definition of a *Segal space*. Imposing a weak equivalence as in (2) leads to the definition of a *complete* Segal space. We look in more detail at this structure in Chapter 5.

But what about the third fact? Since the simplicial set $(NC)_{*,0}$ is just the nerve of C, it is, in particular, a quasi-category. This property will continue to hold for more general complete Segal spaces, as we shall see in Chapter 7.

3.4 Higher Categories

We now consider some motivating questions from a categorical perspective. Let us begin with the idea of higher categories.

The structure of a category, as we have seen, consists of objects and morphisms between them, satisfying some compatibility conditions. However, many examples within mathematics have further structure, in that morphisms themselves have some kind of functions between them. One can depict such a *2-morphism* as follows:

This kind of data can be assembled into a *2-category*, consisting of objects, 1-morphisms between objects, and 2-morphisms between 1-morphisms. One can imagine inductively continuing this process of considering n-morphisms between $(n-1)$-morphisms with common source and target to obtain n-categories for any $n \geq 1$, or even adding arbitrarily high n-morphisms to obtain an ∞-category.

But one can ask also what structure these higher morphisms possess. We should be able to compose morphisms at any level, and there should be an identity n-morphism from an $(n-1)$-morphism to itself. If we want composition to be defined strictly, and identities to possess the properties that we expect from ordinary categories, then we can describe n-categories concretely in terms of enriched categories, as given in Definition 1.1.40.

Definition 3.4.1 A *(strict) n-category* is a category enriched in (strict) $(n-1)$-categories.

Let us look at the case of 2-categories. Using this definition, a 2-category should be a category enriched in categories. Therefore, it should have objects, and for any two objects x and y, $\mathrm{Hom}(x, y)$ should form a category. The objects of $\mathrm{Hom}(x, y)$ can be regarded as the 1-morphisms, and the morphisms of $\mathrm{Hom}(x, y)$ are the 2-morphisms.

Example 3.4.2 There is a 2-category consisting of categories, functors, and natural transformations. The objects are all categories, and the 1-morphisms are the functors between them. We could stop here, at least if the categories were assumed to be small, and get the ordinary category *Cat*. We could, however, include natural transformations of functors as 2-morphisms and define a 2-category of categories, even without the smallness restriction. Observe that we can think of this 2-category as a category enriched in *Cat* as follows. Given categories C and $\mathcal{D}$, there is a category $\mathcal{D}^C$ whose objects are the functors $C \to \mathcal{D}$ and whose morphisms are the natural transformations between those functors.

Unfortunately, many examples of n-categories that arise in mathematics are not strict in the sense described above. Often the higher structure, such as associativity of higher morphisms, only holds up to isomorphism, and these isomorphisms must satisfy some coherence conditions. These structures are called *weak n-categories*. In low dimensions, it is possible to write down all these conditions; weak 2-categories (or *bicategories*) are well-understood [22], as are weak 3-categories (or *tricategories*) [63, 64], but even at this level the coherence conditions become unwieldy.

There have been many proposed definitions of what a weak n-category for general n, or even a weak ∞-category, should be. These definitions use a variety of methods: some are defined via operads [15, 75, 83, 84], others from simplicial objects [65, 111, 113, 117], and others via opetopes [8, 41, 42, 67, 77] or blobs [95]; for a survey see Leinster [85]. Unfortunately, establishing relationships between these definitions, in particular equivalences, has proved to be very difficult.

However, there are properties that any good definition of higher categories is expected to possess. A weak n-category should still be, even if in some weak sense, a category enriched in $(n-1)$-categories. In the special case of groupoids, a weak n-groupoid should be a model for homotopy n-types of spaces. This idea has been explored from many perspectives; for example, see [34, 38, 39, 44, 86, 96, 98, 110].

It is from this second desired property that the notion of homotopical higher categories arises. A weak ∞-groupoid should just be a topological space. So, one starting point is to take this principle as a definition; see [2] or [89].

Definition 3.4.3 A *weak ∞-groupoid* is a topological space.

Let us consider further what a weak ∞-groupoid should be. It should have objects, n-morphisms for any $n \geq 1$, and all those morphisms at all levels should be weakly invertible. If we look at a topological space, we can see this kind of structure arising naturally. The points of the space are objects, and the 1-morphisms between objects are paths between them. While paths are not strictly invertible (in fact, composition of paths is not even strictly associative), they are invertible up to homotopy, via the reverse path. If we define 2-morphisms to be the endpoint-preserving homotopies between paths, then, again, these homotopies are invertible up to some 3-morphism, or homotopy between homotopies. Iterating this process, we get arbitrarily high homotopies between homotopies, all of which are weakly invertible. Indeed, we have described here the fundamental ∞-groupoid of a topological space, generalizing the fundamental groupoid.

We should comment here that we have chosen one particular approach, but it is not the only one. Many authors take a different definition of weak

∞-groupoid and try to prove an equivalence with topological spaces [34, 38, 44, 96]. In this context, what we have stated as a definition is known as the homotopy hypothesis, and is one desired property of any good theory of higher categories.

Many of these structures can be investigated from the viewpoint of homotopy theory, in particular considering such structures in the framework of model categories. Although we will not investigate this direction here, we refer the reader to [3, 40, 78, 79, 80, 81, 102, 117].

It is still harder to describe exactly what the structure of a weak ∞-category should be. In principle, however, a weak ∞-category could have all morphisms at some level invertible, but possibly noninvertible morphisms at another level.

Definition 3.4.4 For any $n \geq 0$, an (∞, n)-*category* is a weak ∞-category such that all k-morphisms are invertible for $k > n$.

Thus, an $(\infty, 0)$-category is precisely a weak ∞-groupoid. But for higher values of n, it is perhaps less clear how to model this structure with a precise mathematical object. It is our goal in this book to look at the multiple ways to model $(\infty, 1)$-categories. While these structures are not simple, and possess much subtlety, they are all known to be equivalent to one another. Even more general (∞, n)-categories, while more complicated than the case when $n = 1$, are more tractable than their nonhomotopical counterparts; see [12, 13, 32, 33, 70, 97, 102].

We begin our investigation of $(\infty, 1)$-categories by applying our original principle for higher categories, that they should be some kind of enriched category. In particular, they should be categories enriched in $(\infty, 0)$-categories.

Definition 3.4.5 An $(\infty, 1)$-*category* is a category enriched in topological spaces.

This definition is in fact quite precise. However, one might want to consider variations on it. Firstly, we would like to work instead with categories enriched in simplicial sets. Since simplicial sets have an equivalent homotopy theory to that of topological spaces, this change is not too difficult. The motivation for still more models comes down to a problem of realization: many examples simply do not have the fairly rigid structure of an enriched category. As for weak n-categories, associativity is often not defined as strictly as would be necessary, for instance. Thus, we need more models where the structure is defined less strictly. Yet, as we will see, these models are equivalent to simplicially enriched categories.

3.5 Homotopy Theories

While we saw how simplicial categories arise from a categorical viewpoint in the previous section, here we consider their motivation from the perspective of homotopy theory.

Recall from Section 1.2 that a category $\mathcal{M}$ with weak equivalences $\mathcal{W}$ gives rise to $\mathcal{M}[\mathcal{W}^{-1}]$, its localization with respect to the class $\mathcal{W}$ of weak equivalences. If $\mathcal{M}$ is a model category, this localization is given by its homotopy category $\mathrm{Ho}(\mathcal{M})$. The difficulty with this construction is that localization does not preserve limits and colimits from $\mathcal{M}$. For example, consider the pullback diagram in $\mathcal{T}op$:

$$
\begin{array}{ccc}
S^0 & \longrightarrow & S^1 \\
\downarrow & & \downarrow{\scriptstyle \times 2} \\
* & \longrightarrow & S^1.
\end{array}
$$

The diagram

$$
\begin{array}{ccc}
\mathrm{Hom}_{\mathrm{Ho}(\mathcal{T}op)}(D^2, S^0) & \longrightarrow & \mathrm{Hom}_{\mathrm{Ho}(\mathcal{T}op)}(D^2, S^1) \\
\downarrow & & \downarrow{\scriptstyle \times 2} \\
\mathrm{Hom}_{\mathrm{Ho}(\mathcal{T}op)}(D^2, *) & \longrightarrow & \mathrm{Hom}_{\mathrm{Ho}(\mathcal{T}op)}(D^2, S^1)
\end{array}
$$

is not a pullback, since the set in the upper left-hand corner consists of two points, whereas all other sets in the diagram consist of a single point.

We want to correct this construction so that it does respect pullbacks, via taking mapping spaces $\mathrm{Map}_{\mathcal{M}}(-, -)$ in the original category instead of morphism sets $\mathrm{Hom}_{\mathrm{Ho}(\mathcal{M})}(-, -)$ in the homotopy category. For the above example, we get a diagram of simplicial sets

$$
\begin{array}{ccc}
\mathrm{Map}_{\mathcal{T}op}(D^2, S^0) & \longrightarrow & \mathrm{Map}_{\mathcal{T}op}(D^2, S^1) \\
\downarrow & & \downarrow \\
\mathrm{Map}_{\mathcal{T}op}(D^2, *) & \longrightarrow & \mathrm{Map}_{\mathcal{T}op}(D^2, S^1),
\end{array}
$$

which is a homotopy pullback diagram of spaces. Taking homotopy classes gives the previous diagram of morphism sets, but these sets (other than the one in the top left-hand corner) form the beginning of a long exact sequence rather than a pullback.

So, how do we define these mapping spaces in a general setting? In the case where $\mathcal{M}$ is a simplicial model category, these mapping spaces are part of the data. In Section 2.4, we mentioned that they can be defined more generally;

in this section we look at one approach to doing so. The process is called simplicial localization and there are two primary methods. In each case, the output is a simplicial category in the sense of Definition 2.4.1.

The first construction is obtained by iterating a free category construction. We begin with the necessary definitions.

Definition 3.5.1 [57, 2.4] Let C be a category. The *free category on C* is the category FC whose objects are the same as those of C and whose morphisms are freely generated by the nonidentity morphisms of C.

We can think of taking the free category on a given category as given by a forgetful-free adjoint pair. There is a forgetful functor from the category *Cat* of small categories to the category of small directed graphs (with identities), where one forgets the composition structure. The left adjoint to this functor is given by taking the free category on a directed graph, obtained by freely adjoining composites of nonidentity arrows in a directed graph to get a category.

Observe that there are natural functors $\varphi \colon FC \to C$, which take any generating morphism Fc of FC to the morphism c in C from which it was defined, and $\psi \colon FC \to F^2C$, defined by taking a generating morphism Fc of FC to the generating morphism $F(Fc)$ of F^2C.

Definition 3.5.2 [57, 2.5] Let C be a category. The *standard simplicial resolution* of C is the simplicial category F_*C (thought of here as a simplicial object in *Cat*) which in degree k is the category $F^{k+1}C$. Each face map $d_i \colon F^{k+1}C \to F^kC$ is given by the composite functor $F^i\varphi F^{k-i}$, and each degeneracy map $s_i \colon F^{k+1}C \to F^{k+2}C$ is given by the composite functor $F^i\psi F^{k-i}$.

Observe that we are treating F_*C as a simplicial object in *Cat*, as discussed in Remark 2.4.2. However, since the free category construction does not change the objects, it is in fact a category enriched in simplicial sets, or a simplicial category in the sense we want to consider.

We use this definition to define the simplicial localization of a category with weak equivalences $(\mathcal{M}, \mathcal{W})$.

Definition 3.5.3 [57, 4.1] The *simplicial localization* of $\mathcal{M}$ with respect to $\mathcal{W}$ is the localization $(F_*\mathcal{W})^{-1}(F_*\mathcal{M})$, given by levelwise localization of simplicial categories. This simplicial localization is denoted by $L(\mathcal{M}, \mathcal{W})$ or simply $L\mathcal{M}$.

We can recover an ordinary category from a simplicial category via the following construction.

Definition 3.5.4   Let S be a simplicial category. Its *category of components* is the category $\pi_0 S$ with $\mathrm{ob}(\pi_0 S) = \mathrm{ob}(S)$ and, for any objects x and y,

$$\mathrm{Hom}_{\pi_0 S}(x, y) = \pi_0 \mathrm{Map}_S(x, y).$$

The following result shows that the simplicial localization of a category with weak equivalences is a higher-order version of its homotopy category.

Theorem 3.5.5   [57, 4.2] *Let (M, W) be a category with weak equivalences. There is an equivalence of categories*

$$\pi_0 L(M, W) \simeq M[W^{-1}].$$

This simplicial localization functor satisfies a universal property similar to that of the homotopy category; given a simplicial category $\mathcal{D}$ and a functor $T : (M, W) \to \mathcal{D}$, such that $T(w)$ is a weak equivalence for every w in W, then there exists a unique simplicial functor $L(M, W) \to \mathcal{D}$ making the following diagram commute:

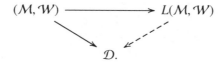

The advantage of this construction for the simplicial localization is that it is easy to describe. However, it suffers from the same problem as ordinary localization of categories with weak equivalences, in that the result can be a category with proper classes of morphisms in each simplicial degree between a fixed pair of objects.

To remedy this difficulty, there is another method for obtaining a simplicial category arising from a category with weak equivalences, also first constructed by Dwyer and Kan [56].

Definition 3.5.6   [56, 3.1] Let (M, W) be a category with weak equivalences. The *hammock localization* of M with respect to W, denoted by $L^H(M, W)$, or simply $L^H M$, is the simplicial category defined as follows.

1   The simplicial category $L^H M$ has the same objects as M.

2   Given objects X and Y of M, the simplicial set $\mathrm{Map}_{L^H M}(X, Y)$ has as k-simplices the *reduced hammocks* of width k and any length between X and

Y, or commutative diagrams of the form

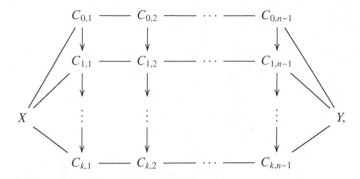

in which

- (i) the length of the hammock is any integer $n \geq 0$,
- (ii) the vertical maps are all in $\mathcal{W}$,
- (iii) in each column all the horizontal maps go in the same direction, and if they go to the left, then they are in $\mathcal{W}$,
- (iv) the maps in adjacent columns go in opposite directions, and
- (v) no column contains only identity maps.

The description of the hammock localization can be greatly simplified if $(\mathcal{M}, \mathcal{W})$ has a model structure. In this case, Dwyer and Kan prove that it suffices to consider hammocks of length 3 such as the following [53, §8]:

$$X \xleftarrow{\;\simeq\;} C_{0,1} \longrightarrow C_{0,2} \xleftarrow{\;\simeq\;} Y.$$

In particular, in the case of a model category, the set-theoretic difficulties of the simplicial localization can be avoided by using the hammock localization instead.

We would like to know that these two notions of simplicial localization are equivalent, in some natural sense. But what should equivalences of simplicial categories be? We consider the following definition which nicely generalizes the definition of equivalence of categories.

Definition 3.5.7 [55, 1.3] Let $F: C \to \mathcal{D}$ be a simplicial functor between simplicial categories. Then F is a *Dwyer–Kan equivalence* if

- (W1) for every $x, y \in \mathrm{ob}(C)$, $\mathrm{Map}_C(x, y) \to \mathrm{Map}_{\mathcal{D}}(Fx, Fy)$ is a weak equivalence of simplicial sets, and
- (W2) the induced functor $\pi_0 C \to \pi_0 \mathcal{D}$ is an equivalence of categories.

Remark 3.5.8   Observe that, in the presence of the first condition, the second condition is equivalent to the functor $\pi_0 C \to \pi_0 \mathcal{D}$ being essentially surjective.

Now we have the following comparison result.

Proposition 3.5.9   [53, 2.2] *For a given model category $\mathcal{M}$, the simplicial categories $L\mathcal{M}$ and $L^H \mathcal{M}$ are Dwyer–Kan equivalent.*

In fact, if $\mathcal{M}$ is a simplicial category, in particular, if it is a simplicial model category, then we recover this simplicial structure from the hammock localization (and hence also by the simplicial localization).

Proposition 3.5.10   [56, 4.8] *If $\mathcal{M}$ is a simplicial model category, then the map of simplicial categories $\mathcal{M} \to L^H \mathcal{M}$ is a Dwyer–Kan equivalence.*

Thus, every model category, and more generally, any category with weak equivalences, gives rise to a simplicial category in a natural way given by either of these two constructions. Perhaps more surprisingly, we also have the following converse statement.

Proposition 3.5.11   [55, 2.5] *Up to Dwyer–Kan equivalence, any simplicial category can be obtained as the simplicial localization of some category with weak equivalences.*

If the basic data of a "homotopy theory" is a category together with a choice of weak equivalences, we can regard simplicial categories as modeling homotopy theories. But then simplicial categories themselves, together with Dwyer–Kan equivalences between them, themselves form a homotopy theory. Thus, the homotopy theory of simplicial categories can be regarded as the "homotopy theory of homotopy theories". Indeed, combining the ideas of the previous sections, we can see that $(\infty, 1)$-categories, which we defined from a categorical perspective, equally describe the structure of a homotopy theory.

4

Simplicial Categories

In this chapter, we develop the homotopy theory of our first model for $(\infty, 1)$-categories, that of simplicial categories. Observe that one could analogously also work with topological categories, and their homotopy theory was developed explicitly by Ilias [72].

Our goal in this chapter is to show that there is a model structure on the category of small simplicial categories.

4.1 The Category of Small Simplicial Categories

Recall that a *simplicial category* is a category enriched in simplicial sets, or a category such that, for any objects x and y, there is a simplicial set $\mathrm{Map}(x, y)$ of morphisms from x to y. A *simplicial functor* $F \colon C \to \mathcal{D}$ is a functor such that, for any objects x and y of C, there is an induced morphism of simplicial sets $\mathrm{Map}(x, y) \to \mathrm{Map}(Fx, Fy)$.

We begin by proving that the first three axioms for a model structure are satisfied.

Proposition 4.1.1 *The category SC has all finite limits and colimits, and its class of weak equivalences is closed under retracts and satisfies the two-out-of-three property.*

Proof One can show that the category of all simplicial categories has all coproducts and all coequalizers, and therefore all finite colimits, and all products and equalizers, and therefore all finite limits, by Proposition 1.1.28. To prove the existence of coequalizers, for example, we use the existence of coequalizers for sets (for the objects) and simplicial sets (for the morphisms). The two properties for the class of weak equivalences are left as an exercise but are not difficult to check. $\qquad\square$

While we ultimately want to consider the category SC of all small simplicial categories and all simplicial functors between them, we begin with subcategories whose objects have a fixed set of objects and whose morphisms are the identity on this object set.

4.2 Fixed-Object Simplicial Categories

Given a set O, we denote by SC_O the category whose objects are simplicial categories with object set O and functors preserving this object set. As a first step in obtaining a model structure on the category of all small simplicial categories, we first establish model structures in this more restricted setting.

The first step is to show that the category SC_O has the necessary limits and colimits. We know the category of all small simplicial categories has all small limits and all small colimits. However, in general, taking limits and colimits in SC does not preserve the object set. For example, suppose C and $\mathcal{D}$ are both simplicial categories with two objects. Then their product is a simplicial category with four objects. Thus, we need to show that a modification which does preserve the object set still retains the necessary universal property.

Proposition 4.2.1 *The category SC_O has all small limits and all small colimits.*

Proof Let $X \colon \mathcal{D} \to SC_O$ be a small diagram of simplicial categories with fixed object set O. Let us first consider the colimit $\operatorname{colim}_{\mathcal{D}} X$. Regard O as a discrete simplicial category with no nonidentity morphisms, and likewise for the set $\operatorname{ob}(\operatorname{colim}_{\mathcal{D}} X)$. Take the pushout in the diagram

$$O \overset{\text{fold}}{\leftarrow} \operatorname{ob}(\operatorname{colim}_D X) \to \operatorname{colim}_D X$$

and denote it by $\operatorname{colim}_{\mathcal{D}}^O X$. Then one can check that this simplicial category has object set O and satisfies the necessary universal property to be the colimit of X in the category SC_O.

For limits, let $\lim_{\mathcal{D}}^O X$ be the full simplicial subcategory of $\lim_{\mathcal{D}} X$ whose objects are in the image of the diagonal map $O \to \lim_{\mathcal{D}} X$. Again, one can check that this simplicial category can be regarded as having object set O and satisfies the necessary universal property. $\square$

Recall from Definition 3.5.1 the definition of a free category. Given any two categories C and $\mathcal{D}$ with the same object set O, their *free product* is the category $C * \mathcal{D}$ whose morphisms are given by words of composable morphisms in C

and $\mathcal{D}$, where composable is taken to mean having compatibility of source and target objects.

In the following definition, we again find it useful to regard a simplicial category as a simplicial object, here in the category of small categories with fixed object set O.

Definition 4.2.2 A functor $f \colon C \to \mathcal{D}$ in SC_O is a *free map* if:

1 the map f is a monomorphism on mapping spaces;
2 in each degree k, the category $\mathcal{D}_k$ can be written as a free product $f(C_k) * F_k$, where F_k is a free category; and
3 for any $k \geq 0$, all degeneracies of generators of the free category F_k are generators of F_{k+1}.

Observe that, if C is the initial object of SC_O, the category with object set O and only identity morphisms, then f is a free map if and only if $\mathcal{D}$ is a free category.

Definition 4.2.3 [57, 7.5] A functor $f \colon C \to \mathcal{D}$ between simplicial categories is a *strong retract* of a map $f' \colon C \to \mathcal{D}'$ if there exists a commutative diagram

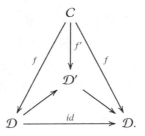

With these definitions, we can describe the model structure on SC_O.

Theorem 4.2.4 [57, 7.2, 7.3, 7.6] *There is a proper simplicial model structure on the category SC_O, where*

1 *weak equivalences are the simplicial functors $f \colon C \to \mathcal{D}$ such that, for any objects x and y of C, the induced map $\mathrm{Map}_C(x, y) \to \mathrm{Map}_\mathcal{D}(fx, fy)$ is a weak equivalence of simplicial sets;*
2 *fibrations are the simplicial functors $f \colon C \to \mathcal{D}$ such that, for any objects x and y of C, the induced map $\mathrm{Map}_C(x, y) \to \mathrm{Map}_\mathcal{D}(fx, fy)$ is a fibration of simplicial sets; and*
3 *cofibrations are the simplicial functors which are strong retracts of free maps.*

Observe that the weak equivalences here are substantially simpler than more general Dwyer–Kan equivalences of simplicial categories. Because the objects are assumed to be fixed, we need only compare mapping spaces to understand whether two simplicial categories are weakly equivalent in this setting.

Example 4.2.5 If $O = \{x\}$, then we recover the classical model structure on simplicial monoids, first proved by Quillen [100, II.3].

We use the following example frequently in what follows.

Example 4.2.6 Suppose that $O = \{x, y\}$. Then $SC_{\{x,y\}}$ is the category of all simplicial categories with two objects, and morphisms which preserve those objects.

Let C be an ordinary category, regarded as a discrete simplicial category. In general, C is not cofibrant in SC_O. A cofibrant replacement of C can be obtained as follows. Treating C as a simplicial object in *Cat*, first take a simplicial resolution at each level C_k as in Definition 3.5.2 to obtain a bisimplicial category F_*C. Applying a diagonal functor, as used in [57, 6.1], results in a simplicial category which is cofibrant.

4.3 The Model Structure

Now, we would like to consider all small simplicial categories as the objects of a model category. In particular, we no longer want to require simplicial functors to preserve the objects.

We first need some definitions. Given a simplicial set K, define the simplicial category UK to have objects x and y and morphism spaces defined as $\mathrm{Map}_{UK}(x, x) = \mathrm{Map}_{UK}(y, y) = \{\mathrm{id}\}$, $\mathrm{Map}_{UK}(x, y) = K$, and $\mathrm{Map}_{UK}(y, x) = \varnothing$. Observe that this construction defines a functor $U \colon SSets \to SC$.

We also need the following notion of what it means for a morphism in a simplicial category to be an equivalence.

Definition 4.3.1 A morphism $e \colon x \to y$ in a simplicial category C is a *homotopy equivalence* if its image in $\pi_0(C)$ is an isomorphism.

Now we are able to describe the model structure on the category of small simplicial categories.

Theorem 4.3.2 [27, 1.1] *There is a cofibrantly generated model structure SC on the category of small simplicial categories in which*

1 the weak equivalences are the Dwyer–Kan equivalences, and

2 *the fibrations are the maps* $f : C \to \mathcal{D}$ *satisfying the following two conditions:*

(F1) *for any objects* a_1 *and* a_2 *in* C, *the map*

$$\mathrm{Map}_C(a_1, a_2) \to \mathrm{Map}_{\mathcal{D}}(fa_1, fa_2)$$

   is a fibration of simplicial sets, and

(F2) *for any object* a_1 *in* C, *b in* $\mathcal{D}$, *and homotopy equivalence* $e : fa_1 \to b$ *in* $\mathcal{D}$, *there is an object* a_2 *in* C *and homotopy equivalence* $d : a_1 \to a_2$ *in* C *such that* $fd = e$.

A set of generating cofibrations consists of

(C1) *the maps* $U\partial\Delta[n] \to U\Delta[n]$ *for* $n \geq 0$, *and*

(C2) *the map* $\varnothing \to \{x\}$, *where* $\varnothing$ *is the empty simplicial category and* $\{x\}$ *denotes the simplicial category with one object* x *and no nonidentity morphisms.*

A set of generating acyclic cofibrations consists of

(A1) *the maps* $UV[n, k] \to U\Delta[n]$ *for* $n \geq 1$, *and*

(A2) *inclusion maps* $\{x\} \to \mathcal{H}$ *which are Dwyer–Kan equivalences, where* $\{x\}$ *is as in (C2) and* $\{\mathcal{H}\}$ *is a set of representatives for the isomorphism classes of simplicial categories with two objects* x *and* y, *weakly contractible mapping spaces, and only countably many simplices in each mapping space. Furthermore, we require that the inclusion map* $\{x\} \amalg \{y\} \to \mathcal{H}$ *be a cofibration in* $\mathcal{SC}_{\{x,y\}}$.

We prove this theorem in the next section, after first establishing a number of preliminary results. We conclude this section with a few comments about the statement of the theorem.

It follows from the definition of Dwyer–Kan equivalence that the functor $\{x\} \to \pi_0\mathcal{H}$ is an equivalence of categories. In particular, all 0-simplices of the mapping spaces of $\mathcal{H}$ are homotopy equivalences.

The idea behind the set (A2) of generating acyclic cofibrations is the fact that two simplicial categories can have a weak equivalence between them which is not a bijection on objects, much as two categories can be equivalent even if they do not have the same objects. We only require that our weak equivalences be surjective on equivalence classes of objects. Thus, we must consider acyclic cofibrations for which the object sets are not isomorphic. The requirement that there be only countably many simplices is included so that we have a set rather than a proper class of such maps.

The purpose of the map in (C2) is to encode the idea that the inclusion of a category into another category with one more object should be a cofibration. Again, this situation does not arise in the fixed-object situation but is important when we allow object sets to vary.

4.4 Proof of the Existence of the Model Structure

In this section we give the main steps of the proof of the existence of the model category SC. We first verify lifting conditions involving the proposed sets of generating cofibrations and generating acyclic cofibrations.

Applying the functor $U: SSets \rightarrow SC$ to each of the generating acyclic cofibrations for $SSets$, we obtain exactly the set (A1) of maps $UV[n, k] \rightarrow U\Delta[n]$ for $n \geq 1$ and $0 \leq k \leq n$. Thus, we can see that a map of simplicial categories has the right lifting property with respect to the maps in (A1) if and only if it satisfies the property (F1).

Remark 4.4.1 Some technicalities aside, essentially what we have just argued is that (A1) is a set of generating acyclic cofibrations for the fixed-object model category SC_O. Similarly, (C1) is a set of generating cofibrations for SC_O. The arguments that follow illustrate the additional complications of including the maps in (A2) to the generating set of acyclic cofibrations.

We want to show that maps with the right lifting property with respect to the maps in both (A1) and (A2) are precisely the maps which satisfy conditions (F1) and (F2).

Proposition 4.4.2 [27, 2.3] *Suppose that a map $f: C \rightarrow D$ of simplicial categories has the right lifting property with respect to the maps in (A1) and (A2). Then f satisfies condition (F2).*

The proof uses the following technical lemma, whose proof can be found in [27, §4].

Lemma 4.4.3 [27, 2.4] *Let $\mathcal{F}$ be the simplicial category with object set $\{x, y\}$ and one nonidentity morphism $g: x \rightarrow y$. Let $\mathcal{E}'$ be any simplicial category with the same object set $\{x, y\}$, and suppose $i: \mathcal{F} \rightarrow \mathcal{E}'$ is a functor which takes g to a homotopy equivalence in $\mathrm{Hom}_{\mathcal{E}'}(x, y)$. Then the map i can be factored as a composite $\mathcal{F} \rightarrow \mathcal{H} \rightarrow \mathcal{E}'$ in such a way that the composite map $\{x\} \rightarrow \mathcal{F} \rightarrow \mathcal{H}$ is isomorphic to a map in (A2).*

Proof of Proposition 4.4.2 Suppose $f: C \rightarrow D$ has the right lifting property with respect to maps in (A1) and (A2). Let $e: x \rightarrow y$ be a homotopy equivalence in D such that $x = fa$ for some object a of C. We need to find a homotopy equivalence $d: a \rightarrow b$ in C such that $fd = e$.

We first consider the case where $x \neq y$. Let $\mathcal{E}'$ be the full simplicial subcategory of D with objects x and y, and let $\mathcal{F}$ be a simplicial category with objects x and y and a single nonidentity morphism $g: a \rightarrow b$, as in Lemma 4.4.3. Suppose $i: \mathcal{F} \rightarrow \mathcal{E}'$ is the identity map on objects and sends g to a homotopy equivalence $e: a \to b$. By Lemma 4.4.3, we can factor this map as $\mathcal{F} \to \mathcal{H} \to \mathcal{E}'$ in such a way that the composite $\{x\} \to \mathcal{F} \to \mathcal{H}$ is isomorphic to a map in (A2).

It follows that the composite $\{a\} \to \{x\} \to \mathcal{H}$ is also isomorphic to a map in (A2). Then consider the composite $\mathcal{H} \to \mathcal{E}' \to \mathcal{D}$, where the functor $\mathcal{E}' \to \mathcal{D}$ is the inclusion. These functors fit into a commutative diagram

The dotted arrow lift exists because we have assumed that the map $f: C \to \mathcal{D}$ has the right lifting property with respect to all maps in (A2). Now, composing the map $\mathcal{F} \to \mathcal{H}$ with the lift sends the map g in $\mathcal{F}$ to a map d in C such that $fd = e$. Since all the morphisms of $\mathcal{H}$ are homotopy equivalences, their images in C must also be homotopy equivalences; in particular, d must be a homotopy equivalence. Thus, f satisfies condition (F2) in the case when $x \neq y$.

Now suppose that $x = y$. Define $\mathcal{E}'$ to be the simplicial category with two objects x and x' such that each mapping space of $\mathcal{E}'$ is given by the simplicial set $\mathrm{Hom}_{\mathcal{D}}(x, x)$ and composition is defined as in $\mathcal{D}$. Consider the map $\mathcal{E}' \to \mathcal{D}$ which sends both objects of $\mathcal{E}'$ to x in $\mathcal{D}$ and is the identity on all mapping spaces. Using this simplicial category $\mathcal{E}'$, the argument proceeds as above. □

We now consider the converse direction, that fibrations have the right lifting property with respect to the maps in (A1) and (A2). As we already know fibrations have the right lifting property with respect to maps in (A1), it suffices to prove the following proposition.

Proposition 4.4.4 [27, 2.5] *Suppose $f: C \to \mathcal{D}$ is a map of simplicial categories which satisfies properties (F1) and (F2). Then f has the right lifting property with respect to the maps in (A2).*

Proof Suppose $f: C \to \mathcal{D}$ satisfies (F1) and (F2). We need to show that there exists a lift in any diagram of the form

$$\begin{array}{ccc} \{x\} & \longrightarrow & C \\ \downarrow & \nearrow & \downarrow f \\ \mathcal{H} & \longrightarrow & \mathcal{D} \end{array}$$

where $\{x\} \to \mathcal{H}$ is a map in (A2).

Let $g: x \to y$ be a homotopy equivalence in $\mathcal{H}$. Let $\mathcal{F}$ denote the subcategory of $\mathcal{H}$ consisting of the objects x and y with g its only nonidentity morphism. Consider the composite $\{x\} \to \mathcal{F} \to \mathcal{H}$ and the resulting diagram

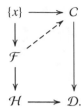

Because the functor $\mathcal{F} \to \mathcal{D}$ factors through $\mathcal{H}$, which consists of homotopy equivalences, the image of g in $\mathcal{D}$ is a homotopy equivalence. Thus, the existence of the lift in the above diagram follows from the fact that f satisfies (F2).

Now, we need to show that the rest of $\mathcal{H}$ lifts to C. For simplicity, let us denote also by x and y the images of the objects x and y of $\mathcal{H}$ in $\mathcal{D}$; and also by x the image of x in C. We begin by assuming that $x \neq y$ in $\mathcal{D}$. Let e be the image of g in $\mathcal{D}$, which is necessarily a homotopy equivalence. Since f satisfies (F2), there is a homotopy equivalence $d\colon x \to z$ in C such that $fd = e$, and since $x \neq y$ in $\mathcal{D}$, we must also have $x \neq z$ in C. Consider the full simplicial subcategory of C with objects x and z, and denote by C' the isomorphic simplicial category with objects x and y. Let $\mathcal{D}'$ be the simplicial subcategory of $\mathcal{D}$ with objects x and y. Now, we can consider $\mathcal{H}$, C', and $\mathcal{D}'$ as objects in the category $SC_{\{x,y\}}$ of simplicial categories with fixed object set $\{x, y\}$. Note that the functor $C' \to \mathcal{D}'$ is a fibration in $SC_{\{x,y\}}$. Now define $\mathcal{E}$ to be the pullback in the diagram

Then the functor $\mathcal{E} \to \mathcal{H}$ is also a fibration in $SC_{\{x,y\}}$ by Proposition 1.4.11.

By Lemma 4.4.3, we can factor $\mathcal{F} \to \mathcal{E}$ as the composite $\mathcal{F} \to \mathcal{H}' \to \mathcal{E}$ for some simplicial category $\mathcal{H}'$ such that the composite $\{x\} \to \mathcal{F} \to \mathcal{H}'$ is isomorphic to a functor in (A2). Then, the composite $\mathcal{H}' \to \mathcal{E} \to \mathcal{H}$ is a weak equivalence in $SC_{\{x,y\}}$ since it is the identity on objects and all the mapping spaces of $\mathcal{H}$ and $\mathcal{H}'$ are weakly contractible.

Now, we can assemble these functors into a commutative diagram

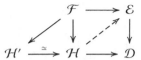

in which the dotted arrow lift exists by Lemma 1.4.12.

If $x = y$ in $\mathcal{D}$, then $\mathcal{D}'$ (and possibly C') as defined above only has one object x. In this case, define the simplicial category $\mathcal{D}''$ with two objects x and y such that each mapping space is the simplicial set $\mathrm{Map}_{\mathcal{D}'}(x, x)$ (as in the proof of Proposition 4.4.2). We can then factor $\mathcal{H} \to \mathcal{D}'$ through the object $\mathcal{D}''$, where $\mathcal{D}'' \to \mathcal{D}$ sends both objects of $\mathcal{D}''$ to a in $\mathcal{D}$ and is the identity on each mapping space. If C' also has one object, then we construct a simplicial category C'' analogously. We can repeat the argument above in the left-hand square of the diagram

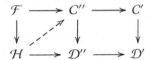

to obtain a lift $\mathcal{H} \to C''$, and hence a lift $\mathcal{H} \to C'$ via composition. □

We now consider the sets (C1) and (C2). Suppose a functor $f\colon C \to \mathcal{D}$ is a fibration and a weak equivalence. Using the model structure on $\mathcal{SSets}$, we can see that a functor satisfies conditions (F1) and (W1) if and only if it has the right lifting property with respect to the maps $U\partial\Delta[n] \to U\Delta[n]$ for $n \geq 0$, i.e., those in the set (C1).

Proposition 4.4.5 [27, 3.2] *A map in SC is a fibration and a weak equivalence if and only if it has the right lifting property with respect to the maps in (C1) and (C2).*

Proof First suppose that $f\colon C \to \mathcal{D}$ is both a fibration and a weak equivalence. By conditions (F1) and (W1), $\mathrm{Map}_C(a, b) \to \mathrm{Map}_{\mathcal{D}}(fa, fb)$ is an acyclic fibration of simplicial sets for any choice of objects a and b in C. In other words, there is a lift in any diagram of the form

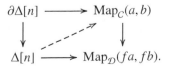

However, having this lift is equivalent to having a lift in the diagram

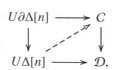

where we assume the objects x and y of $U\partial\Delta[n]$ map to a and b in C, and analogously the objects x and y of $U\Delta[n]$ map to fx and fy in $\mathcal{D}$. Hence, f has the right lifting property with respect to the maps in (C1).

It remains only to show that f has the right lifting property with respect to the map $\varnothing \to \{x\}$. Equivalently, we need to show that f is surjective on objects. Being surjective on homotopy equivalence classes of objects follows from condition (W2). So suppose that $e\colon x \to y$ is an isomorphism in $\mathcal{D}$ and there is an object a in C such that $fa = x$. Since e is a homotopy equivalence, by property (F2) there is a homotopy equivalence in C with domain a and which maps to e under f. In particular, there is an object b in C mapping to y. Thus, f is surjective on objects.

Conversely, suppose that f has the right lifting property with respect to the maps in (C1) and (C2). Again, using the model structure on simplicial sets, we have that

$$\mathrm{Map}_C(a, b) \to \mathrm{Map}_{\mathcal{D}}(fa, fb)$$

is both a fibration and a weak equivalence, and hence f satisfies both (F1) and (W1). It follows that $\mathrm{Hom}_{\pi_0 C}(a, b) \to \mathrm{Hom}_{\pi_0 \mathcal{D}}(fa, fb)$ is an isomorphism. As above, having the right lifting property with respect to the map $\varnothing \to \{x\}$ is equivalent to being surjective on objects. These two facts show then that $\pi_0 C \to \pi_0 \mathcal{D}$ is an equivalence of categories, establishing condition (W2).

It remains to show that f satisfies property (F2). By Proposition 4.4.2 and the fact that satisfying (F1) is equivalent to having the right lifting property with respect to maps in (A1), it suffices to show that f has the right lifting property with respect to the maps in (A2). But, a map $\{x\} \to \mathcal{H}$ in (A2) can be written as a (possibly infinite) composition of pushouts along $\varnothing \to \{x\}$ followed by pushouts along maps of the form $U\partial\Delta[n] \to U\Delta[n]$, and f has the right lifting property with respect to all such maps since they are precisely those in (C1) and (C2). □

Proposition 4.4.6 [27, 3.3] *A map in SC is an acyclic cofibration if and only if it has the left lifting property with respect to the fibrations.*

The proof requires the use of the following lemma:

Lemma 4.4.7 [27, 3.4] *Let $\mathcal{A} \to \mathcal{B}$ be a map in (A1) or (A2) and $i \colon \mathcal{A} \to C$ any map in SC. Then in the pushout diagram*

$$
\begin{array}{ccc}
\mathcal{A} & \xrightarrow{\ i\ } & C \\
\downarrow & & \downarrow \\
\mathcal{B} & \longrightarrow & \mathcal{D}
\end{array}
$$

the map $C \to \mathcal{D}$ is a weak equivalence.

Proof First suppose that the map $\mathcal{A} \to \mathcal{B}$ is in (A2), so we can instead write it as $\{x\} \to \mathcal{H}$, with x and y the objects of $\mathcal{H}$. Let O be the set of objects of C and define O' to be the set $O \backslash \{x\}$. (For simplicity of notation, we assume that $ix = x$.) We denote also by O and O' the respective discrete (simplicial) categories with no nonidentity morphisms. Consider the diagram

$$
\begin{array}{ccc}
X = O \amalg \{y\} & \longrightarrow & C \amalg \{y\} = C' \\
\downarrow & & \downarrow \\
\mathcal{H}' = O' \amalg \mathcal{H} & \longrightarrow & \mathcal{D}
\end{array}
$$

and notice that $\mathcal{D}$ is also the pushout of this diagram. Since X (regarded as a set) is the object set of any of these categories, note that the left-hand vertical arrow is a cofibration in SC_X.

We factor the map $X \to C'$ as the composite of a cofibration and an acyclic fibration in SC_X:

$$
X \lhook\joinrel\longrightarrow C'' \overset{\sim}{\longrightarrow\mkern-14mu\rightarrow} C'.
$$

Since SC_X is proper, it follows from Proposition 1.7.4 that the pushouts of each row in the diagram

$$
\begin{array}{ccc}
\mathcal{H}' & \longleftarrow\joinrel\rhook \ X \ \lhook\joinrel\longrightarrow & C' \\
\uparrow{\scriptstyle =} & \uparrow{\scriptstyle =} & \uparrow{\scriptstyle \sim} \\
\mathcal{H}' & \longleftarrow\joinrel\rhook \ X \ \lhook\joinrel\longrightarrow & C'' \\
\downarrow{\scriptstyle \sim} & \downarrow{\scriptstyle =} & \downarrow{\scriptstyle =} \\
\pi_0 \mathcal{H}' & \longleftarrow \ X \ \lhook\joinrel\longrightarrow & C''
\end{array}
$$

are weakly equivalent to one another. In particular, the pushout of the bottom row is weakly equivalent to $\mathcal{D}$. It remains to show that there is a weak equivalence of pushouts of the rows of the diagram

$$
\begin{array}{ccccc}
\pi_0 \mathcal{H}' & \longleftarrow & X & \lhook\joinrel\longrightarrow & C'' \\
\downarrow & & \downarrow & & \downarrow \\
\pi_0 \mathcal{H}' & \longleftarrow & X & \longrightarrow & C'.
\end{array}
$$

One can check that the pushout of this bottom row is weakly equivalent in SC to the pushout of the diagram

$$
\pi_0 \mathcal{H} \longleftarrow \{x\} \longrightarrow C
$$

and therefore that the pushout of the top row is weakly equivalent to the pushout of the bottom row. It follows that the map $C \to \mathcal{D}$ is a weak equivalence in SC.

Now let us suppose that $\mathcal{A} \to \mathcal{B}$ is in (A1). Thus, we consider pushout diagrams of the form

$$
\begin{array}{ccc}
UV[n,k] & \xrightarrow{\ j\ } & C \\
\downarrow & & \downarrow \\
U\Delta[n] & \longrightarrow & \mathcal{D}
\end{array}
$$

for $n \geq 1$ and $0 \leq k \leq n$. As before, take $\{x, y\}$ to be the object set of $UV[n,k]$ and of $U\Delta[n]$, and take O to be the object set of C. Let $O'' = O\backslash\{x, y\}$. (Again, for notational simplicity we assume that $jx = x$ and $jy = y$.) Now we consider the diagram

$$
\begin{array}{ccc}
O'' \amalg UV[n,k] & \longrightarrow & C \\
\downarrow & & \downarrow \\
O'' \amalg U\Delta[n] & \longrightarrow & \mathcal{D}
\end{array}
$$

in SC_O. However, since the left vertical map is a weak equivalence and assuming that the top map is a cofibration (factoring if necessary as above), we can again use the fact that SC_O is proper to show that $C \to \mathcal{D}$ is a weak equivalence in SC_O and thus also in SC.   □

Now we can bring these results together to prove that acyclic cofibrations are exactly the maps with the left lifting property with respect to the fibrations.

Proof of Proposition 4.4.6   First suppose that a map $C \to \mathcal{D}$ is an acyclic cofibration. By Proposition 1.7.11, there is a factorization of the map $C \to \mathcal{D}$ as the composite $C \to C' \to \mathcal{D}$ where C' is obtained from C by a directed colimit of iterated pushouts along the maps in (A1) and (A2). Thus, by Lemma 4.4.7

and Proposition 1.7.6, this map $C \to C'$ is a weak equivalence. Furthermore, the map $C' \to \mathcal{D}$ has the right lifting property with respect to the maps in (A1) and (A2). Thus, by Proposition 4.4.2, it is a fibration. It is also a weak equivalence since the maps $C \to \mathcal{D}$ and $C \to C'$ are, by the two-out-of-three property. In particular, by the definition of cofibration, it has the right lifting property with respect to the cofibrations. Therefore, there exists a dotted arrow lift in the diagram

Hence the map $C \to \mathcal{D}$ is a retract of the map $C \to C'$ and therefore also has the left lifting property with respect to fibrations.

Conversely, suppose that $C \to \mathcal{D}$ has the left lifting property with respect to fibrations. In particular, it has the left lifting property with respect to the acyclic fibrations, so it is a cofibration by definition. We again obtain a factorization of this functor as the composite $C \to C' \to \mathcal{D}$ where C' is obtained from C by iterated pushouts of the functors in (A1) and (A2). Once again, $C' \to \mathcal{D}$ has the right lifting property with respect to the functors in (A1) and (A2) and thus is a fibration by Proposition 4.4.2. Therefore there is a lift in the diagram

Again using Lemma 4.4.7, the map $C \to \mathcal{D}$ is a weak equivalence because it is a retract of $C \to C'$. $\qquad\square$

We have now proved everything we need for the existence of the model category structure on $S\mathcal{C}$.

Proof of Theorem 4.3.2 We show that the conditions of Theorem 1.7.15 are satisfied. The category of small simplicial categories has all small limits and colimits, and the weak equivalences satisfy the two-out-of-three property by Proposition 4.1.1. It can be shown that both $\varnothing$ and $\{x\}$ are small, and using the smallness of $V[n, k]$ and $\Delta[n]$ in $S\mathit{Sets}$, it can be shown that each $U\partial\Delta[n]$ is small relative to the set (C1) and each $UV[n, k]$ is small relative to the set (A1). Therefore, condition (1) of that theorem holds. Propositions 4.4.2 and 4.4.4, together with what we know from the fixed-object model structure, prove that the fibrations are precisely the J-fibrations, where J is the union of the maps in (A1) and (A2). Then conditions (2) and (4)(i) of Theorem 1.7.15 are satisfied by Proposition 6.5.3, and conditions (3) and (4)(ii) (of which the latter is unnecessary) are satisfied by Proposition 4.4.6.   □

4.5 Properties of the Model Structure

Now that we have a model structure on the category of small simplicial categories, we investigate some of its additional properties.

To start, we give characterizations of the fibrant and cofibrant objects.

Proposition 4.5.1 *The fibrant objects of SC are precisely the simplicial categories whose mapping spaces are all Kan complexes.*

Proof Let C be a simplicial category whose mapping spaces are all Kan complexes. Then, for any objects x and y of C, any $n \geq 1$, and any $0 \leq k \leq n$, a lift exists in any diagram of the form

But the existence of such lifts is equivalent to lifts in all diagrams

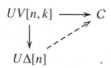

Thus, such simplicial categories are exactly those which have the lifting property with respect to the maps in (A1).

Now we must consider the maps in (A2). But if $\{x\} \to \mathcal{H}$ is a map in (A2), then the unique section $\mathcal{H} \to \{x\}$ gives a lift in any diagram

for any C. Thus the lifting condition with respect to maps in (A2) is immediate when we consider maps from C to a terminal object.   □

Next, we give a description of the cofibrant objects, which Riehl [104] attributes to Verity. Here, we make use of the fact that a simplicial category

C can be regarded as a simplicial object in the category *Cat* of small categories, and in particular has associated categories C_n for each $n \geq 0$. We start with some preliminary definitions.

Definition 4.5.2 A morphism in a category $\mathcal{D}$ is *atomic* if it admits no non-trivial factorizations.

Definition 4.5.3 Let C be a simplicial category. An *n-arrow* is a morphism in the category C_n. Alternatively, an n-arrow $f : a \to b$ is an element of the set $\mathrm{Map}_C(a,b)_n$ of n-simplices of the mapping space $\mathrm{Map}_C(a,b)$.

Recall from properties of simplicial sets that any n-arrow f of a simplicial category C can be uniquely factored as $f = f' \circ \alpha$, where $\alpha : [n] \to [m]$ is surjective in Δ and f' is a nondegenerate m-arrow of C.

Definition 4.5.4 [104, 16.2.1] A *simplicial computad* is a simplicial category C such that

1 each category C_n is a free category on a directed graph of atomic n-arrows, and
2 given any surjective $\alpha : [n] \to [m]$ in Δ and any atomic arrow f in C_m, the n-arrow $f \circ \alpha$ is atomic in C_n.

The second condition can be restated less formally as saying that degenerate images of atomic morphisms are atomic.

Proposition 4.5.5 [104, 16.2.2] *A simplicial category is cofibrant in SC if and only if it is a simplicial computad.*

Proof First, we show that simplicial computads are cellular cofibrant in *SC*, in that they can be obtained by taking pushouts along the generating cofibrations. Let C be a simplicial computad. Then the map $\varnothing \to C$ can be obtained as a pushout along maps $\varnothing \to \{x\}$, one for each object of C, and along maps $U\partial\Delta[n] \to U\Delta[n]$ for each atomic n-arrow and for all $n \geq 0$.

Next, we show that cellular cofibrant simplicial categories are simplicial computads. The objects of such simplicial categories are obtained by the pushouts along the copies of the map $\varnothing \to \{x\}$, and, for any $n \geq 0$, the n-arrows obtained by taking pushouts along the map $U\partial\Delta[n] \to U\Delta[n]$ are atomic n-arrows. One can check that the necessary properties hold for degeneracies of atomic arrows.

Finally, we prove that simplicial computads are closed under retract. Working on each simplicial level, we begin by proving that a retract of a free category is free. Suppose $\mathcal{A}$ is a retract of a free category $\mathcal{B}$. Suppose we have a composite $h = g \circ f$ in $\mathcal{B}$. If two of the maps f, g, and h are in $\mathcal{A}$, then so is the third. By induction, it follows that any morphism of $\mathcal{A}$ can be factored uniquely into arrows of $\mathcal{A}$ which are either atomic in $\mathcal{B}$ (and hence also in $\mathcal{A}$) or atomic in $\mathcal{A}$; in the latter case, these morphisms can be further factored into atomic morphisms in $\mathcal{B}$, but the component morphisms are not in $\mathcal{A}$. Therefore, the category $\mathcal{A}$ is also free.

Now suppose that $\mathcal{D}$ is a retract of a simplicial computad C. We know that each $\mathcal{D}_n$ is free, but it remains to show that degenerate images of atomic n-arrows are atomic $(n+1)$-arrows. If a morphism is atomic in $\mathcal{D}_n$ and also atomic in C_n, then its degenerate image is atomic since C is a simplicial computad. So suppose h is the degenerate image of an atomic morphism of $\mathcal{D}_n$ which is not atomic in C_n, and consider a factorization $h = g \circ f$ in $\mathcal{D}_{n+1}$. We can also factor $h = g' \circ f'$ in C_n where $g' \mapsto g$ and $f' \mapsto f$ under the retraction map. Using a face map which is a retraction of the degeneracy map being used, either g or f must be an identity map in $\mathcal{D}_n$. $\qquad\qquad\square$

Next we turn to properness of the model category SC.

Proposition 4.5.6 [27, 3.5] *The model category SC is right proper.*

Proof Suppose that

$$\mathcal{A} = \mathcal{B} \times_{\mathcal{D}} C \xrightarrow{\ f\ } B$$

$$\begin{array}{ccc} & & \downarrow{g} \\ C & \xrightarrow{\ h\ } & \mathcal{D} \end{array}$$

is a pullback diagram, where $g \colon \mathcal{B} \to \mathcal{D}$ is a fibration and $h \colon C \to \mathcal{D}$ is a Dwyer–Kan equivalence. We need to show that $f \colon \mathcal{A} \to \mathcal{B}$ is a Dwyer–Kan equivalence.

We first need to show that $\mathrm{Map}_{\mathcal{A}}(x, y) \to \mathrm{Map}_{\mathcal{B}}(fx, fy)$ is a weak equivalence of simplicial sets for any objects x and y of $\mathcal{A}$. However, this fact follows since the model structure on simplicial sets is right proper.

It remains to prove that $\pi_0 A \to \pi_0 B$ is an equivalence of categories. Given that the mapping spaces are weakly equivalent, it suffices to show that $\mathcal{A} \to \mathcal{B}$ is essentially surjective on objects.

Consider an object b of $\mathcal{B}$ and its image $g(b)$ in $\mathcal{D}$. Since $C \to \mathcal{D}$ is a Dwyer–Kan equivalence, there exists some object c of C together with a homotopy equivalence $g(b) \to h(c)$ in $\mathcal{D}$. Since $\mathcal{B} \to \mathcal{D}$ is a fibration, there exists an object b' and homotopy equivalence $b \to b'$ in $\mathcal{B}$ such that $g(b') = h(c)$. Using the fact that $\mathcal{A}$ is a pullback, we have a homotopy equivalence $b \to f((b', c))$, completing the proof. $\qquad\qquad\square$

The following result also holds, but its proof is substantially more difficult, and we do not include it here. It is proved as part of a much more general result by Lurie [88].

Proposition 4.5.7   [88, A.3.2.4] *The model structure SC is left proper.*

Remark 4.5.8   Aside from the rigidity of the structure of simplicial categories, from a homotopy-theoretic point of view there are reasons to look for equivalent models. Although, as we have just seen, the model structure SC is proper, many of the other desirable structures on model categories are not present. For example, the model structure SC is not cartesian, nor does it seem to be a simplicial model category for any natural simplicial structure.

The fact that it is not cartesian can be seen via the following argument. The discrete simplicial category [1] is cofibrant, but taking the pushout product along two copies of the cofibration $\varnothing \rightarrow [1]$ results in the map $\varnothing \rightarrow [1] \times [1]$. But, using the description of the cofibrant simplicial categories above, $[1] \times [1]$ is not cofibrant. Thus the first condition for a cartesian model category fails.

To attempt a simplicial structure, one might attempt to define $C \otimes \Delta[n]$ to be $C \times [n]$ or $C \times F_*[n]$, but one can check that either definition fails axiom (MC7). As a potential remedy one could instead take $C \times I[n]$, where $I[n]$ is the groupoid with $n + 1$ objects and a single isomorphism between any two objects, but this definition leads to problems with axiom (MC6).

4.6 Nerves of Simplicial Categories

There are two ways to consider the nerve of a simplicial category, one which results in a simplicial space and one which results in a simplicial set. Both are significant in defining weak versions of simplicial categories as models for $(\infty, 1)$-categories.

Definition 4.6.1   Let C be a simplicial category, thought of as a functor $\Delta^{op} \rightarrow$ *Cat*. Its *simplicial nerve* is the simplicial space snerve(C) defined by

$$\text{snerve}(C)_{*,m} = \text{nerve}(C_m).$$

Observe that, for any $m \geq 0$,

$$\text{snerve}(C)_{0,m} = \text{ob}(C_m) = \text{ob}(C).$$

In particular, the simplicial set snerve(C)$_0$ is discrete for any simplicial category C. Furthermore, for any $n \geq 0$, the composition in C induces an isomorphism

$$\text{snerve}(C)_n \cong \underbrace{\text{snerve}(C)_1 \times_{\text{snerve}(C)_0} \cdots \times_{\text{snerve}(C)_0} \text{snerve}(C)_1}_{n}.$$

Alternatively, we can compress the same essential information into the structure of a simplicial set. The idea here is to modify the definition of the nerve of a category by replacing the category $[n]$ by a free resolution as described in Definition 3.5.2.

Definition 4.6.2 Let C be a simplicial category. Its *coherent nerve* is the simplicial set $\widetilde{N}(C)$ defined by

$$\widetilde{N}(C)_n = \text{Hom}_{SC}(F_*[n], C)$$

where $F_*[n]$ denotes the free resolution of the category $[n]$.

These two ways to think of the nerve of a simplicial category will be important as we consider the models of Segal categories and quasi-categories, respectively.

5

Complete Segal Spaces

In this chapter we introduce our first model for an up-to-homotopy version of simplicial categories, the complete Segal spaces first defined by Rezk [103]. We saw in Chapter 3 that one motivation for complete Segal spaces is given by the classifying diagram, which is a refinement of the nerve construction which separates out information about isomorphisms. Its output is a simplicial space whose spaces in degrees zero and one determine the rest of the simplicial structure, using composition in the original category. For more general Segal spaces, we only ask that this composition structure be given up to homotopy. Furthermore, rather than isomorphisms we instead distinguish information about homotopy equivalences. In a complete Segal space, the space in degree zero is weakly equivalent to the subspace of the degree one space consisting of homotopy equivalences. A major point in understanding complete Segal spaces is making sense of what these "homotopy equivalences" should be in the context of simplicial spaces.

As we will see in all our models for $(\infty, 1)$-categories with weak composition, in order to have a model structure we must work in a larger underlying category; typically the homotopical properties that we require are not closed under limits and colimits. In the case of complete Segal spaces, the underlying category is that of all simplicial spaces. The model structure for complete Segal spaces is a localization of the Reedy model structure, in such a way that the fibrant objects are precisely the complete Segal spaces. Therefore, once we determine what the localization should be, the existence of the model structure is automatic. What is more difficult to prove is that this model structure is cartesian.

5.1 Segal Spaces

The idea behind Segal spaces goes back to the work of Segal on Γ-spaces and Δ-spaces [109]. We want to have a simplicial diagram of simplicial sets such that the spaces in degree zero ("objects") and degree one ("morphisms") determine all other spaces in the diagram, in that the space in degree n is weakly equivalent to a pullback of n copies of the space in degree one along the space in degree zero. The precise weak equivalences are given by the Segal maps; while we have mentioned these maps previously, we now give an explicit construction.

In the category Δ, consider the maps $\alpha^i \colon [1] \to [n]$, where $0 \le i < n$, given by $\alpha^i(0) = i$ and $\alpha^i(1) = i + 1$. Define the simplicial set

$$G(n) = \bigcup_{i=0}^{n-1} \alpha^i \Delta[1] \subseteq \Delta[n].$$

Alternatively, we can write

$$G(n) = \underbrace{\Delta[1] \amalg_{\Delta[0]} \cdots \amalg_{\Delta[0]} \Delta[1]}_{n}$$

where the right-hand side is the colimit of representables induced by the diagram

$$[1] \xrightarrow{d^0} [0] \xleftarrow{d^1} \cdots \xrightarrow{d^0} [0] \xleftarrow{d^1} [1]$$

in the category Δ.

Let us look at some examples for small values of n. While $G(0) = \Delta[0]$ and $G(1) = \Delta[1]$, when $n = 2$, the simplicial set $G(2) = \Delta[1] \amalg_{\Delta[0]} \Delta[1]$ can be depicted as

In this case, $G(2)$ coincides with the horn $V[2, 1]$, but in higher dimensions these simplicial sets continue to be strings of 1-simplices and are no longer horns; we look further at the relationship between the two in Proposition 7.1.2.

Observe that, for any simplicial set K and any $n \ge 2$, there is an isomorphism of sets

$$\operatorname{Hom}_{SSets}(G(n), K) \cong \underbrace{K_1 \times_{K_0} \cdots \times_{K_0} K_1}_{n},$$

where the right-hand side is the limit of the diagram

$$K_1 \xrightarrow{d_0} K_0 \xleftarrow{d_1} K_1 \xrightarrow{d_0} \cdots \xleftarrow{d_1} K_1.$$

However, we want to work instead in the setting of simplicial spaces. Recall that we can regard a simplicial set K as a discrete simplicial set K^t, where each K_n^t is the discrete simplicial set given by the set K_n. Thus let us consider the inclusion of simplicial spaces $G(n)^t \to \Delta[n]^t$. Given any simplicial space W, it induces a map of simplicial sets

$$\mathrm{Map}(\Delta[n]^t, W) \to \mathrm{Map}(G(n)^t, W)$$

which can be rewritten simply as

$$W_n \to \underbrace{W_1 \times_{W_0} \cdots \times_{W_0} W_1}_{n}.$$

We call such maps *Segal maps* and denote them by φ_n.

As we saw at the end of the previous chapter, if W is the simplicial nerve of a simplicial category, then the Segal maps are all isomorphisms of simplicial sets. We are interested in a weaker situation, where the Segal maps are instead weak equivalences of simplicial sets for all $n \geq 2$. We call such a requirement the *Segal condition*.

Observe also that we have taken honest mapping spaces here. In order to make our constructions homotopy invariant, we should work in the context of a model category and take homotopy mapping spaces, in which we replace W by a fibrant replacement. However, we instead remedy this situation by restricting to simplicial spaces W which are Reedy fibrant. Since all simplicial spaces, and in particular $G(n)^t$ and $\Delta[n]^t$, are Reedy cofibrant, we can continue to take mapping spaces (and a limit, rather than homotopy limit, in the codomains of the Segal maps) and still expect pullback constructions to be homotopy invariant.

Definition 5.1.1 [103, 4.1] A simplicial space W is a *Segal space* if it is Reedy fibrant and the Segal maps

$$\varphi_n \colon W_n \to \underbrace{W_1 \times_{W_0} \cdots \times_{W_0} W_1}_{n}$$

are weak equivalences for all $n \geq 2$.

Example 5.1.2

1 A discrete simplicial space is a Segal space if and only if it is of the form $\mathrm{nerve}(C)^t$ for some category C.
2 A Segal space is the simplicial nerve of a simplicial category if and only if its Segal maps are isomorphisms and its degree zero space is discrete.

3 The classifying diagram of a category, as described in Section 3.3, is a Segal space.

Theorem 5.1.3   [103, 7.1]   *There is a simplicial, left proper, combinatorial model structure SeSp on the category of simplicial spaces such that*

1 the cofibrations are the monomorphisms, so that every object is cofibrant;
2 the fibrant objects are exactly the Segal spaces; and
3 the weak equivalences are the maps f such that $\text{Map}(f, W)$ *is a weak equivalence of simplicial sets for every Segal space W.*

Proof   We localize the Reedy model structure with respect to the set of inclusions

$$\{G(n)^t \to \Delta[n]^t \mid n \geq 2\}.$$

We have seen in the arguments above that a Reedy fibrant simplicial space W is local with respect to these maps if and only if its Segal maps are all weak equivalences. Then the existence of the simplicial model structure follows from Theorem 2.8.2.   □

However, we can say more about this model structure.

Theorem 5.1.4   [103, 7.1]   *The model structure SeSp is cartesian.*

To prove this result, by Proposition 2.9.5, it suffices to prove that, if W is a Segal space, then so is the internal hom object $W^{\Delta[1]^t}$. We follow the approach of Rezk using covers [103, §10].

The morphisms $\alpha^i \colon [1] \to [n]$ in Δ used to define the Segal maps can be generalized to more general $\alpha^i \colon [k] \to [n]$ for any $k \leq n$ and $0 \leq i \leq n - k$, defined by $\alpha^i(j) = i + j$ for each $0 \leq j \leq k$. Each such map gives rise to a corresponding map of simplicial spaces $\alpha^i \colon \Delta[k]^t \to \Delta[n]^t$.

Definition 5.1.5   [103, §10] Let $n \geq 1$. A simplicial space $G \subseteq \Delta[n]^t$ is a *cover* of $\Delta[n]^t$ if:

1 $G_0 = (\Delta[n]^t)_0$, and
2 the simplicial space G has the form

$$G = \bigcup_\lambda \alpha^{i_\lambda} \Delta[k_\lambda]^t$$

where each $k_\lambda \geq 1$ and $0 \leq i_\lambda \leq k_\lambda - 1$.

Observe that $G(n)^t$ is the smallest cover of $\Delta[n]^t$, since each $k_\lambda = 1$ and hence takes the smallest possible value, and $\Delta[n]^t$ is its own largest cover.

Lemma 5.1.6 [103, 10.1] *Let G be a cover of $\Delta[n]^t$. Then the inclusion maps*

$$G(n)^t \xrightarrow{i} G \xrightarrow{j} \Delta[n]^t$$

are weak equivalences in $SeSp$.

Proof We know by construction that the composite map ji is a weak equivalence in $SeSp$, since it is one of the maps with respect to which we localize to obtain the model structure $SeSp$. Using the two-out-of-three property, it thus suffices to prove that the map i is a weak equivalence.

Consider $\alpha^{i_1}\Delta[k_1]^t, \alpha^{i_2}\Delta[k_2]^t \subseteq \Delta[n]^t$. Observe that their intersection is either empty or of the form $\alpha^{i_3}\Delta[k_3]^t$ for some i_3 and k_3. Thus, the cover G, which is a union of such simplicial spaces by definition, can more precisely be written as a colimit of partially ordered sets of the form $\alpha^i\Delta[k]^t$.

Intersecting with the minimal cover $G(n)^t$, we observe that

$$G(n)^t \cap \alpha^i\Delta[k]^t = \alpha^iG(k)^t,$$

so in particular $G(k)^t$ can also be written as a colimit, indexed by the same partially ordered set, of simplicial spaces of the form $\alpha^iG(k)^t$.

Let W be any Segal space. Then, for any $k \geq 0$, the map

$$\mathrm{Map}(\alpha^i\Delta[k]^t, W) \to \mathrm{Map}(\alpha^iG(k)^t, W)$$

is a weak equivalence of simplicial sets. Thus we have, using part (2) of Proposition 2.4.10,

$$\begin{aligned}
\mathrm{Map}(G, W) &= \mathrm{Map}(\mathrm{colim}\,\alpha^i\Delta[k]^t, W) \\
&\simeq \lim \mathrm{Map}(\alpha^i\Delta[k]^t, W) \\
&\simeq \lim \mathrm{Map}(\alpha^iG(k)^t, W) \\
&\simeq \mathrm{Map}(\mathrm{colim}\,\alpha^iG(k)^t, W) \\
&= \mathrm{Map}(G(n)^t, W).
\end{aligned}$$

Thus, the map i is a weak equivalence in $SeSp$, as we wished to show. □

With this result about covers in place, we can prove the following lemma. The decomposition we use here is analogous to the canonical decomposition of $\Delta[1] \times \Delta[n]$ into $(n + 1)$-simplices, used for example in proving homotopy invariance of singular homology [66, 2.10].

Lemma 5.1.7 [103, 10.3] *Let $n \geq 2$. The inclusion map $\Delta[1]^t \times G(n)^t \to \Delta[1]^t \times \Delta[n]^t$ is a weak equivalence in $SeSp$.*

Proof We first define, for any $0 \le i \le n$, the functor $\gamma^i : [n+1] \to [1] \times [n]$ by

$$\gamma^i(j) = \begin{cases} (0, j) & j \le i \\ (1, j-1) & j > i. \end{cases}$$

Similarly, for any $0 \le i \le n$, define the functor $\delta^i : [n] \to [1] \times [n]$ by

$$\delta^i(j) = \begin{cases} (0, j) & j \le i \\ (1, j) & j > i. \end{cases}$$

Then the simplicial space $\Delta[1]^t \times \Delta[n]^t$ can be written as a colimit of the diagram

$$\gamma^0 \Delta[n+1]^t \leftarrow \delta^0 \Delta[n]^t \to \gamma^1 \Delta[n+1]^t \leftarrow \delta^1 \Delta[n]^t \to \cdots \to \gamma^n \Delta[n+1]^t. \quad (5.1)$$

Observe that, for each i, the domains of the maps

$$\gamma^i \Delta[n+1]^t \cap (\Delta[1]^t \times G(n)^t) \to \gamma^i \Delta[n+1]^t$$

and

$$\delta^i \Delta[n]^t \cap (\Delta[1]^t \times G(n)^t) \to \delta^i \Delta[n]^t$$

are covers of their codomains and are hence weak equivalences by Lemma 5.1.6.

However, intersecting each object in the diagram (5.1) with the appropriate minimal cover produces the domain of one of these two kinds of maps; taking the colimit of all these intersections is precisely $\Delta[1]^t \times G[n]^t$. The inclusion $\Delta[1]^t \times G(n)^t \to \Delta[1]^t \times \Delta[n]^t$ is the induced map of colimits, which must be a weak equivalence since the maps between objects in the colimit diagrams are all weak equivalences. $\square$

Proof of Theorem 5.1.4 To prove that $SeSp$ is cartesian, it suffices to prove that, for any Segal space W, the simplicial space $W^{\Delta[1]^t}$ is also a Segal space, by Proposition 2.9.5. Recall from Example 2.9.4 that if W is Reedy fibrant, then so is $W^{\Delta[1]^t}$. It remains to show that, for any $n \ge 2$, the map

$$\mathrm{Map}(\Delta[n]^t, W^{\Delta[1]^t}) \to \mathrm{Map}(G(n)^t, W^{\Delta[1]^t})$$

is a weak equivalence of simplicial sets.

Using Lemma 5.1.7, we can see, however, that

$$\mathrm{Map}(\Delta[n]^t, W^{\Delta[1]^t}) \simeq \mathrm{Map}(\Delta[1]^t \times \Delta[n]^t, W)$$

$$\simeq \mathrm{Map}(\Delta[1]^t \times G(n)^t, W)$$

$$\simeq \mathrm{Map}(G(n)^t, W^{\Delta[1]^t}).$$

Hence, $W^{\Delta[1]'}$ is a Segal space, as we needed to show. $\square$

For later use, we note the following property of Segal spaces.

Lemma 5.1.8 [103, 12.4] *Let W be a Segal space. Then the two diagrams*

$$
\begin{array}{ccc}
W_1 & \xrightarrow{\ d_1\ } & W_0 \\
{\scriptstyle s_0}\downarrow & & \downarrow{\scriptstyle s_0} \\
W_2 & \xrightarrow{\ d_2\ } & W_1
\end{array}
\qquad
\begin{array}{ccc}
W_1 & \xrightarrow{\ d_0\ } & W_0 \\
{\scriptstyle s_1}\downarrow & & \downarrow{\scriptstyle s_0} \\
W_2 & \xrightarrow{\ d_0\ } & W_1
\end{array}
$$

are homotopy pullback squares.

Proof Since W is a Segal space, the map $(d_2, d_0)\colon W_2 \to W_1 \times_{W_0} W_1$ is a weak equivalence. Hence we also have weak equivalences

$$
W_2 \times_{W_1} W_0 \to (W_1 \times_{W_0} W_1) \times_{W_1} W_0 \simeq W_1
$$

and

$$
W_0 \times_{W_1} W_2 \to W_0 \times_{W_1} (W_1 \times_{W_0} W_1) \simeq W_1.
$$

The result follows. $\square$

5.2 Segal Spaces as Categories Up to Homotopy

In this section, we look more closely at the ways in which a Segal space mimics the structure of a simplicial category. In particular, we apply much of the language of simplicial categories in this setting. These ideas were developed by Rezk [103, §5].

Definition 5.2.1 [103, 5.1] Let W be a Segal space.

1 The set of *objects* of W is $\mathrm{ob}(W) = W_{0,0}$.
2 Given $x, y \in \mathrm{ob}(W)$, the *mapping space* between them is the pullback

$$
\begin{array}{ccc}
\mathrm{map}_W(x, y) & \longrightarrow & W_1 \\
\downarrow & & \downarrow \\
\{(x, y)\} & \longrightarrow & W_0 \times W_0.
\end{array}
$$

3 Given $x \in \mathrm{ob}(W)$, its *identity map* is $\mathrm{id}_x = s_0(x) \in \mathrm{map}_W(x, x)_0$.

In the definition of mapping space, the fact that W is Reedy fibrant implies that the right-hand vertical map is a fibration, so that the mapping space is actually given by a homotopy pullback. In particular, the homotopy type of $\text{map}_W(x, y)$ depends only on the equivalence classes of x and y in $\pi_0 W_0$.

More generally, given any $x_0, \ldots, x_n \in \text{ob}(W)$, consider the map $W_n \to W_0^{n+1}$ which is the degree n component of the map $W \to \text{cosk}_0 W$. If W is Reedy fibrant, then this map is a fibration. Define $\text{map}_W(x_0, \ldots, x_n)$ to be the (homotopy) fiber of this map over the point $(x_0, \ldots, x_n)$.

Notice that the map $W_n \to W_0^{n+1}$ factors through the Segal map φ_n, so for a Segal space W we have a commutative diagram

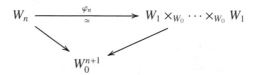

which induces acyclic fibrations

$$\varphi_n \colon \text{map}_W(x_0, \ldots, x_n) \to \text{map}_W(x_0, x_1) \times \cdots \times \text{map}_W(x_{n-1}, x_n)$$

between the fibers of the downward maps. These maps enable us to define composition in a Segal space.

Definition 5.2.2   Let x, y, and z be objects of a Segal space W. Given $(f, g) \in \text{map}_W(x, y) \times \text{map}_W(y, z)$, a *composite* of f and g is a lift of φ_2 to some $k \in \text{map}_W(x, y, z)$; a *result* of this composition is $d_1(k) \in \text{map}_W(x, z)$.

Since φ_2 is only a weak equivalence, not necessarily an isomorphism, we cannot expect that k is unique, nor even that its result is unique. However, we would like to know that it is unique up to homotopy in an appropriate sense.

Definition 5.2.3   Given objects x and y of a Segal space W, two maps $f, g \in \text{map}_W(x, y)_0$ are *homotopic* if they lie in the same component of $\text{map}_W(x, y)$, denoted by $f \simeq g$.

Proposition 5.2.4   [103, 5.4] *Let W be a complete Segal space. Suppose that $f \in \text{map}_W(w, x)_0$, $g \in \text{map}_W(x, y)_0$, and $h \in \text{map}_W(y, z)_0$. Then*

1 $(h \circ g) \circ f \simeq h \circ (g \circ f)$, and
2 $f \circ \text{id}_w \simeq f \simeq \text{id}_x \circ f$.

Definition 5.2.5   [103, 5.5] Given a Segal space W, its *homotopy category* $\text{Ho}(W)$ has objects $W_{0,0}$ and morphisms given by

$$\text{Hom}_{\text{Ho}(W)}(x, y) = \pi_0 \text{map}_W(x, y).$$

Given any $f \in \mathrm{map}_W(x, y)_0$, let $[f]$ be its homotopy class in $\mathrm{Hom}_{\mathrm{Ho}(W)}(x, y)$. Observe that maps $f, g \in \mathrm{map}_W(x, y)_0$ are homotopic if and only if

$$[f] = [g] \in \mathrm{Hom}_{\mathrm{Ho}(W)}(x, y).$$

Definition 5.2.6   [103, 5.5] Let W be a Segal space and x, y objects of W. If $f \in \mathrm{map}_W(x, y)_0$ and $[f]$ is an isomorphism in $\mathrm{Ho}(W)$, then f is a *homotopy equivalence*.

Notice in particular that for any object x, $\mathrm{id}_x = s_0(x)$ is a homotopy equivalence.

Let us look at this notion more closely. If $[f]$ is an isomorphism in $\mathrm{Ho}(W)$, then there exist $g, h \in \mathrm{map}_W(y, x)$ such that $f \circ g \simeq \mathrm{id}_y$ and $h \circ g \simeq \mathrm{id}_x$. Using Proposition 5.2.4, we can conclude that $g \simeq h$, since $g \simeq h \circ f \circ g \simeq h$.

Consider the simplicial set $Z(3) \subseteq \Delta[3]$ given by the nerve of the category $(0 \to 2 \leftarrow 1 \to 3)$. Given any Segal space W, the inclusion $Z(3)^t \to \Delta[3]^t$ induces a fibration of simplicial sets

$$W_3 \cong \mathrm{Map}(\Delta[3]^t, W) \to \mathrm{Map}(Z(3)^t, W).$$

However, we can use the definition of $Z(3)$ to write

$$\mathrm{Map}(Z(3)^t, W) \cong \lim(W_1 \xrightarrow{d_0} W_0 \xleftarrow{d_0} W_1 \xrightarrow{d_1} W_0 \xleftarrow{d_1} W_1) \simeq W_1 \times_{W_0} W_1 \times_{W_0} W_1.$$

Then we can see that $f \in \mathrm{map}_W(x, y)$ is a homotopy equivalence if and only if any element $(\mathrm{id}_x, f, \mathrm{id}_y)$ of $\mathrm{Map}(Z(3)^t, W)$ can be lifted to an element of W_3.

We can use this description to show that homotopy equivalences comprise components of mapping spaces.

Proposition 5.2.7   [103, 5.8] *If* $[f] = [g] \in \pi_0 \mathrm{map}_W(x, y)$, *then* f *is a homotopy equivalence if and only if* g *is a homotopy equivalence.*

Proof   Suppose f is a homotopy equivalence. Let $\gamma: \Delta[1] \to W_1$ be a path from f to g in $\mathrm{map}_W(x, y)$. We know that the point $(\mathrm{id}_x, f, \mathrm{id}_y) \in \mathrm{Map}(Z(3)^t, W)$ admits a lift to W_3, which is given by a map $\Delta[0] \to W_3$. Thus we can consider the diagram

$$
\begin{array}{ccc}
\Delta[0] & \longrightarrow & W_3 \\
{\scriptstyle 0}\big\downarrow & \nearrow & \big\downarrow \\
\Delta[1] & \longrightarrow & \mathrm{Map}(Z(3)^t, W).
\end{array}
$$

Since the left-hand vertical map is an acyclic cofibration and the right-hand vertical map is a fibration, the dotted arrow lift exists, thus defining a lift of $(\mathrm{id}_x, g, \mathrm{id}_y)$ to W_3, making g a homotopy equivalence. $\square$

Definition 5.2.8   Let W be a Segal space. The *space of homotopy equiva-lences* $W_{\text{heq}} \subseteq W_1$ consists of the components whose 0-simplices are homotopy equivalences.

Notice that the degeneracy map s_0 factors through W_{heq}, so we can consider the map $s_0 \colon W_0 \to W_{\text{heq}}$.

5.3 Complete Segal Spaces

The main problem with considering Segal spaces as up-to-homotopy models for simplicial categories is that they have a space, rather than a set, of objects. However, if we return to the properties of the classifying diagram that we observed in Section 3.3, we observe that we have not yet imposed the condition that the space of objects be weakly equivalent to the space of homotopy equiv-alences. This condition turns out to remedy the difficulty of having a space of objects, rather than a set; this idea is made explicit in the comparison with Segal categories in the next chapter.

Since we have defined the space W_{heq} of homotopy equivalences in a Segal space W, we can use it to make the following definition.

Definition 5.3.1   [103, §6] A Segal space W is *complete* if the map $s_0 \colon W_0 \to W_{\text{heq}}$ is a weak equivalence.

Consider the category I with two objects and a single isomorphism between them. Define the simplicial set $E = \text{nerve}(I)$. There is an inclusion map $\Delta[1] \to E$ which, after passing to a map of simplicial spaces $\Delta[1]^t \to E^t$, induces a map of simplicial sets

$$\text{Map}(E^t, W) \to \text{Map}(\Delta[1]^t, W) = W_1$$

whose image is in W_{heq}. In fact, $\text{Map}(E^t, W) \to W_{\text{heq}}$ is a weak equivalence of simplicial sets.

The proof of the following result is technical, but key to understanding the model structure for complete Segal spaces. Its proof can be found in [103, §11].

Proposition 5.3.2   [103, 6.2] *The collapse map $E \to \Delta[0]$ induces a map $W_0 \to \text{Map}(E^t, W)$ which is a weak equivalence if and only if $s_0 \colon W_0 \to W_{\text{heq}}$ is a weak equivalence.*

Thus, we can localize the model category $SeSp$ with respect to the map $E^t \to \Delta[0]^t$ to get a model category whose fibrant objects are the complete Segal spaces.

Theorem 5.3.3 [103, 7.2] *There is a simplicial, left proper, combinatorial model structure on the category of simplicial spaces such that:*

1 *the cofibrations are the monomorphisms, so that every object is cofibrant;*
2 *the fibrant objects are exactly the complete Segal spaces; and*
3 *the weak equivalences are the maps f such that* $\mathrm{Map}(f, W)$ *is a weak equivalence of simplicial sets for every complete Segal space W.*

We denote this model structure by CSS.

It follows that a map $f: X \to Y$ of complete Segal spaces is a weak equivalence in CSS if and only if f is a levelwise weak equivalence of simplicial sets, using Proposition 2.8.4.

As for the model structure $SeSp$, the proof of the following result requires more work.

Theorem 5.3.4 [103, 7.2, 12.1] *The model structure CSS is cartesian.*

For the proof, we need the following definition.

Definition 5.3.5 [103, 12.2] A map $f: X \to Y$ of simplicial sets is a *homotopy monomorphism* if the diagram

$$
\begin{array}{ccc}
X & \xrightarrow{\ \mathrm{id}\ } & X \\
{\scriptstyle \mathrm{id}}\downarrow & & \downarrow{\scriptstyle f} \\
X & \xrightarrow{\ f\ } & Y
\end{array}
$$

is a homotopy pullback square.

The next result essentially follows from the fact that, for a Segal space W, the space W_{heq} is a subspace of W_1.

Proposition 5.3.6 *If W is a complete Segal space, then the degeneracy map $s_0: W_0 \to W_1$ is a homotopy monomorphism.*

We need one general result about homotopy monomorphisms.

Lemma 5.3.7 *Let $X, Y: \mathcal{D} \to SSets$ be diagrams and $X \to Y$ a natural transformation such that, for every object d of $\mathcal{D}$, the map $X_d \to Y_d$ is a homotopy monomorphism. Then $\mathrm{holim}_{\mathcal{D}} X_d \to \mathrm{holim}_{\mathcal{D}} Y_d$ is also a homotopy monomorphism.*

Proof This result is a consequence of the fact that homotopy limits commute with one another, by Proposition 2.3.7.   □

Proof of Theorem 5.3.4 By Proposition 2.9.5, it suffices to prove that if W is a complete Segal space, then so is $W^{\Delta[1]^t}$. We have already proved that $W^{\Delta[1]^t}$ is a Segal space, so we must prove that the map

$$(W^{\Delta[1]^t})_0 \to (W^{\Delta[1]^t})_{\text{heq}}$$

is a weak equivalence of simplicial sets.

First recall that $(W^{\Delta[1]^t})_0 = \text{Map}(\Delta[1]^t, W) = W_1$. Similarly, we have

$$(W^{\Delta[1]^t})_1 = \text{Map}(\Delta[1]^t \times \Delta[1]^t, W) \simeq W_2 \times_{W_1} W_2.$$

Thus, the degeneracy map $s_0 \colon (W^{\Delta[1]^t})_0 \to (W^{\Delta[1]^t})_1$ is given by taking limits of the rows in the diagram

$$
\begin{array}{ccccc}
W_1 & \xrightarrow{\ =\ } & W_1 & \xleftarrow{\ =\ } & W_1 \\
{\scriptstyle s_0}\downarrow & & {\scriptstyle \text{id}}\downarrow & & \downarrow{\scriptstyle s_1} \\
W_2 & \xrightarrow{\ d_1\ } & W_1 & \xleftarrow{\ d_1\ } & W_2.
\end{array}
$$

Since W is a complete Segal space, by Proposition 5.3.6, the map $s_0 \colon W_0 \to W_1$ is a homotopy monomorphism. It follows that the maps $s_0, s_1 \colon W_1 \to W_2$ are also homotopy monomorphisms, since they are weakly equivalent to the maps

$$W_1 \times_{W_0} s_0 \colon W_1 \times_{W_0} W_0 \to W_1 \times_{W_0} W_1$$

and

$$s_0 \times_{W_0} W_1 \colon W_0 \times_{W_0} W_1 \to W_1 \times_{W_0} W_1,$$

respectively. Applying Lemma 5.3.7 we see that the map

$$s_0 \colon (W^{\Delta[1]^t})_0 \to (W^{\Delta[1]^t})_1$$

is a homotopy monomorphism.

Since inclusion maps are homotopy monomophisms, we also have that

$$(W^{\Delta[1]^t})_{\text{heq}} \to (W^{\Delta[1]^t})_1$$

is also a homotopy monomorphism, and the map

$$s_0 \colon (W^{\Delta[1]^t})_0 \to (W^{\Delta[1]^t})_1$$

factors through it. Thus, it suffices to prove that the map

$$(W^{\Delta[1]^t})_0 \to (W^{\Delta[1]^t})_{\text{heq}}$$

is surjective on components.

Let $f, g \colon W^{\Delta[1]^t} \to W^{\Delta[0]^t}$ be the maps induced by the two face inclusions

$d^0, d^1 : \Delta[0]^t \to \Delta[1]^t$. Using Lemma 5.1.8, notice that a point $x \in (W^{\Delta[1]^t})_{\text{heq}}$ lies in a component in the image of $(W^{\Delta[1]^t})_0$ if and only if its images fx and gx in $(W^{\Delta[0]^t})_1 \simeq W_1$ are homotopy equivalences in W. But if x is a homotopy equivalence, then so are its images under f and g. □

5.4 Categorical Equivalences

Although we now have the model category for complete Segal spaces, we want to gain a deeper understanding of how morphisms between Segal spaces can behave sufficiently like equivalences of categories. In this section, we discuss one approach.

Definition 5.4.1 Let $f, g : U \to V$ be maps between Segal spaces. A *categorical homotopy* between f and g is given by a map $H : U \times E \to V$ such that the diagram

$$
\begin{array}{c}
U \\
{\scriptstyle \text{id} \times i_0} \downarrow \qquad \searrow^{f} \\
U \times E \xrightarrow{\ H\ } V \\
{\scriptstyle \text{id} \times i_1} \uparrow \qquad \nearrow_{g} \\
U
\end{array}
$$

commutes. Equivalently, a categorical homotopy is given by a map H' or H'' as given in its respective commutative diagram

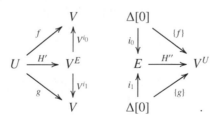

Proposition 5.4.2 [103, 13.2] *Suppose that U is a Segal space and W is a complete Segal space. Then maps $f, g : U \to W$ are categorically homotopic if and only if there exists a homotopy $K : U \to W^{\Delta[1]^t}$ such that the two composites*

$$U \to W^{\Delta[1]^t} \to W^{\Delta[0]^t} = W,$$

induced by the two inclusions $\Delta[0] \to \Delta[1]$, are the maps f and g.

Proof   Consider the diagram

$$
\begin{array}{ccc}
W & \longrightarrow & W^{E^t} \\
\downarrow & & \downarrow \\
W^{\Delta[1]^t} & \longrightarrow & W \times W
\end{array}
$$

in the category of functors $\Delta^{op} \to SSets$ over $W \times W$, where the maps out of
$W = W^{\Delta[0]^t}$ are induced by the maps $\Delta[1] \to \Delta[0]$ and $E \to \Delta[0]$, respectively.
The map $W \to W^{\Delta[1]^t}$ is a levelwise weak equivalence since it is the inclusion
of the subspace of constant paths in the path space. The map $W \to W^{E^t}$ is a
levelwise weak equivalence because W is a complete Segal space. Further, the
maps $W^{\Delta[1]^t} \to W \times W$ and $W^{E^t} \to W \times W$ are both fibrations in the Reedy
model structure.

If f and g are simplicially homotopic, then there is a map $K \colon U \to W^{\Delta[1]^t}$
such that the composite with the map to $W \times W$ is precisely (f, g). To obtain
a categorical homotopy, we need this composite map to factor through W^{E^t}. If
the maps out of W were fibrations (in particular if W were a pullback), then we
could factor K through W. However, they are not.

To remedy this situation, take the pullback

$$
P = W^{\Delta[1]^t} \times_{W \times W} W^{E^t}
$$

and then factor the natural map $W \to P$ as a levelwise acyclic cofibration
followed by a fibration: $W \to Q \to P$. Now the maps out of P are fibrations,
since P is a pullback along fibrations, and by composition so are the maps
$Q \to W^{\Delta[1]^t}$ and $Q \to W^{E^t}$. But we also know that these two maps are levelwise
weak equivalences, by the two-out-of-three property. Therefore, we can lift to
obtain a map $U \to Q$ which we can compose with $Q \to W^{E^t}$ to obtain the
desired categorical homotopy.

To prove the converse, observe that this argument is completely symmetric;
if we begin instead with a categorical homotopy $U \to W^{E^t}$, we can produce a
simplicial homotopy $U \to W^{\Delta[1]^t}$.   □

Definition 5.4.3   A map $g \colon U \to V$ between Segal spaces is a *categorical
equivalence* if there exist maps $f, h \colon V \to U$ together with categorical homotopies $gf \simeq \mathrm{id}_V$ and $hg \simeq \mathrm{id}_U$.

Example 5.4.4   If $U = \mathrm{nerve}(C)^t$ and $V = \mathrm{nerve}(\mathcal{D})^t$ are discrete nerves of
categories, then a categorical homotopy between maps $U \to V$ corresponds
exactly to a natural isomorphism of functors between C and $\mathcal{D}$. Then a categorical homotopy is precisely defined by an equivalence of categories.

Proposition 5.4.5   [103, 13.4] *The following are equivalent for a map $g: U \to$*
V between complete Segal spaces:

1 the map g is a categorical equivalence,
2 the map g is a simplicial homotopy equivalence, and
3 the map g is a Reedy weak equivalence.

Proof   We know already that (1) is equivalent to (2) by Proposition 5.4.2.
Then (2) is equivalent to (3) since complete Segal spaces are Reedy fibrant and
cofibrant; thus weak equivalences between them agree with simplicial homotopy equivalences.   □

Proposition 5.4.6   [103, 13.5] *Suppose that A, B, and W are Segal spaces.*
If $f, g: A \to B$ are categorically homotopic maps, then the induced maps
$W^f, W^g: W^B \to W^A$ are categorically homotopic.

Proof   Let $H: A \times E \to B$ be a categorical homotopy between f and g. Then
there is a categorical homotopy

$$W^H: W^B \to W^{A \times E^t} \cong (W^A)^{E^t}$$

between W^f and W^g.   □

Corollary 5.4.7   *Let A, B, and W be Segal spaces. If $f: A \to B$ is a categori-*
cal equivalence, then so is the induced map $W^f: W^B \to W^A$.

Now we can use these results to understand categorical equivalences in the
model category CSS.

Proposition 5.4.8   [103, 13.6] *If $f: U \to V$ is a categorical equivalence*
between Segal spaces, then it is a weak equivalence in CSS.

Proof   By definition of the model structure CSS as a localization, f is a weak
equivalence in CSS if and only if

$$\mathrm{Map}(f, W): \mathrm{Map}(V, W) \to \mathrm{Map}(U, W)$$

is a weak equivalence of simplicial sets for any complete Segal space W. However, each such map is a weak equivalence if and only if each map $W^f: W^V \to$
W^U is a Reedy weak equivalence for any complete Segal space W, since for
each $n \geq 0$, $(W^f)_n \cong \mathrm{Map}(f, W^{\Delta[n]^t})$ and we know that $W^{\Delta[n]^t}$ is a complete
Segal space. But then by Corollary 5.4.7, we know that W^f must be a categor-
ical equivalence, and it is such between complete Segal spaces, so it is hence a
Reedy weak equivalence by Proposition 5.4.5.   □

Finally, we record the following result for later reference.

Lemma 5.4.9 *A categorical equivalence between Segal spaces induces an equivalence of homotopy categories.*

Proof Suppose U is a Segal space. Using the isomorphisms

$$\mathrm{Ho}(U \times E^t) \cong \mathrm{Ho}(U) \times \mathrm{Ho}(E^t) \cong \mathrm{Ho}(U) \times I[1],$$

one can verify that categorically homotopic maps between Segal spaces induce isomorphic functors on homotopy categories. In particular, categorical equivalences induce equivalences of homotopy categories.   □

5.5 Dwyer–Kan Equivalences

In this section, we consider a notion of equivalence of Segal spaces which mimics the definition of Dwyer–Kan equivalence between simplicial categories, so much so that we use the same name.

Definition 5.5.1 A map $f \colon W \to Z$ of Segal spaces is a *Dwyer–Kan equivalence* if:

1 the map $\mathrm{map}_W(x, y) \to \mathrm{map}_Z(fx, fy)$ is a weak equivalence of simplicial sets for any objects x and y of W, and
2 the functor $\mathrm{Ho}(W) \to \mathrm{Ho}(Z)$ is an equivalence of categories.

The first part of the following lemma follows from the definition, using the fact that both weak equivalences of simplicial sets and equivalences of categories satisfy the two-out-of-three property. The second part is also not difficult to check.

Lemma 5.5.2 [103, 7.5] *Dwyer–Kan equivalences satisfy the two-out-of-three property. Furthermore, if we have composable maps of Segal spaces*

$$U \xrightarrow{f} V \xrightarrow{g} W \xrightarrow{h} X$$

such that gf and hg are Dwyer–Kan equivalences, then the maps f, g, and h are all Dwyer–Kan equivalences.

We want to understand how Dwyer–Kan equivalences relate to the several other notions of equivalence that we have introduced. We begin by understanding when Dwyer–Kan equivalences are actually levelwise weak equivalences of simplicial sets.

Proposition 5.5.3 [103, 7.6] *A map $f \colon U \to V$ between complete Segal spaces is a Dwyer–Kan equivalence if and only if it is a levelwise weak equivalence of simplicial sets.*

Proof A levelwise weak equivalence is always a Dwyer–Kan equivalence, so we need only prove the reverse implication. Suppose that U and V are complete Segal spaces and that $f : U \to V$ is a Dwyer–Kan equivalence. Then for any $x, y \in U_{0,0}$, we have that $\mathrm{map}_U(x, y) \to \mathrm{map}_V(fx, fy)$ is a weak equivalence of simplicial sets, and that U_0 and V_0 have the same set of components.

Recall that $\mathrm{map}_U(x, y)$ can be written as a (homotopy) pullback

$$
\begin{array}{ccc}
\mathrm{map}_U(x, y) & \longrightarrow & U_1 \\
\downarrow & & \downarrow \\
\{(x, y)\} & \longrightarrow & U_0 \times U_0
\end{array}
$$

and that U_{heq} consists of components of U_1. Therefore, we can describe the mapping space of equivalences $h\,\mathrm{map}_U(x, y)$ by restricting U_1 to U_{heq} and taking the pullback

$$
\begin{array}{ccc}
h\,\mathrm{map}_U(x, y) & \longrightarrow & U_{\mathrm{heq}} \\
\downarrow & & \downarrow \\
\{(x, y)\} & \longrightarrow & U_0 \times U_0.
\end{array}
$$

We obtain that for any pair of objects (x, y), $h\,\mathrm{map}_U(x, y) \to h\,\mathrm{map}_V(fx, fy)$ is a weak equivalence.

By precomposing with the degeneracy map s_0, we obtain a commutative diagram

$$
\begin{array}{ccc}
U_0 & \longrightarrow & U_0 \times U_0 \\
\downarrow & & \downarrow \\
V_0 & \longrightarrow & V_0 \times V_0,
\end{array}
$$

which is a homotopy pullback diagram since taking the horizontal fibers results in a weak equivalence $h\,\mathrm{map}_U(x, y) \to h\,\mathrm{map}_V(fx, fy)$. It follows that the map $U_0 \to V_0$ is a weak equivalence. Then $U_1 \to V_1$ must also be a weak equivalence, since

$$
\begin{array}{ccc}
U_1 & \longrightarrow & U_0 \times U_0 \\
\downarrow & & \downarrow \\
V_1 & \longrightarrow & V_0 \times V_0
\end{array}
$$

is a pullback diagram with horizontal maps fibrations, since U and V are Reedy fibrant. Using the Segal condition, we have established that f is a levelwise weak equivalence. $\qquad\square$

Next, we show how Dwyer–Kan equivalences can arise from categorical equivalences between Segal spaces. We first give a lemma.

Lemma 5.5.4   [103, 13.9] *If W is a Segal space, then the map $W \to W^{E^t}$ induced from the map $E \to \Delta[0]$ and both maps $W^{E^t} \to W$ induced from the two maps $\Delta[0] \to E$ are Dwyer–Kan equivalences.*

Proof   Using the two-out-of-three property for Dwyer–Kan equivalences and the fact that either composite $W \to W^{E^t} \to W$ is the identity, it suffices to prove that the map $W \to W^{E^t}$ is a Dwyer–Kan equivalence. Since $\Delta[1]^t \to E^t$ is a categorical equivalence, by Corollary 5.4.7 and Lemma 5.4.9, we know that $\mathrm{Ho}(W) \to \mathrm{Ho}(W^{E^t})$ is an equivalence of categories. If we let $j: E \to \Delta[0]$, so that $j^*: W \to W^{E^t}$ is the map in question, it remains to show that, for any objects $x, y \in \mathrm{ob}(W)$, the induced map $\mathrm{map}_W(x, y) \to \mathrm{map}_{W^{E^t}}(j^*x, j^*y)$ is a weak equivalence of simplicial sets.

Let $i: \Delta[1] \to E$ be the inclusion and $i^*: W^{E^t} \to W^{\Delta[1]^t}$ be the induced map. Observe that $s^0 = j \circ i: \Delta[1] \to \Delta[0]$. Consider the diagram

$$
\begin{array}{ccccc}
W_1 & \xrightarrow{\;\;j^*\;\;} & (W^{E^t})_1 & \xrightarrow{\;\;i^*\;\;} & (W^{\Delta[1]^t})_1 \\
{\scriptstyle (d_1,d_0)}\downarrow & & \downarrow{\scriptstyle (d_1,d_0)} & & \downarrow{\scriptstyle (d_1,d_0)} \\
W_0 \times W_0 & \xrightarrow{\;\;j^*\;\;} & (W^{E^t})_0 \times (W^{E^t})_0 & \xrightarrow{\;\;i^*\;\;} & (W^{\Delta[1]^t})_0 \times (W^{\Delta[1]^t})_0.
\end{array}
$$

Notice that the two maps labeled by i^* are homotopy monomorphisms, since, for any $n \geq 0$,

$$(W^{E^t})_1 \simeq (W^{\Delta[1]^t})_{\mathrm{heq}} \simeq (W^{\Delta[1]^t})_0$$

and

$$(W^{\Delta[1]^t})_0 \to (W^{\Delta[1]^t})_1$$

is a homotopy monomorphism by Lemma 5.3.6. Then observe that the outer rectangle of the diagram is isomorphic to the diagram

$$
\begin{array}{ccc}
W_1 & \xrightarrow{\;\;(s_0,s_1)\;\;} & W_2 \times_{W_1} W_2 \\
{\scriptstyle (d_1,d_0)}\downarrow & & \downarrow{\scriptstyle (d_2 p_1, d_0 p_2)} \\
W_0 \times W_0 & \xrightarrow{\;\;s_0 \times s_0\;\;} & W_1 \times W_1
\end{array}
$$

where $W_2 \times_{W_1} W_2$ is the pullback along two copies of the map $d_1: W_2 \to W_1$, and p_i denotes the ith projection. But this square can be checked to be a homotopy pullback diagram.

Thus, taking homotopy fibers over any $x, y \in \mathrm{ob}(W)$, the composite

$$\mathrm{map}_W(x, y) \xrightarrow{j^*} \mathrm{map}_{W^{E^t}}(j^*x, j^*y) \to \mathrm{map}_{W^{\Delta[1]^t}}(s_0 x, s_0 y)$$

is a weak equivalence. Since we have shown that i^* is a homotopy monomor-phism, it follows that j^* is a weak equivalence.   □

Proposition 5.5.5   [103, 13.8] *If* $g \colon U \to V$ *is a categorical equivalence between Segal spaces, then it is a Dwyer–Kan equivalence.*

Proof   Suppose $g \colon U \to V$ is a categorical equivalence, with U and V Segal spaces. Then there exist maps $f, h \colon V \to U$ and categorical homotopies H from $g \circ f$ to id_V and K from $h \circ g$ to id_U. Consider the diagram

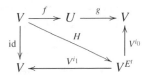

where $i_0, i_1 \colon \Delta[0] \to E$ are the two inclusions. By Lemma 5.5.4, the maps U^{i_0} and U^{i_1} are Dwyer–Kan equivalences. By the two-out-of-three property, K is thus a Dwyer–Kan equivalence, and hence also $h \circ g$. Applying the analogous argument to the diagram

$$
\begin{array}{ccccc}
V & \xrightarrow{\ f\ } & U & \xrightarrow{\ g\ } & V \\
{\scriptstyle \mathrm{id}}\downarrow & \searrow{\scriptstyle H} & & & \uparrow{\scriptstyle V^{i_0}} \\
V & \xleftarrow{\ V^{i_1}\ } & & V^{E^t} &
\end{array}
$$

we obtain that $g \circ f$ is a Dwyer–Kan equivalence. Then we can apply the second statement of Lemma 5.5.2 to conclude that g is a Dwyer–Kan equivalence.   □

Theorem 5.5.6   [103, 7.7] *A map* $f \colon W \to Z$ *of Segal spaces is a Dwyer–Kan equivalence if and only if it is a weak equivalence in CSS.*

To prove this theorem, we need a completion functor taking a Segal space to a complete Segal space in a particularly nice way. Suppose W is a Segal space. Let $I(m)$ be the groupoid with $m + 1$ objects and a single isomorphism between any two objects, and $E(m) = \mathrm{nerve}(I(m))$. For a fixed $n \geq 0$, we can define a simplicial space given by

$$[m] \mapsto \mathrm{Map}(E(m)^t, W^{\Delta[n]^t}).$$

As we range over all values of n, we obtain a bisimplicial space (or trisimpli-cial set), i.e., a functor $\Delta^{op} \to SSets^{\Delta^{op}}$; we first need to take an appropriate diagonal to recover a simplicial space. For each $n \geq 0$, define

$$\widetilde{W}_n = \mathrm{diag}([m] \mapsto \mathrm{Map}(E(m)^t, W^{\Delta[n]^t})).$$

Using the equivalences

$$\mathrm{Map}(E(m)^t, W^{\Delta[n]^t}) \simeq \mathrm{Map}(E(m)^t \times \Delta[n]^t, W)$$
$$\simeq \mathrm{Map}(\Delta[n]^t, W^{E(m)^t})$$
$$\simeq (W^{E(m)^t})_n,$$

we can alternatively write

$$\widetilde{W}_n = \mathrm{diag}([m] \mapsto (W^{E(m)^t})_n).$$

These simplicial sets assemble to form a simplicial space $\widetilde{W}$ equipped with a natural map $W \to \widetilde{W}$ induced by the collapse maps $E(m) \to E(0) = \Delta[0]$. Since this simplicial space may not be Reedy fibrant, we apply a functorial Reedy fibrant replacement $\widetilde{W} \to \widehat{W}$. We consider the composite $W \to \widetilde{W} \to \widehat{W}$.

Definition 5.5.7 [103, §14] The assignment $W \mapsto \widehat{W}$ defines a *completion functor* taking any Segal space to a complete Segal space. Each component map $i_W \colon W \to \widehat{W}$ is called a *completion map*.

We consider some of the nice properties of this completion procedure. First, we consider the relationship between the completion and the classifying diagram functor N from Section 3.3.

Lemma 5.5.8 [103, 14.2] *If C is a category, then $\widehat{\mathrm{nerve}(C)^t} \cong NC$. In particular, $NI(1) \cong \widehat{E^t}$ and $\widehat{E^t}$ are weakly equivalent to $\Delta[0]$.*

Proof Using definitions, we can show that there are isomorphisms

$$(\widehat{\mathrm{nerve}(C)^t})_{n,m} = \mathrm{Hom}_{SSets^{\Delta^{op}}}(E(m) \times \Delta[n]^t, \mathrm{nerve}(C)^t)$$
$$\cong \mathrm{Hom}_{Cat}(I(m) \times [n], C)$$
$$= (NC)_{n,m}.$$

Since the category $I(1)$ is equivalent to the terminal category and the classifying diagram functor N takes equivalences of categories to weak equivalences by Proposition 3.3.4, the second statement follows.   □

Lemma 5.5.9 [103, 14.3] *If $U \to V$ is a categorical equivalence between Segal spaces, then $\widehat{U} \to \widehat{V}$ is a levelwise weak equivalence.*

Proof Observe that, for any Segal space U, we have $\widetilde{U \times E} \simeq \widetilde{U} \times \widetilde{E}$, and

$\widetilde{E}$ is contractible by Lemma 5.5.8. It follows that the completion of a categorical homotopy is a simplicial homotopy. In particular, the completion of a categorical equivalence is a Dwyer–Kan equivalence. □

The following result gives the most important properties of the completion map.

Proposition 5.5.10 [103, §14] *Given any Segal space W,*

1 the completion $\widehat{W}$ is a complete Segal space;

2 the completion map i_W is a weak equivalence in CSS; and

3 the completion map i_W is a Dwyer–Kan equivalence.

Proof By Proposition 5.4.6, the maps $W \to W^{E(m)^t}$ are categorical equivalences and hence weak equivalences in *CSS* by Proposition 5.4.8. Since the map $W \to \widetilde{W}$ is given by the induced map on homotopy colimits of such maps, it is also a weak equivalence in *CSS*. It follows that the composite $W \to \widetilde{W} \to \widehat{W}$ is a weak equivalence in *CSS*.

Let $\delta \colon [n] \to [m]$ be a map in Δ, and consider the induced diagram

$$
\begin{array}{ccc}
(W^{E(m)^t})_k & \longrightarrow & (W^{E(n)^t})_k \\
\downarrow & & \downarrow \\
(W^{E(m)^t})_0^{k+1} & \longrightarrow & (W^{E(n)^t})_0^{k+1}.
\end{array}
$$

By Proposition 5.5.5, the map $W^{E(m)^t} \to W^{E(n)^t}$ is a Dwyer–Kan equivalence, so for any $x_0, \ldots, x_k \in \mathrm{ob}(W)$, the map

$$
\mathrm{map}_{W^{E(m)^t}}(x_0, \ldots, x_k) \to \mathrm{map}_{W^{E(n)^t}}(\delta x_0, \ldots, \delta x_k)
$$

between the fibers of the vertical maps in the diagram is a weak equivalence. It follows that the induced map of diagonals has homotopy fibers weakly equivalent to the appropriate k-fold product of mapping spaces. Thus $\widehat{W}$ is a Segal space and $\mathrm{map}_W(x, y) \to \mathrm{map}_{\widehat{W}}(i_W x, i_W y)$ is a weak equivalence for every $x, y \in \mathrm{ob}(W)$. By construction, $\pi_0 W_0 \to \pi_0 \widehat{W}$ is surjective, so the functor $\mathrm{Ho}(W) \to \mathrm{Ho}(\widehat{W})$ is essentially surjective. Hence, $W \to \widehat{W}$ is a Dwyer–Kan equivalence.

It remains to prove that $\widehat{W}$ is complete. Let $j \colon E(n) \to E(m)$ be any map, and consider the induced square

$$
\begin{array}{ccc}
(W^{E(m)^t})_{\text{heq}} & \xrightarrow{\ (W^j)_{\text{heq}}\ } & (W^{E(n)^t})_{\text{heq}} \\
\downarrow & & \downarrow \\
(W^{E(m)^t})_0 \times (W^{E(m)^t})_0 & \xrightarrow{(W^j)_0 \times (W^j)_0} & (W^{E(n)^t})_0 \times (W^{E(n)^t})_0.
\end{array}
$$

Since the map $W^{E(m)^t} \to W^{E(n)^t}$ is a categorical equivalence by Proposition 5.4.6 and thus a Dwyer–Kan equvialence by Proposition 5.5.5, we may conclude that

$$
h \operatorname{map}_{W^{E(m)^t}}(x, y) \to h \operatorname{map}_{W^{E(n)^t}}(jx, jy)
$$

is a weak equivalence for all $x, y \in (W^{E(m)^t})_0$. It follows that the above diagram is a homotopy pullback square and hence the induced map on diagonals has homotopy fiber weakly equivalent to $h \operatorname{map}_W(x, y)$. In other words, we have

$$
\widehat{W}_{\text{heq}} \simeq \operatorname{diag}([m] \mapsto (W^{E(m)^t})_{\text{heq}}).
$$

Since $(W^{E(m)^t})_{\text{heq}} \simeq (W^{E(m)^t \times E(1)^t})_0$ by Proposition 5.3.2, there is an equivalence $\widehat{W}_{\text{heq}} \simeq (\widehat{W^{E(1)^t}})_0$. But Lemma 5.5.9 shows that since $W^{E(1)^t}$ is categorically equivalent to W, $\widehat{W}_{\text{heq}} \simeq (\widehat{W^{E(1)^t}})_0 \simeq \widehat{W}_0$, so $\widehat{W}$ is complete. $\square$

Proof of Theorem 5.5.6 Let $f : U \to V$ be a map of Segal spaces. Using the properties of the completion functor from Proposition 5.5.10, we have that f is a weak equivalence in CSS if and only if $\widehat{f} : \widehat{U} \to \widehat{V}$ is a weak equivalence. But $\widehat{f}$ is a map of complete Segal spaces, which are the fibrant objects in a localized model structure, so it is a weak equivalence if and only if it is a levelwise weak equivalence. But then by Proposition 5.5.3, $\widehat{f}$ is a levelwise weak equivalence if and only if $\widehat{f}$ is a Dwyer–Kan equivalence. Finally, since Dwyer–Kan equivalences satisfy the two-out-of-three property, we also know that f is a Dwyer–Kan equivalence if and only if $\widehat{f}$ is. Therefore, f is a weak equivalence if and only if it is a Dwyer–Kan equivalence. $\square$

The completion functor should be regarded as a simplicial analogue of the classifying diagram construction, in the following sense. If C is a simplicial category, its simplicial nerve is a Segal space, at least up to Reedy fibrant replacement. We can then apply the completion functor to this Segal space to obtain a complete Segal space. In Chapter 9, we explore this relationship further.

Observe that, thus far, we have only defined complete Segal spaces and established an appropriate model structure in which the fibrant objects are the complete Segal spaces. In order to show that they give another good model for

(∞, 1)-categories, by showing that this model structure is Quillen equivalent to the model category for simplicial categories, we need to consider a third intermediate model, that of Segal categories.

6

Segal Categories

Segal categories can be seen as an intermediate model between simplicial categories and complete Segal spaces. Like complete Segal spaces, they are simplicial spaces with a notion of up-to-homotopy composition encoded by a Segal condition, but rather than imposing the completeness condition, we simply ask that the space in degree zero be discrete. Thus, Segal categories more closely resemble simplicial nerves of simplicial categories. Because this discreteness assumption is simpler than completeness, as objects Segal categories are more straightforward to define than complete Segal spaces. However, asking that a given simplicial set be discrete (rather than homotopy discrete) is awkward from a homotopical point of view, so doing homotopy theory with these structures is substantially more difficult. In particular, neither of the two model structures for Segal categories can be obtained as a localization of a model category with levelwise weak equivalences.

Segal categories first appeared in a paper of Dwyer, Kan, and Smith [58], although not under that name. They were developed extensively by Hirschowitz and Simpson [70], and their homotopy theory was investigated by Pellissier [97]. For the purposes of establishing Quillen equivalences, we consider two different model structures on the category of Segal precategories; $SeCat_c$ resembles an injective model structure, that its cofibrations are precisely the monomorphisms, and is more easily compared to CSS, whereas $SeCat_f$ more closely resembles a projective model structure and is more easily compared to SC. The notation is further justified by the fact that the fibrant objects in these model categories are Segal categories which are Reedy fibrant and projective fibrant, respectively, as simplicial spaces. However, these two model structures have the same weak equivalences, and they are Quillen equivalent to one another.

6.1 Basic Definitions and Constructions

As in the case of complete Segal spaces, the category of Segal categories alone does not have the necessary categorical properties to have a model structure. Thus, we need to work in a larger category to get a model structure for Segal categories. However, here we work in a more restricted category than that of all simplicial spaces.

Definition 6.1.1 A *Segal precategory* is a simplicial space X such that X_0 is a discrete simplicial set.

We denote the category of all Segal precategories by $\mathcal{SeCat}$, and we use it as the underlying category for both the model structures for Segal categories. We now give a formal definition of Segal category.

Definition 6.1.2 A Segal precategory X is a *Segal category* if the Segal maps

$$\varphi_n \colon X_n \to \underbrace{X_1 \times_{X_0} \cdots \times_{X_0} X_1}_{n}$$

are weak equivalences of simplicial sets for every $n \geq 2$.

Observe that the definition of a Segal category is similar to that of a Segal space, with the additional requirement that the degree zero space be discrete. Since we follow the convention that Segal categories are not required to be Reedy fibrant, they are not necessarily Segal spaces. However, the discreteness of the space in X_0 makes the face maps $X_1 \to X_0$ fibrations, so again the iterated pullback that appears in the Segal map is actually a homotopy limit.

However, unlike for the category of simplicial spaces, there is no model structure on the category of Segal precategories with levelwise weak equivalences if we want cofibrations to be monomorphisms, as is the case for all the model structures that we consider here. For example, the map of doubly constant simplicial spaces $\Delta[0] \amalg \Delta[0] \to \Delta[0]$ cannot be factored as a monomorphism followed by a weak equivalence in the category of Segal precategories; in the context of simplicial spaces the factorization must necessarily be through a simplicial space which is not discrete in degree zero. Therefore, we cannot obtain an appropriate model structure for Segal categories by localizing such a model structure, as we did for complete Segal spaces.

Thus, we must develop the model structures $\mathcal{SeCat}_c$ and $\mathcal{SeCat}_f$ on the category of Segal precategories directly. However, we use features of the Reedy and projective model structures, respectively, on simplicial spaces, as well as the Segal space model structure $\mathcal{SeSp}$, to give model structures which have the kinds of properties we want. Specifically, we want the weak equivalences to be analogous to Dwyer–Kan equivalences of simplicial categories or Segal spaces and for the fibrant objects to be Segal categories.

A primary strategy that we use is to modify various simplicial spaces and maps between them so that they reside in the category of Segal precategories, and then to adapt various constructions to this context. The first task in this direction is to define sets of generating cofibrations and generating acyclic cofibrations which are similar to those of the Reedy and projective model category structures on simplicial spaces. To obtain these sets, we start with the generating sets for the Reedy and projective model structures on $SSets^{\Delta^{op}}$ and then modify them to be maps of Segal precategories. We use slightly different methods for the two structures.

The first method we call reduction, and we use it to define the generating cofibrations in $SeCat_c$. Consider the inclusion functor

$$SeCat \to SSets^{\Delta^{op}}$$

from the category of Segal precategories to the category of simplicial spaces. This functor has a left adjoint, which we call the *reduction functor*. Given a simplicial space X, we denote its reduction by X_r. The reduction is obtained from X by collapsing the space X_0 to its set of components and making the subsequent changes to degenerate simplices in higher degrees.

Recall that the cofibrations in the Reedy model category structure on simplicial spaces are monomorphisms and that the Reedy generating cofibrations are of the form

$$\partial\Delta[m] \times \Delta[n]^t \cup \Delta[m] \times \partial\Delta[n]^t \to \Delta[m] \times \Delta[n]^t$$

for all $n, m \geq 0$. In general, these maps are not in $SeCat$ because the constant simplicial spaces $\partial\Delta[m]$ and $\Delta[m]$ are not Segal precategories except for $\partial\Delta[0]$, $\Delta[0]$, and $\partial\Delta[1]$. Therefore, we apply the reduction functor and hence consider the maps

$$(\partial\Delta[m] \times \Delta[n]^t \cup \Delta[m] \times \partial\Delta[n]^t)_r \to (\Delta[m] \times \Delta[n]^t)_r.$$

However, we want cofibrations to be monomorphisms, and some of these reduced maps are not. Specifically, when $n = 0$, we have the map

$$\partial\Delta[m]_r \to \Delta[m]_r.$$

If $m = 1$, then $\partial\Delta[1]$ is already reduced, and consists of the constant simplicial space on two points, but the reduction of $\Delta[1]$ is the constant simplicial space consisting of a single point. Thus, we want to omit this map from our generating set because it is not a monomorphism. If $m \geq 2$, then the above map is an isomorphism between simplicial spaces consisting of a single point, since both $\partial\Delta[m]$ and $\Delta[m]$ are constant on a connected simplicial set. Thus, there is no need to include these maps in the generating set. On the other hand, if $n = m = 0$ we obtain the map $\varnothing \to \Delta[0]$, which is a monomorphism, and we want to include it.

Therefore, we define the set

$$I_c = \{(\partial\Delta[m] \times \Delta[n]^t \cup \Delta[m] \times \partial\Delta[n]^t)_r \to (\Delta[m] \times \Delta[n]^t)_r\}$$

for all $m \geq 0$ when $n \geq 1$ and for $n = m = 0$. While this point should be more clear when we establish the model structure using these generating cofibrations, we have not lost any important information by discarding the maps for which $n = 0$ and $m > 0$. Such maps are used to generate higher simplices in the space in degree zero, which are necessarily degenerate when we work with Segal precategories.

However, reduction is not quite the right procedure for modifying the generating cofibrations in the projective model structure on the category of simplicial spaces, so we turn to our second method. Recall that these generating cofibrations are the maps

$$\partial\Delta[m] \times \Delta[n]^t \to \Delta[m] \times \Delta[n]^t$$

for $m, n \geq 0$. Consider the case where $m = 1$, where we have maps

$$\partial\Delta[1] \times \Delta[n]^t \to \Delta[1] \times \Delta[n]^t$$

for any $n \geq 0$. The reduction of $\Delta[1] \times \Delta[n]^t$ is a Segal precategory with $n + 1$ points in degree zero, but the object $\partial\Delta[1] \times \Delta[n]^t$ reduces to a Segal precategory with $2(n + 1)$ points in degree zero. In other words, the reduction of such a map is no longer a monomorphism. Unlike the problematic reduced Reedy cofibrations, we cannot simply exclude these maps; when $n > 0$, we need maps of this flavor to generate nondegenerate 1-simplices in the space in degree n. We need an alternative method of obtaining a Segal precategory version of these maps which are monomorphisms.

Consider the set $\Delta[n]_0$ and denote by $\Delta[n]_0^t$ the doubly constant simplicial space defined by it. If $m = 0$, define $P_{0,n}$ to be the empty simplicial space. For $m \geq 1$ and $n \geq 0$, define $P_{m,n}$ to be the pushout of the diagram

$$
\begin{array}{ccc}
\partial\Delta[m] \times \Delta[n]_0^t & \longrightarrow & \partial\Delta[m] \times \Delta[n]^t \\
\downarrow & & \downarrow \\
\Delta[n]_0^t & \longrightarrow & P_{m,n}.
\end{array}
$$

For all $m \geq 0$ and $n \geq 1$, define $Q_{m,n}$ to be the pushout of the diagram

$$
\begin{array}{ccc}
\Delta[m] \times \Delta[n]_0^t & \longrightarrow & \Delta[m] \times \Delta[n]^t \\
\downarrow & & \downarrow \\
\Delta[n]_0^t & \longrightarrow & Q_{m,n}.
\end{array}
$$

For each m and n, the map $\partial\Delta[m] \times \Delta[n]^t \to \Delta[m] \times \Delta[n]^t$ induces a map $i_{m,n}\colon P_{m,n} \to Q_{m,n}$. We then define the set

$$
I_f = \{i_{m,n}\colon P_{m,n} \to Q_{m,n} \mid m, n \geq 0\}.
$$

Remark 6.1.3 Observe that when $m \geq 2$ this construction gives exactly the same objects as those given by reduction. Namely, the Segal precategory $P_{m,n}$ is precisely $(\partial\Delta[m] \times \Delta[n]^t)_r$ and likewise $Q_{m,n}$ is precisely $(\Delta[m] \times \Delta[n]^t)_r$.

Remark 6.1.4 Just as we defined in Section 5.2 for a Segal space, given a Segal precategory X, we denote by $\mathrm{map}_X(v_0, \ldots, v_n)$ the fiber of the map $X_n \to X_0^{n+1}$ over $(v_0, \ldots, v_n) \in X_0^{n+1}$, where the map $X_n \to X_0^{n+1}$ is given by the degree n component of the canonical map $X \to \mathrm{cosk}_0 X$. We can use the pushout diagrams defining the objects $P_{m,n}$ and $Q_{m,n}$ to see that

$$
\mathrm{Hom}_{\mathcal{SeCat}}(P_{m,n}, X) \cong \coprod_{v_0,\ldots,v_n} \mathrm{Hom}_{\mathcal{SSets}}(\partial\Delta[m], \mathrm{map}_X(v_0, \ldots, v_n))
$$

and

$$
\mathrm{Hom}_{\mathcal{SeCat}}(Q_{m,n}, X) \cong \coprod_{v_0,\ldots,v_n} \mathrm{Hom}_{\mathcal{SSets}}(\Delta[m], \mathrm{map}_X(v_0, \ldots, v_n)).
$$

These isomorphisms enable us to establish the following lifting property.

Lemma 6.1.5 [30, 4.1] *Suppose a map $f\colon X \to Y$ of Segal precategories has the right lifting property with respect to the maps in I_f. Then the map $X_0 \to Y_0$ of discrete simplicial sets is surjective and each map*

$$
\mathrm{map}_X(v_0, \ldots, v_n) \to \mathrm{map}_Y(fv_0, \ldots, fv_n)
$$

is an acyclic fibration of simplicial sets for $n \geq 1$ and $(v_0, \ldots, v_n) \in X_0^{n+1}$.

Proof The surjectivity of $X_0 \to Y_0$ follows from the fact that f has the right lifting property with respect to the map $P_{0,0} \to Q_{0,0}$, which is just the inclusion $\varnothing \to \Delta[0]$.

In order to prove the remaining statement, it suffices to show that there is a dotted arrow lift in any diagram of the form

$$
\begin{array}{ccc}
\partial\Delta[m] & \longrightarrow & \mathrm{map}_X(v_0,\ldots,v_n) \\
\downarrow & \nearrow & \downarrow \\
\Delta[m] & \longrightarrow & \mathrm{map}_Y(fv_0,\ldots,fv_n)
\end{array}
\tag{6.1}
$$

for $m \geq 0$ and $n \geq 1$.

By our hypothesis, there is a dotted arrow lift in any diagram of the form

$$
\begin{array}{ccc}
P_{m,n} & \longrightarrow & X \\
\downarrow & \nearrow & \downarrow \\
Q_{m,n} & \longrightarrow & Y
\end{array}
\tag{6.2}
$$

with $m, n \geq 0$. The existence of the lift in diagram (6.2) is equivalent to the surjectivity of the map $\mathrm{Hom}(Q_{m,n}, X) \to P$, where P is the pullback in the square of the diagram

$$
\begin{array}{ccccc}
\mathrm{Hom}(Q_{m,n}, X) & \longrightarrow & P & \longrightarrow & \mathrm{Hom}(P_{m,n}, X) \\
& & \downarrow & & \downarrow \\
& & \mathrm{Hom}(Q_{m,n}, Y) & \longrightarrow & \mathrm{Hom}(P_{m,n}, Y).
\end{array}
$$

Now, as noted above we have that

$$
\mathrm{Hom}(Q_{m,n}, X) \cong \coprod_{v_0,\ldots,v_n} \mathrm{Hom}(\Delta[m], \mathrm{map}_X(v_0,\ldots,v_n))
$$

and

$$
\mathrm{Hom}(P_{m,n}, X) \cong \coprod_{v_0,\ldots,v_n} \mathrm{Hom}(\partial\Delta[m], \mathrm{map}_X(v_0,\ldots,v_n)).
$$

Similarly, if $(w_0,\ldots,w_n) \in Y_0^{n+1}$, we have

$$
\mathrm{Hom}(Q_{m,n}, Y) \cong \coprod_{w_0,\ldots,w_n} \mathrm{Hom}(\Delta[m], \mathrm{map}_Y(w_0,\ldots,w_n))
$$

and

$$
\mathrm{Hom}(P_{m,n}, Y) \cong \coprod_{w_0,\ldots,w_n} \mathrm{Hom}(\partial\Delta[m], \mathrm{map}_Y(w_0,\ldots,w_n)).
$$

Restricting to a fixed $(v_0,\ldots,v_n) \in X_0^{n+1}$, we have surjectivity of the map from $\mathrm{Hom}(\Delta[m], \mathrm{map}_X(v_0,\ldots,v_n))$ to the pullback

$$
\mathrm{Hom}(\partial\Delta[m], \mathrm{map}_X(v_0,\ldots,v_n)) \times_{\mathrm{Hom}(\partial\Delta[m], \mathrm{map}_Y(fv_0,\ldots,fv_n))} \mathrm{Hom}(\Delta[m], \mathrm{map}_Y(fv_0,\ldots,fv_n)).
$$

An element of this pullback consists of maps $g\colon \partial\Delta[m] \to \mathrm{map}_X(v_0,\ldots,v_n)$ and $h\colon \Delta[m] \to \mathrm{map}_Y(fv_0,\ldots,fv_n)$ such that the composites

$$\partial\Delta[m] \xrightarrow{g} \mathrm{map}_X(v_0,\ldots,v_n) \xrightarrow{f} \mathrm{map}_Y(fv_0,\ldots,fv_n)$$

and

$$\partial\Delta[m] \to \Delta[m] \xrightarrow{h} \mathrm{map}_Y(fv_0,\ldots,fv_n)$$

coincide. The fact that the map from $\mathrm{Hom}(\Delta[m],\mathrm{map}_X(v_0,\ldots,v_n))$ to this pullback is surjective implies exactly that there is a lift in the diagram (6.1).   □

6.2 Fixed-Object Segal Categories

As with simplicial categories, it is helpful to consider a model structure on Segal categories with a fixed object set, or, more precisely, a fixed discrete simplicial set in degree zero. As we saw in the case of simplicial categories, establishing such a model structure is substantially easier than the general case, although here the reasons for the relative simplicity are somewhat different. When the space in degree zero is fixed, we can define a model structure defined via levelwise weak equivalences of simplicial sets. The problem in the general case arises when trying to factor a map from a Segal precategory with a larger object set to one with a smaller object set. When the object set is fixed, this difficulty does not arise. Then we can obtain the desired model structure, whose fibrant objects are Segal categories, via localization. Most of the results here are taken from the present author's work [29].

Consider the category $\mathcal{SS}p_O$ whose objects are Segal precategories with a fixed set O in degree zero and whose morphisms are maps of Segal precategories which are the identity map on O. We begin by establishing a model structure whose weak equivalences are levelwise weak equivalences of simplicial sets. We have two choices, since we can ask that either the fibrations or the cofibrations are defined levelwise. We will establish both, but first we turn to showing that $\mathcal{SS}p_O$ has all small limits and colimits. We remark that this proof is not the one that appears in [29, 3.5], which is incorrect.

Lemma 6.2.1   *The category $\mathcal{SS}p_O$ has all small limits.*

Proof   Let $\mathcal{D}$ be a small category, and consider a functor $\mathcal{D} \to \mathcal{SS}p_O$. Let d be an object of $\mathcal{D}$. In the category of all simplicial spaces, the limit $\lim_{\mathcal{D}} X_d$ exists. However, if $\mathcal{D}$ is not connected, then this limit is not an object of $\mathcal{SS}p_O$, since $\lim_{\mathcal{D}}(X_d)_0 = O^{\pi_0\mathcal{D}}$. However, if $\mathrm{diag}\colon O \to \lim_{\mathcal{D}}(X_d)_0$ is the diagonal map, we can define the limit in $\mathcal{SS}p_O$, denoted by $\lim_{\mathcal{D}}^O X_d$, as the sub-Segal precategory of $\lim_{\mathcal{D}} X_d$ whose discrete set in degree zero is precisely the image of the diagonal map $O \to (\lim_{\mathcal{D}} X_d)_0$, and whose higher simplices are all those lying above these objects. This Segal precategory now satisfies the universal property of limits when we require the maps involved to be the identity on degree zero and hence in the category $\mathcal{SS}p_O$. $\qquad \square$

Lemma 6.2.2 [29, 3.6] *The category $\mathcal{SS}p_O$ has all small colimits.*

Proof As for the proof for limits, begin by considering the colimit of a small diagram $\mathcal{D} \to \mathcal{SS}p_O$. Note that again we have a problem in degree zero if our index category $\mathcal{D}$ has more than one component, since in this case, for any object d of $\mathcal{D}$, $\text{colim}_{\mathcal{D}}(X_d)_0 = \coprod_{\pi_0 \mathcal{D}} O$. If we consider the fold map

$$\text{fold}: \text{colim}_{\mathcal{D}}(X_d)_0 \to O,$$

we can define the colimit in $\mathcal{SS}p_O$, denoted by $\text{colim}_{\mathcal{D}}^{O} X_d$, as the pushout in the diagram

$$
\begin{array}{ccc}
\text{colim}_{\mathcal{D}}(X_d)_0 & \xrightarrow{\text{fold}} & O \\
\downarrow & & \downarrow \\
\text{colim}_{\mathcal{D}} X_d & \longrightarrow & \text{colim}_{\mathcal{D}}^{O} X_d
\end{array}
$$

where the left-hand vertical map is the inclusion map. Similarly to the case for limits, this new simplicial space is a Segal precategory with discrete space O in degree zero and satisfies the universal property for colimits. $\qquad \square$

We first consider the projective model structure on the category of Segal precategories with a fixed object set.

Theorem 6.2.3 [29, 3.7] *There is a cofibrantly generated model structure on the category $\mathcal{SS}p_O$, which we denote by $\mathcal{SS}p_{O,f}$, in which:*

1 *the weak equivalences are levelwise weak equivalences of simplicial sets; and*
2 *the fibrations are the levelwise fibrations of simplicial sets.*

The first step toward defining this model structure is finding candidates for the sets of generating cofibrations and generating acyclic cofibrations. Recall that for the projective model structure on the category $\mathcal{SS}ets^{\Delta^{op}}$ of simplicial spaces, as described in Theorem 2.5.2, the generating acyclic cofibrations are of the form

$$V[m, k] \times \Delta[n]^t \to \Delta[m] \times \Delta[n]^t$$

for $n \geq 0$, $m \geq 1$ and $0 \leq k \leq m$.

The first problem with working instead in $SSpo_{,f}$ is that $\Delta[n]^t$ is not an object of $SSpo$ without further clarifying what its discrete space of 0-simplices is. In order to have this set preserved by all necessary maps, we need to define a separate n-simplex for any $(x_0, \ldots, x_n) \in O^n$. We denote such an n-simplex with specified 0-simplices by $\Delta[n]^t_{x_0, \ldots, x_n}$; setting $\underline{x} = (x_0, \ldots, x_n)$, we can write this simplex as $\Delta[n]^t_{\underline{x}}$. Furthermore, $\Delta[n]^t_{\underline{x}}$ must have any remaining elements of O as disjoint 0-simplices.

With this adaptation, we then utilize the strategy from the previous section to obtain maps of Segal precategories. For a fixed object set O, consider the maps

$$V[m, k] \times \Delta[n]^t_{\underline{x}} \rightarrow \Delta[m] \times \Delta[n]^t_{\underline{x}}$$

where m, n, k, and $\underline{x}$ are as above. We want to modify them so they are maps of Segal precategories.

As before, define the object $(Q_{m,n})_{\underline{x}}$ to be the pushout of the diagram

$$
\begin{array}{ccc}
\Delta[m] \times (\Delta[n]^t_{\underline{x}})_0 & \longrightarrow & \Delta[m] \times \Delta[n]^t_{\underline{x}} \\
\downarrow & & \downarrow \\
(\Delta[n]^t_{\underline{x}})_0 & \longrightarrow & (Q_{m,n})_{\underline{x}}.
\end{array}
$$

Also define the object $(R_{m,n,k})_{\underline{x}}$ to be the pushout of the diagram

$$
\begin{array}{ccc}
V[m, k] \times (\Delta[n]^t_{\underline{x}})_0 & \longrightarrow & V[m, k] \times \Delta[n]^t_{\underline{x}} \\
\downarrow & & \downarrow \\
(\Delta[n]^t_{\underline{x}})_0 & \longrightarrow & (R_{m,n,k})_{\underline{x}}.
\end{array}
$$

Consider the set

$$J_{f,O} = \{(R_{m,n,k})_{\underline{x}} \rightarrow (Q_{m,n})_{\underline{x}}\}$$

where $m, n \geq 1$, $0 \leq k \leq m$, and $\underline{x} = (x_0, \ldots, x_n) \in O^{n+1}$. This set $J_{f,O}$ is our candidate for a set of generating acyclic cofibrations for $SSpo_{,f}$.

Similarly, we can define the set

$$I_{f,O} = \{(P_{m,n})_{\underline{x}} \rightarrow (Q_{m,n})_{\underline{x}}\}$$

for all $m, n \geq 0$ and $\underline{x} \in O^{n+1}$, where $(P_{m,n})_{\underline{x}}$ is the pushout of the diagram

$$\begin{array}{ccc} \partial\Delta[m] \times (\Delta[n]^t_{\underline{x}})_0 & \longrightarrow & \partial\Delta[m] \times \Delta[n]^t_{\underline{x}} \\ \downarrow & & \downarrow \\ (\Delta[n]^t_{\underline{x}})_0 & \longrightarrow & (P_{m,n})_{\underline{x}}. \end{array}$$

This set is our candidate for a set of generating cofibrations for $\mathcal{S}\mathcal{S}p_{O,f}$.

To prove Theorem 6.2.3, we need only show that the adjustments that we have made to the projective model structure, namely restricting the space in degree zero to a fixed discrete space and requiring all maps to be the identity on that space, do not alter the essential features of the original model structure.

Proof of Theorem 6.2.3 Lemmas 6.2.1 and 6.2.2 show that our category has small limits and colimits. The two-out-of-three property and the retract axiom for weak equivalences hold because they hold in the projective model structure for simplicial spaces and are not altered by the restriction on spaces in degree zero. It now suffices to check the conditions of Theorem 1.7.15. To prove statement (1), notice that the maps in the sets $I_{f,O}$ and $J_{f,O}$ are modified versions of the generating cofibrations in the projective model structure on simplicial spaces, which permit the small object argument. One can check that these modifications do also.

Notice that the $I_{f,O}$-injectives are precisely the levelwise acyclic fibrations, and that the $J_{f,O}$-injectives are precisely the levelwise fibrations. Thus, we have satisfied conditions (3) and (4)(ii) of Theorem 1.7.15.

Now notice that the $I_{f,O}$-cofibrations are precisely the cofibrations, by our definition of cofibration. Furthermore, the $J_{f,O}$-cofibrations are the maps with the left lifting property with respect to the fibrations. It follows that a $J_{f,O}$-cofibration is an $I_{f,O}$-cofibration. Using the definition of the maps in $J_{f,O}$ and the model structure for simplicial sets, we can see that a $J_{f,O}$-cofibration is also a weak equivalence, establishing condition (2). □

We now want to localize this model category so that the fibrant objects are Segal categories with the set O in degree zero. The localization process is similar to that for Segal spaces; we need only modify the maps with respect to which we localize so that they are in the category $\mathcal{S}\mathcal{S}p_O$.

Recall the inclusion maps of simplicial sets $G(n) \to \Delta[n]$, and observe that the simplicial spaces $G(n)^t$ and $\Delta[n]^t$ are Segal precategories. As before, we can replace $\Delta[n]^t$ with the objects $\Delta[n]^t_{\underline{x}}$, where $\underline{x} - (x_0, \ldots, x_n) \in O^{n+1}$. Then, define

$$G(n)^t_{\underline{x}} = \bigcup_{i=0}^{n-1} \alpha^i \Delta[1]^t_{x_i, x_{i+1}}.$$

Now, we take all inclusion maps

$$G(n)^t_{\underline{x}} \to \Delta[n]^t_{\underline{x}},$$

where $n \geq 0$ and $\underline{x} = (x_0, \ldots, x_n) \in O^n$.

Thus, we have defined maps with which to localize the fixed-object model structure to obtain Segal categories as fibrant objects. The following result is immediate from Theorem 2.8.2.

Proposition 6.2.4 [29, 3.8] *There is a model structure $SeCat_{O,f}$ on the category of Segal precategories with a fixed set O in degree zero in which:*

1 the weak equivalences are the local equivalences with respect to the maps $G(n)^t_{\underline{x}} \to \Delta[n]^t_{\underline{x}}$ for $n \geq 0$ and $\underline{x} \in O^n$; and
2 the cofibrations are those of $SSp_{O,f}$.

For making some of our calculations, we find it convenient to work in a model structure $SSp_{O,c}$ in which the weak equivalences are again given by levelwise weak equivalences of simplicial sets, but in which the cofibrations, rather than the fibrations, are defined levelwise.

Theorem 6.2.5 [29, 3.9] *There is a model structure $SSp_{O,c}$ on the category of Segal precategories with a fixed set O in degree zero in which the weak equivalences and cofibrations are defined levelwise.*

To define candidate sets $I_{c,O}$ and $J_{c,O}$ for generating cofibrations and generating acyclic cofibrations, respectively, we apply the fixed-object techniques used for the previous model structure and the reduction functor to the generating set of cofibrations and acyclic cofibrations in the Reedy model structure on simplicial spaces.

We start by reducing the objects defining the Reedy generating cofibrations and generating acyclic cofibrations to obtain maps of the form

$$(\partial \Delta[m] \times \Delta[n]^t \cup \Delta[m] \times \partial \Delta[n]^t)_r \to (\Delta[m] \times \Delta[n])_r$$

and

$$(V[m,k] \times \Delta[n]^t \cup \Delta[m] \times \partial \Delta[n]^t)_r \to (\Delta[m] \times \Delta[n]^t)_r.$$

Then, in order to have our maps in SSp_O, we define a separate such map for each choice of vertices $\underline{x}$ in degree zero and adding in the remaining elements of O if necessary. As above, we use $\Delta[n]^t_{\underline{x}}$ to denote the object $\Delta[n]^t$ with the $(n + 1)$-tuple $\underline{x}$ of vertices. We then define sets

$$I_{c,O} = \{(\partial\Delta[m] \times \Delta[n]^t_{\underline{x}} \cup \Delta[m] \times \partial\Delta[n]^t_{\underline{x}})_r \to (\Delta[m] \times \Delta[n]^t_{\underline{x}})_r\}$$

for all $m \geq 0$ and $n \geq 1$, and

$$J_{c,O} = \{(V[m,k] \times \Delta[n]^t_{\underline{x}} \cup \Delta[m] \times \partial\Delta[n]^t_{\underline{x}})_r \to (\Delta[m] \times \Delta[n]^t_{\underline{x}})_r\}$$

for all $m \geq 1$, $n \geq 1$, $0 \leq k \leq m$, and $\underline{x} \in O^{n+1}$.

Given these maps, we are now able to prove the existence of the model structure $SSp_{O,c}$.

Proof of Theorem 6.2.5 The proofs that $SSp_{O,c}$ has finite limits and colimits, satisfies the two-out-of-three and retract axioms, as well as the smallness for the domains of $I_{c,O}$ and $J_{c,O}$, are the same as for $SSp_{O,f}$.

From the definitions of $I_{c,O}$ and $J_{c,O}$, it follows that the $I_{c,O}$-fibrations are the maps of Segal precategories with O in degree zero which are Reedy fibrations and that the $J_{c,O}$-fibrations are the maps of Segal precategories with O in degree zero which are Reedy acyclic fibrations. Furthermore, it follows from these facts that the $I_{c,O}$-cofibrations are precisely the cofibrations and that the $J_{c,O}$-cofibrations are precisely the acyclic cofibrations. □

We can then localize $SSp_{O,c}$ just as we localized $SSp_{O,f}$.

Proposition 6.2.6 [29, §3] *There is a model structure $SeCat_{O,c}$ on the category of Segal precategories with object set O in which:*

1 *the weak equivalences are the local equivalences with respect to the maps $G(n)^t_{\underline{x}} \to \Delta[n]^t_{\underline{x}}$ for $n \geq 0$ and $\underline{x} \in O^n$; and*
2 *the cofibrations are those of $SSp_{O,c}$; in particular, all objects are cofibrant.*

Recall from Proposition 2.8.6 that the weak equivalences in a localized model structure depend only on the weak equivalences in the original model structure. Therefore, the weak equivalences in $SeCat_{f,O}$ and $SeCat_{c,O}$ are the same. Hence, we can prove the following.

Proposition 6.2.7 [29, 3.10] *The adjoint pair given by the identity functor induces a Quillen equivalence of model categories*

$$SeCat_{O,f} \rightleftarrows SeCat_{O,c}.$$

Proof Since the cofibrations in $SeCat_{O,f}$ are monomorphisms, the identity functor

$$SeCat_{O,f} \to SeCat_{O,c}$$

preserves both cofibrations and acyclic cofibrations, so this adjoint pair is a Quillen pair. It remains to show that for any cofibrant X in $SeCat_{O,f}$ and fibrant Y in $SeCat_{O,c}$, the map $FX \to Y$ is a weak equivalence if and only if the map $X \to RY$ is a weak equivalence. However, this fact follows since the weak equivalences are the same in each category.   □

We conclude this section with characterizations of the fibrant objects in these localized model structures.

Proposition 6.2.8   [24, 3.1] *The fibrant objects in $SSp_{O,c}$ are fibrant in the Reedy model structure on simplicial spaces. In particular, the fibrant objects of $SeCat_{O,c}$ are the Segal categories which are Reedy fibrant.*

Proof   Let X be a fibrant object in $SSp_{O,c}$. We need to show that the map $X \to \Delta[0]^I$ has the right lifting property with respect to all levelwise acyclic cofibrations in the Reedy model structure, not just the ones in $SSp_{O,c}$. We first consider the special case where $A \to B$ is an acyclic cofibration in $SSp_{O',c}$ for some set $O' \neq O$.

Using the 0-skeleta $sk_0(A)$ and $sk_0(X)$, we define a simplicial space A' as the pushout

$$
\begin{array}{ccc}
sk_0(A) & \longrightarrow & sk_0(X) \\
\downarrow & & \downarrow \\
A & \longrightarrow & A'.
\end{array}
$$

Note in particular that the induced map $A'_0 \to X_0$ is an isomorphism. Now, define B' as the pushout

$$
\begin{array}{ccc}
A & \longrightarrow & A' \\
\downarrow & & \downarrow \\
B & \longrightarrow & B'.
\end{array}
$$

Because it is defined as a pushout along an acyclic cofibration in the Reedy structure, the map $A' \to B'$ is also a Reedy acyclic cofibration, and $A'_0 \cong B'_0 \cong \amalg_O \Delta[0]^I$. We make use of $\amalg_O \Delta[0]^I$ here as the terminal object in the category

$SSp_{O,c}$ and consider the commutative diagram

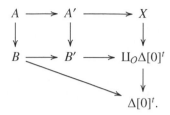

Therefore there exists a lift $B' \to X$, from which there exists a lift $B \to X$, using the universal property of pushouts.

Now, suppose that $A \to B$ is a Reedy acyclic cofibration between simplicial spaces which are not necessarily Segal precategories. Apply the reduction functor to get a map $A_r \to B_r$, which must be an isomorphism in degree zero. Since X and $\Delta[0]'$ are Segal precategories, we can factor our diagram as

$$
\begin{array}{ccccc}
A & \longrightarrow & A_r & \longrightarrow & X \\
\downarrow & & \downarrow & & \downarrow \\
B & \longrightarrow & B_r & \longrightarrow & \Delta[0]'.
\end{array}
$$

Then, we obtain a lift $B_r \to X$ from the previous argument, from which we obtain a lift $B \to X$.

The second statement of the proposition follows from the fact that fibrant objects in $SeCat_{O,c}$ are necessarily fibrant in $SSp_{O,c}$, and that they are Segal categories by construction of the localization. □

We can prove the following result using the same techniques.

Proposition 6.2.9 [24, 4.1] *The fibrant objects in $SeCat_{O,f}$ are precisely the Segal categories which have the set O in degree zero and are fibrant in the projective model structure on simplicial spaces.*

We conclude this section with a comparison result between simplicial categories with a fixed object set and Segal precategories with that same fixed set in degree zero.

Recall from Proposition 4.2.4 that there is a model structure SC_O on the category whose objects are the simplicial categories with a fixed set O of objects and whose morphisms are the functors which are the identity on the objects, and that there is a model structure $SeCat_{O,f}$ on the category whose objects are the Segal precategories with the set O in degree zero and whose morphisms are the identity on degree zero.

Theorem 6.2.10   [29, 5.5] *There is an adjoint pair*

$$F_O \colon SeCat_{O,f} \rightleftarrows SC_O \colon R_O$$

which is a Quillen equivalence.

The proof of this theorem uses a generalization of a result by Badzioch [7, 6.5] which relates strict and homotopy algebras over an algebraic theory to multi-sorted algebraic theories [28]. Since this argument would require an extensive treatment of these methods, we do not include it here. However, the theorem is an important stepping-stone to the comparison between simplicial categories and Segal categories without fixed objects.

6.3 The First Model Structure

In this section, we prove the existence of the model structure $SeCat_c$.

To begin, we would like to define a functorial "localization" functor on *SeCat* which assigns to any Segal precategory a Segal category which is weakly equivalent to it in the Segal space model structure *SeSp* on the category of all simplicial spaces.

Proposition 6.3.1   [30, §5] *Given any Segal precategory X, there is a functor* $L_c \colon SeCat \to SeCat$ *such that* $L_c X$ *is a Segal space which is a Segal category weakly equivalent to X in* $SeSp_c$.

Proof We begin by considering a functorial fibrant replacement functor in $SeSp_c$ and then modifying it so that it takes values in *SeCat*. Using Proposition 2.8.7, a choice of generating acyclic cofibrations for $SeSp_c$ is the set of maps

$$\{\partial\Delta[m] \times \Delta[n]^t \cup \Delta[m] \times G(n)^t \to \Delta[m] \times \Delta[n]^t\}$$

for $n, m \geq 0$. Therefore, by Proposition 1.7.11 we can use the small object argument to construct a functorial fibrant replacement functor by taking a colimit of pushouts, each of which is along the coproduct of all these maps.

If we apply this functor to a Segal precategory, the maps for which $n = 0$ are problematic because taking a pushout along one of them does not result in a Segal precategory. We claim that we can obtain a functorial localization functor L_c on the category *SeCat* by omitting these maps and simply taking a colimit of iterated pushouts along the maps

$$\partial\Delta[m] \times \Delta[n]^t \cup \Delta[m] \times G(n)^t \to \Delta[m] \times \Delta[n]^t$$

for $n \geq 1$ and $m \geq 0$.

To see that this restricted set of maps is sufficient, consider a Segal precategory X and the Segal category L_cX we obtain from taking such a colimit. Then for any $m \geq 0$, consider the diagram

$$
\begin{array}{ccc}
\partial\Delta[m] & \longrightarrow & \mathrm{Map}^h(G(0)^t, L_cX) \\
\downarrow & \nearrow & \downarrow \\
\Delta[m] & \longrightarrow & \mathrm{Map}^h(\Delta[0]^t, L_cX).
\end{array}
$$

Since $\Delta[0]^t$ is isomorphic to $G(0)^t$, and since L_cX is discrete in degree zero, the right-hand vertical map is an isomorphism of discrete simplicial sets. Therefore, a dotted arrow lift exists in this diagram. It follows that the map $L_cX \to \Delta[0]^t$ has the right lifting property with respect to the maps

$$
\partial\Delta[m] \times \Delta[n]^t \cup \Delta[m] \times G(n)^t \to \Delta[m] \times \Delta[n]^t
$$

for all $n, m \geq 0$. Therefore, L_cX is fibrant in $SeSp_c$, namely, a Segal space. $\quad\square$

Since L_cX is a Segal space, we saw in Section 5.2 that it has mapping spaces $\mathrm{map}_{L_cX}(x, y)$ and an associated homotopy category $\mathrm{Ho}(L_cX)$.

Definition 6.3.2 A map $X \to Y$ of Segal precategories is a *Dwyer–Kan equivalence* if the induced map of Segal categories $L_cX \to L_cY$ is a Dwyer–Kan equivalence of Segal spaces, in the sense of Definition 5.5.1.

In other words, $X \to Y$ is a Dwyer–Kan equivalence if $L_cX \to L_cY$ induces a weak equivalence on all mapping spaces and an equivalence of homotopy categories.

Theorem 6.3.3 [30, 5.1] *There is a cofibrantly generated model structure $SeCat_c$ on the category of Segal precategories such that*

1 *weak equivalences are the Dwyer–Kan equivalences; and*
2 *cofibrations are the monomorphisms, and in particular, every Segal precategory is cofibrant.*

We take as generating cofibrations the set

$$
I_c = \{(\partial\Delta[m] \times \Delta[n]^t \cup \Delta[m] \times \partial\Delta[n]^t)_r \to (\Delta[m] \times \Delta[n]^t)_r\}
$$

for all $m \geq 0$ when $n \geq 1$ and for $n = m = 0$. Notice that since taking a pushout along such a map amounts to attaching an m-simplex to the space in degree n, any cofibration can be written as a directed colimit of pushouts along the maps of I_c.

We then define the set $J_c = \{i\colon A \to B\}$ to be a set of representatives of isomorphism classes of maps in *SeCat* satisfying two conditions:

1 the map $i\colon A \to B$ is a monomorphism and a weak equivalence; and
2 for all $n \geq 0$, the spaces A_n and B_n have countably many simplices.

Given these proposed generating acyclic cofibrations, we need the following result. Its proof is highly technical, so we do not include it here, but it can be found in [30].

Proposition 6.3.4   [30, 5.7] *Any acyclic cofibration $j\colon C \to D$ in $SeCat_c$ can be written as a directed colimit of pushouts along the maps in J_c.*

Now, we have two definitions of acyclic fibration that we need to show coincide: the fibrations which are weak equivalences, and the maps with the right lifting property with respect to the maps in I_c.

Proposition 6.3.5   [30, 5.8] *The maps with the right lifting property with respect to the maps in I_c are fibrations and weak equivalences.*

Before giving a proof of this proposition, we begin by looking at the maps in I_c and determining what an I_c-fibration looks like. If $f\colon X \to Y$ has the right lifting property with respect to the maps in I_c, then for each $n \geq 1$, the map $X_n \to P_n$ is an acyclic fibration of simplicial sets, where P_n is the pullback in the diagram

$$
\begin{array}{ccc}
P_n & \longrightarrow & Y_n \\
\downarrow & & \downarrow \\
(\mathrm{cosk}_{n-1}\, X)_n & \longrightarrow & (\mathrm{cosk}_{n-1}\, Y)_n.
\end{array}
$$

In the case that $n = 0$, the restrictions on m and n give us that the map $X_0 \to Y_0$ is a surjection rather than the isomorphism we get in the Reedy case. Notice that by the same argument given for the Reedy model structure, the simplicial sets P_n are actually homotopy pullbacks and are therefore homotopy invariant.

We have the following technical result.

Lemma 6.3.6   [30, 5.9] *Suppose that $f\colon X \to Y$ is a map of Segal precategories which is an I_c-fibration. Then f is a Dwyer–Kan equivalence.*

Proof of Proposition 6.3.5   Suppose that $f\colon X \to Y$ is an I_c-fibration. Then f has the right lifting property with respect to all cofibrations. Since, in particular, it has the right lifting property with respect to the acyclic cofibrations, it is a fibration by definition. It remains to show that f is a weak equivalence. However, this fact follows from Lemma 6.3.6, proving the proposition.   □

We now state the converse, whose proof is again technical, and indeed builds on the proof of Lemma 6.3.6.

Proposition 6.3.7   [30, 5.10] *The maps in $SeCat_c$ which are both fibrations and weak equivalences are I_c-fibrations.*

Now we prove a lemma which we need to check the last condition for our model structure.

Lemma 6.3.8   [30, 5.11] *A pushout along a map of J_c is an acyclic cofibration in $SeCat_c$.*

Proof   Let $j: A \to B$ be a map in J_c. Notice that j is an acyclic cofibration in CSS. Since CSS is a model category, by Proposition 1.4.11 a pushout along an acyclic cofibration is again an acyclic cofibration. If all the objects involved are Segal precategories, then the pushout is again a Segal precategory and therefore the pushout map is an acyclic cofibration in $SeCat_c$.   □

Proposition 6.3.9   [30, 5.12] *If a map of Segal precategories is a J_c-cofibration, then it is an I_c-cofibration and a weak equivalence.*

Proof   By definition and Proposition 6.3.4, a J_c-cofibration is a map with the left lifting property with respect to the maps which have the right lifting property with respect to the acyclic cofibrations. Using the definition of fibration, the J_c-cofibrations are hence the maps with the left lifting property with respect to the fibrations.

Similarly, using Propositions 6.3.5 and 6.3.7, an I_c-cofibration is a map with the left lifting property with respect to the acyclic fibrations. Thus, we need to show that a map with the left lifting property with respect to the fibrations has the left lifting property with respect to the acyclic fibrations and is a weak equivalence. Since the acyclic fibrations are fibrations, it remains to show that the maps with the left lifting property with respect to the fibrations are weak equivalences.

Let $f: A \to B$ be a map with the left lifting property with respect to all fibrations. By Lemma 6.3.8 above, we know that a pushout along maps of J_c is an acyclic cofibration. Therefore, we can use the small object argument to factor the map $f: A \to B$ as the composite of an acyclic cofibration $A \to A'$ and a fibration $A' \to B$. Then there exists a dotted arrow lift in the diagram showing that the map $A \to B$ is a retract of the map $A \to A'$ and therefore a weak equivalence.   □

Proof of Theorem 6.3.3   Limits and colimits of Segal precategories (computed as simplicial spaces) still have discrete 0-space and are therefore Segal precategories. The two-out-of-three and retract properties hold similarly to the case of Dwyer–Kan equivalences of simplicial categories.

It remains to show that the four conditions of Theorem 1.7.15 are satisfied. The set I_c permits the small object argument because the generating cofibrations in the Reedy model structure do. We can show that the objects A which appear as the sources of the maps in J_c are small using an analogous argument to the one for simplicial sets, so the set J_c permits the small object argument. Thus, condition (1) of that theorem is satisfied.

Condition (2) is precisely the statement of Proposition 6.3.9. Condition (3) and condition (4)(ii) of Theorem 1.7.15 are precisely the statements of Propositions 6.3.5 and 6.3.7, respectively.   □

Corollary 6.3.10   [24, 3.2], [30, 5.3] *The fibrant objects in SeCat$_c$ are precisely the Reedy fibrant Segal categories.*

Proof   Suppose that X is fibrant in *SeCat$_c$*. Let $A \to B$ be any generating acyclic cofibration in the Reedy model structure. As in the proof of Proposition 6.2.8, we can reduce this map and consider the diagram

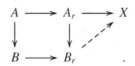

If we can prove that $A_r \to B_r$ is an acyclic cofibration in *SeCat$_c$*, then the existence of the dotted arrow lift, and hence a lift $B \to X$, will follow from fibrancy of X in *SeCat$_c$*.

Recall that generating acyclic cofibrations in the Reedy model structure are of the form

$$V[m,k] \times \Delta[n]^t \cup \Delta[m] \times \partial\Delta[n]^t \to \Delta[m] \times \Delta[n]^t.$$

Recall from when we reduced the generating cofibrations in the Reedy structure that potential problems arose when $n = 0$. However, in this case here we simply get the maps $V[m,k] \to \Delta[m]$, which always reduce to an isomorphism of a single point. Although generating acyclic cofibrations for *SeCat$_c$* are not defined via these reduced maps, it is not difficult to check that they are still acyclic cofibrations in *SeCat$_c$*.

Thus, we have established that X is Reedy fibrant. Further, since the maps

$$(\Delta[m] \times G(n)^t)_r \to (\Delta[m] \times \Delta[n]^t)_r$$

for all $m, n \geq 0$ are acyclic cofibrations in $SeCat_c$, it follows that X is a Segal category.

Conversely, let X be a Reedy fibrant Segal category and suppose that $f : A \to B$ is a generating acyclic cofibration in $SeCat_c$. We need to show that the map $X \to \Delta[0]$ has the right lifting property with respect to the map f. We know that it has the right lifting property with respect to any such f which preserves a fixed discrete space O in degree zero, by Proposition 6.2.8. Therefore, we assume that f is an inclusion but is not surjective in degree zero.

Choose $b \in B_0$ which is not in the image of $f : A \to B$. Since f is a weak equivalence, we know that b is equivalent in $L_c B$ to $f(a)$ for some $a \in (L_c A)_0 = A_0$. Define $(L_c B)_a$ to be the full Segal subcategory of $L_c B$ whose objects are b and $f(a)$. Let B_a be the subsimplicial space of B whose image is in $(L_c B)_a$. Note that $(B_a)_0 = ((L_c B)_a)_0 = \{b, f(a)\}$.

Letting a also denote the doubly constant simplicial space given by a, we define A_1 to be a pushout given by

Notice that the map $A \to A_1$ has a section and that we can factor f as the composite $A \to A_1 \to B$.

We now repeat this process by choosing a b' which is not in the image of the map $A_1 \to B$ and a corresponding a', and continue to do so, perhaps infinitely many times, and take a colimit to obtain a Segal precategory $\widehat{A}$ such that the map f factors as $A \to \widehat{A} \to B$ and there is a section $\widehat{A} \to A$. Furthermore, notice that $\widehat{A}_0 = B_0$ and that the map $\widehat{A} \to B$ is an object-preserving acyclic cofibration. Therefore, the dotted arrow lift exists in the following diagram:

$$\begin{array}{ccc} \widehat{A} & \longrightarrow & X \\ \downarrow & \nearrow & \downarrow \\ B & \longrightarrow & \Delta[0], \end{array}$$

which implies, using the section $\widehat{A} \to A$, that there is a dotted arrow lift in the

diagram

$$\square$$

We further prove that *SeCat_c* has additional structure. First, we have the following immediate consequence of the fact that all objects are cofibrant.

Corollary 6.3.11   *The model category SeCat_c is left proper.*

Proposition 6.3.12   [26, 6.3] *The model category SeCat_c has the structure of a simplicial model category.*

Proof   We need to show that the axioms (MC6) and (MC7) for a simplicial model category hold, in Definition 2.4.3. We begin with (MC6). Suppose that X and Y are Segal precategories and K is a simplicial set. Define $X \otimes K = (X \times K)_r$, where the product is taken in the category of simplicial spaces. Then define $\mathrm{Map}(X, Y)$ and Y^K just as for simplicial spaces; for the latter, observe that if Y_0 is discrete, then so is $(Y^K)_0 = (Y_0)^K$. Then one can verify the necessary isomorphisms to verify (MC6).

To check axiom (MC7), we need to show that if $i \colon A \to B$ is a cofibration and $p \colon X \to Y$ is a fibration in *SeCat_c*, then the pullback-corner map

$$\mathrm{Map}(B, X) \to \mathrm{Map}(A, X) \times_{\mathrm{Map}(A,Y)} \mathrm{Map}(B, Y)$$

is a fibration of simplicial sets which is a weak equivalence if either i or p is. Since we have defined mapping spaces to be the same as in the category of simplicial spaces, we need only verify that a cofibration or fibration in *SeCat_c* is still a cofibration or fibration, respectively, in the Reedy model structure on simplicial spaces. Since cofibrations are exactly the monomorphisms in both categories, the case of cofibrations is immediate.

Suppose, then, that p is a fibration in *SeCat_c*, so that it has the right lifting property with respect to monomorphisms between Segal precategories which are also Dwyer–Kan equivalences. In particular, it has the right lifting property with respect to monomorphisms which are levelwise weak equivalences of simplicial sets. Suppose that $A \to B$ is an acyclic cofibration in the Reedy model structure. Then $\pi_0(A_0) \cong \pi_0(B_0)$, so $(A_r)_0 \cong (B_r)_0$. Therefore the map $A_r \to B_r$ is still a levelwise weak equivalence and monomorphism, and in particular a weak equivalence in $SeCat_c$. Therefore we obtain a lift

$$
\begin{array}{ccc}
A & \longrightarrow & A_r & \longrightarrow & X \\
{\scriptstyle \simeq}\downarrow & & \downarrow & \nearrow & \downarrow{\scriptstyle p} \\
B & \longrightarrow & B_r & \longrightarrow & Y.
\end{array}
$$

Therefore, p is also a fibration in the Reedy model structure on simplicial spaces. It follows that $SeCat_c$ satisfies axiom (MC7) and is a simplicial model category. $\qquad\square$

While we do not give a proof of the following result, we record it for comparison with other models.

Theorem 6.3.13 [110, 19.3.3] *The model category $SeCat_c$ is cartesian.*

6.4 The Equivalence With Complete Segal Spaces

In this section, we show that there is a Quillen equivalence between the model categories $SeCat_c$ and CSS. We first need to show that we have an adjoint pair of functors between the two categories.

Let $I\colon SeCat_c \to CSS$ be the inclusion functor of Segal precategories into the category of all simplicial spaces. We want to show that it has a right adjoint functor $R\colon CSS \to SeCat_c$ which "discretizes" the degree zero space.

Let W be a simplicial space. Define simplicial spaces $U = \mathrm{cosk}_0(W_0)$ and $V = \mathrm{cosk}_0(W_{0,0})$. There exist maps $W \to U \leftarrow V$, using the universal property of coskeleta and the inclusion, respectively, from which we take the pullback

$$
\begin{array}{ccc}
RW & \longrightarrow & V \\
\downarrow & & \downarrow \\
W & \longrightarrow & U.
\end{array}
$$

Note that RW is a Segal precategory since V is. It is not hard to check that, if W is a complete Segal space, then so are U and V. Although in general RW is not expected to be complete, we claim that it is still a Segal space. If we restrict to the simplicial sets in degree one in the above pullback square, we obtain

$$
\begin{array}{ccc}
(RW)_1 & \longrightarrow & W_{0,0} \times W_{0,0} \\
\downarrow & & \downarrow \\
W_1 & \longrightarrow & W_0 \times W_0,
\end{array} \tag{6.3}
$$

and in degree two we get

$$(RW)_1 \times_{(RW)_0} (RW)_1 \longrightarrow (W_{0,0})^3$$
$$\downarrow \qquad\qquad\qquad\qquad \downarrow$$
$$W_2 \simeq W_1 \times_{W_0} W_1 \longrightarrow W_0 \times W_0 \times W_0.$$

Looking at these pullbacks, and the analogous ones for higher n, we notice that RW satisfies the Segal condition. Since we have already established that $(RW)_0$ is discrete, we have shown that RW is a Segal category.

Observe that this construction is functorial, since coskeleta and pullbacks are. We have thus defined a functor $R\colon CSS \to SeCat_c$.

Proposition 6.4.1 [30, 6.1] *The functor $R\colon CSS \to SeCat_c$ is right adjoint to the inclusion $I\colon SeCat_c \to CSS$.*

Proof We need to show that there is an isomorphism

$$\mathrm{Hom}_{SeCat_c}(Y, RW) \cong \mathrm{Hom}_{CSS}(IY, W)$$

for any Segal precategory Y and simplicial space W.

Suppose that we have a map $Y = IY \to W$. Since Y is a Segal precategory, Y_0 is equal to $Y_{0,0}$ viewed as a constant simplicial set. Therefore, we can restrict the map $Y \to W$ to a unique map $Y \to V$, where $V = \mathrm{cosk}_0(W_{0,0})$ as above. Then, the universal property of pullbacks provides a unique map $Y \to RW$. Hence, we obtain a map

$$\psi\colon \mathrm{Hom}_{CSS}(IY, W) \to \mathrm{Hom}_{SeCat_c}(Y, RW).$$

The map ψ is surjective because, given any map $Y \to RW$, we can compose it with the map $RW \to W$ to obtain a map $Y \to W$.

Now for any Segal precategory Y, consider the diagram

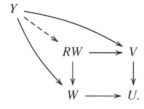

The image of the map $Y_0 \to W_0$ is contained in $W_{0,0}$ since Y is a Segal precategory, so it uniquely determines the map $Y \to V$. Therefore, given a map $Y \to RW$, it could only have come from one map $Y \to W$. Thus, ψ is injective. $\qquad\square$

Now we need to show that this adjoint pair respects the model structures.

Proposition 6.4.2 [30, 6.2] *The adjoint pair of functors*

$$I: SeCat_c \rightleftarrows CSS: R$$

defines a Quillen pair.

Proof It suffices to show that the inclusion functor I preserves cofibrations and acyclic cofibrations. It preserves cofibrations because they are defined to be the monomorphisms in each category. In each of the two model categories, a map is a weak equivalence if it is a Dwyer–Kan equivalence after localizing to obtain a Segal space, as given in Theorem 5.5.6. Thus, in each case an acyclic cofibration is a monomorphism satisfying this property. Therefore, the inclusion map I preserves acyclic cofibrations. □

Theorem 6.4.3 [30, 6.3] *The Quillen pair*

$$I: SeCat_c \rightleftarrows CSS: R$$

defines a Quillen equivalence.

Proof Here, we apply Proposition 1.6.4(2) and show that I reflects weak equivalences between cofibrant objects and that, for any complete Segal space W in CSS, the map $I((RW)^c) = IRW \rightarrow W$ is a weak equivalence in $SeCat_c$.

The fact that I reflects weak equivalences between cofibrant objects follows from the same argument as from the proof of the Quillen pair. To prove the second part, it remains to show that the map $j: RW \rightarrow W$ in the pullback diagram

$$
\begin{array}{ccc}
RW & \longrightarrow & V \\
j\downarrow & & \downarrow \\
W & \longrightarrow & U
\end{array}
$$

is a Dwyer–Kan equivalence. It suffices to show that the map of objects $ob(RW) \rightarrow ob(W)$ is surjective and that the map $map_{RW}(x, y) \rightarrow map_W(jx, jy)$ is a weak equivalence. By the definition of RW, we know that $ob(RW) = ob(W)$. In particular, $jx = x$ and $jy = y$. It remains to show that $map_{RW}(x, y) \simeq map_W(x, y)$. However, this map is precisely the induced map on fibers of the horizontal maps in the diagram (6.3) over points in (the image of) $W_{0,0} \times W_{0,0}$. Therefore, the map $RW \rightarrow W$ is a Dwyer–Kan equivalence. □

6.5 The Second Model Structure

The model structure $SeCat_c$ is good for establishing the Quillen equivalence with the complete Segal space model structure, but a comparison with SC is more difficult. The difficulty is that $SeCat_c$ has all monomorphisms as cofibrations, and not all of them are preserved by the left adjoint to the simplicial nerve functor. Therefore, we need another model structure $SeCat_f$, with fewer cofibrations, to obtain such a Quillen equivalence.

For the model structure $SeCat_c$, we started with the generating cofibrations in the Reedy model category structure and adapted them to be generating cofibrations of Segal precategories. For this second model structure, we use modified generating cofibrations from the projective model structure on the category of simplicial spaces so that the objects involved are Segal precategories.

We want the weak equivalences in this model structure to be the same as those of $SeCat_c$, namely, the Dwyer–Kan equivalences. However, there is one technicality we need to address.

To define the weak equivalences for the new model structure, we want to use a functorial localization in $SeSp_f$ rather than $SeSp_c$. We can define a localization functor L_f in the same way that we defined L_c but making necessary changes in light of the fact that we are starting from the model structure $SeSp_f$. However, just as in the fixed-object situation, we can use Proposition 2.8.6 to see that the weak equivalences in the two model structures agree.

Recall the set of maps I_f, obtained by modifying the generating cofibrations in the projective model structure, from Section 6.1.

Theorem 6.5.1 [30, 7.1] *There is a cofibrantly generated model structure $SeCat_f$ on the category of Segal precategories in which*

1 *the weak equivalences are the Dwyer–Kan equivalences; and*
2 *the cofibrations are the maps which can be formed by taking iterated pushouts along the maps of the set I_f.*

We define the set J_f to be a set of isomorphism classes of maps $\{i \colon A \to B\}$ such that

1 the map $i \colon A \to B$ is an acyclic cofibration, and
2 for all $n \geq 0$, the spaces A_n and B_n have countably many simplices.

We would like to show that I_f is a set of generating cofibrations and that J_f is a set of generating acyclic cofibrations for $SeCat_f$.

We begin with the following lemma, whose proof is similar to that of Proposition 6.3.4.

Lemma 6.5.2   [30, 7.2] *Any acyclic cofibration* $j: C \to D$ *in* $SeCat_f$ *can be obtained as a directed colimit of pushouts along the maps in* J_f.

We have the following characterization of the acyclic fibrations.

Proposition 6.5.3   [30, 7.3] *A map* $f: X \to Y$ *is an acyclic fibration in* $SeCat_f$ *if and only if it is an* I_f-*fibration.*

Proof   First suppose that f has the right lifting property with respect to the maps in I_f. Then we claim that for each $n \geq 0$ and $(v_0, \ldots, v_n) \in X_0^{n+1}$, the map $X_n(v_0, \ldots, v_n) \to Y_n(fv_0, \ldots, fv_n)$ is an acyclic fibration of simplicial sets. This fact, however, follows from Lemma 6.1.5. In particular, it is a weak equivalence, and therefore a proof similar to that of Lemma 6.3.6 can be used to show that the map $X \to Y$ is a Dwyer–Kan equivalence, completing the proof of the first implication.

To prove the converse, assume that f is a fibration and a weak equivalence. Then we can apply the proof of Proposition 6.3.7, making the factorizations in the projective model structure rather than in the Reedy model structure. The argument follows analogously.   □

Proposition 6.5.4   [30, 7.4] *A map in* $SeCat_f$ *is a* J_f-*cofibration if and only if it is an* I_f-*cofibration and a weak equivalence.*

Proof   This proof follows just as the proof of Proposition 6.3.9, again using the projective structure rather than the Reedy structure.   □

Proof of Theorem 6.5.1   As before, we must check the conditions of Theorem 1.7.15. Condition (1) follows just as in the proof of Theorem 6.3.3. Condition (2) is precisely the statement of Proposition 6.5.4. Condition (3) and condition (4)(ii) follow from Proposition 6.5.3 after applying Lemma 6.5.2.   □

We have the following properties of the model category $SeCat_f$.

Theorem 6.5.5   [24, 4.2] *The fibrant objects in* $SeCat_f$ *are precisely the Segal categories which are fibrant in the projective model category structure on simplicial spaces.*

Proof   The argument given for $SeCat_c$ can be applied in this case.   □

Proposition 6.5.6   *The model category* $SeCat_f$ *is left proper.*

Proof　Consider a pushout diagram

$$
\begin{array}{ccc}
A & \xrightarrow{\ g\ }_{\simeq} & C \\
{\scriptstyle f}\downarrow & & \downarrow \\
B & \xrightarrow{\ h\ } & D
\end{array}
$$

in which f is a cofibration and g is a weak equivalence. We want to show that h is a weak equivalence. Since f is a cofibration in $SeCat_f$, it is also a cofibration in $SeCat_c$. Since we know by Corollary 6.3.11 that $SeCat_c$ is left proper, and since weak equivalences are the same in both model structures, we can conclude that h is a weak equivalence in $SeCat_f$.　　　　　□

In general, the model category $SeCat_f$ is not as nice as $SeCat_c$. We have already seen that neither its fibrations nor its cofibrations seem to have a clean description, and it is known not to be cartesian [110, 19.5.1]. Nonetheless, it will be useful in the next section. Furthermore, the two model structures are Quillen equivalent.

Theorem 6.5.7　[30, 7.5] *The identity functor induces a Quillen equivalence*

$$
SeCat_f \rightleftarrows SeCat_c.
$$

Proof　Since both maps are the identity functor, they form an adjoint pair. Notice that the cofibrations of $SeCat_f$ form a subclass of the cofibrations of $SeCat_c$ since they are monomorphisms. Similarly, the acyclic cofibrations of $SeCat_f$ form a subclass of the acyclic cofibrations of $SeCat_c$. In particular, these observations imply that the identify functor $SeCat_f \to SeCat_c$ preserves cofibrations and acyclic cofibrations. Hence, we have a Quillen pair.

It remains to show that this pair is a Quillen equivalence. To do so, we must show that, given any cofibrant object X in $SeCat_f$ and fibrant object Y in $SeCat_c$, a map $X \to Y$ is a weak equivalence in $SeCat_f$ if and only if its corresponding adjoint map, which is just $X \to Y$ again, is a weak equivalence in $SeCat_c$. But, we have already established that the weak equivalences are the same in each category, completing the proof.　　　　　□

Remark 6.5.8　One might ask at this point why we could not just use the $SeCat_f$ model category structure and show a Quillen equivalence between it and the model category structure CSS_f where we localize the projective model category structure (rather than the Reedy) with respect to the same maps.

However, if we work with "complete Segal spaces" which are fibrant in the projective model structure rather than in the Reedy structure, then for a fibrant object W, the map $W \to U$ used in defining the right adjoint $CSS \to SeCat_c$

need not be a fibration. Therefore, the pullback RW is no longer a homotopy pullback and in particular not homotopy invariant. If RW is not homotopy invariant, then there is no guarantee that the map $RW \to W$ is a Dwyer–Kan equivalence, and the argument for a Quillen equivalence fails. Thus, the $SeCat_c$ and CSS model structures as we have defined them are necessary for the arguments that we have used. In any case, it is useful in practice to have model structures in which all objects are cofibrant.

6.6 The Equivalence With Simplicial Categories

We begin by defining an adjoint pair of functors between the two categories SC and $SeCat_f$. The simplicial nerve functor $R \colon SC \to SeCat_f$ is the right adjoint in this pair. Via this nerve functor, we regard simplicial categories as strictly local Segal spaces, or strictly local objects in the sense of Definition 2.8.9, with respect to the maps with which we localize to obtain the Segal space model structure.

Although we are actually working in the subcategory of Segal precategories, rather than the category of all simplicial spaces, we can still use Lemma 2.8.10 to obtain a left adjoint functor F to our inclusion map R, since the construction defining F preserves the property of having a discrete space in degree zero, being defined by iterated pushouts.

The following result gives a formal justification for our overlap of terminology between simplicial categories and Segal spaces or Segal categories.

Proposition 6.6.1 *A simplicial functor $C \to D$ is a Dwyer–Kan equivalence of simplicial categories if and only if its induced map on simplicial nerves $RC \to RD$ is a Dwyer–Kan equivalence of Segal precategories.*

Proof Since the simplicial nerve of a simplicial category is a strict Segal category, its localization is the identity, perhaps up to a fibrant replacement. However, the simplicial nerve preserves objects, and can be shown to preserve mapping spaces. The result follows. □

In the proof of the following result, we see the benefit of having made the comparison in the fixed object set case.

Proposition 6.6.2 [30, 8.3] *The adjoint pair*

$$F \colon SeCat_f \rightleftarrows SC \colon R$$

is a Quillen pair.

Proof We want to show that the left adjoint F preserves cofibrations and acyclic cofibrations. We begin by considering cofibrations.

Since F is a left adjoint functor, it preserves colimits. Therefore, using Proposition 1.7.14, it suffices to show that F preserves the set I_f of generating cofibrations in $SeCat_f$. Recall that the elements of this set are the maps $P_{m,n} \to Q_{m,n}$ for $m, n \geq 0$. We begin by considering the maps $P_{n,1} \to Q_{n,1}$ for any $n \geq 0$. The strict localization of such a map is precisely the map of simplicial categories $U\partial\Delta[n] \to U\Delta[n]$ which is a generating cofibration in SC. Then one can check that the strict localization of any $P_{m,n} \to Q_{m,n}$ can be obtained as the colimit of iterated pushouts along the generating cofibrations of SC. Therefore, F preserves cofibrations.

We now need to show that F preserves acyclic cofibrations. To do so, first consider the model structure $SeCat_{O,f}$ on Segal precategories with a fixed set O in degree zero and the model structure SC_O of simplicial categories with a fixed object set O. Recall from Theorem 6.2.10 that there is a Quillen equivalence

$$F_O\colon SeCat_{O,f} \rightleftarrows SC_O \colon R_O.$$

In particular, if X is a cofibrant object of $SeCat_{O,f}$, then there is a weak equivalence $X \to R_O((F_O X)^f)$. Notice that F_O agrees with F on Segal precategories with the set O in degree zero, and similarly R_O agrees with R.

Suppose first that $X \to Y$ is an acyclic cofibration between cofibrant objects in $SeCat_f$, and let $O = X_0$ and $O' = Y_0$. Consider the commutative diagram

$$
\begin{array}{ccc}
X & \xrightarrow{\simeq} & L_f X \\
\downarrow & & \downarrow \\
Y & \xrightarrow{\simeq} & L_f Y
\end{array}
$$

where the upper and lower horizontal maps are weak equivalences not only in $SeCat_f$, but also in $SeCat_{O,f}$ and $SeCat_{O',f}$, respectively. However, the functors F_O and $F_{O'}$ (and hence also F) preserve these weak equivalences, giving a diagram

$$
\begin{array}{ccc}
FX & \xrightarrow{\simeq} & FL_f X \\
\downarrow & & \downarrow \\
FY & \xrightarrow{\simeq} & FL_f Y
\end{array}
$$

in which the indicated maps are weak equivalences. Thus, to prove that $FX \to FY$ is a weak equivalence, it suffices to prove that $FL_f X \to FL_f Y$ is a weak equivalence. Applying the adjunction (F, R), we can instead consider the diagram

$$
\begin{array}{ccc}
L_f X & \xrightarrow{\simeq} & RFL_f X \\
\downarrow & & \downarrow \\
L_f Y & \xrightarrow{\simeq} & RFL_f Y
\end{array}
$$

in which the indicated maps are weak equivalences from the fixed-object Quillen equivalence. However, we know that the left-hand vertical map is a weak equivalence, so the right-hand vertical map must be also. Using Proposition 6.6.1, it follows that $FL_f X \to FL_f Y$ is a weak equivalence. Thus, we have shown that F preserves acyclic cofibrations between cofibrant objects.

It remains to show that F preserves all acyclic cofibrations. Suppose that $f : X \to Y$ is an acyclic cofibration in $SeCat_f$. Applying the cofibrant replacement functor to the map $X \to Y$, we obtain an acyclic cofibration $X^c \to Y^c$, and in the resulting commutative diagram

$$
\begin{array}{ccc}
X^c & \longrightarrow & Y^c \\
\downarrow & & \downarrow \\
X & \longrightarrow & Y
\end{array}
$$

the vertical arrows are levelwise weak equivalences.

Now consider the following diagram, where the top square is a pushout diagram:

$$
\begin{array}{ccc}
X^c & \xrightarrow{\simeq} & Y^c \\
\downarrow & & \downarrow \\
X & \xrightarrow{\simeq} & Y' \\
\Big\| & & \downarrow \\
X & \xrightarrow{\simeq} & Y.
\end{array}
$$

Notice that all three of the horizontal arrows are acyclic cofibrations in $SeCat_f$, the upper and lower by assumption and the middle one because pushouts preserve acyclic cofibrations by Proposition 1.4.11. Now we apply the functor F

to this diagram to obtain the diagram

$$
\begin{array}{ccc}
FX^c & \xrightarrow{\simeq} & FY^c \\
\downarrow & & \downarrow \\
FX & \xrightarrow{\simeq} & FY' \\
{\scriptstyle =}\downarrow & & \downarrow \\
FX & \longrightarrow & FY.
\end{array}
\tag{6.4}
$$

The top horizontal arrow is an acyclic cofibration since F preserves acyclic cofibrations between cofibrant objects. Furthermore, since F is a left adjoint and hence preserves colimits, the middle horizontal arrow is also an acyclic cofibration because the top square is a pushout square.

Now we work in the category of Segal precategories under X, whose objects are maps $X \rightarrow Y$ in $SeCat_c$. To show that the bottom horizontal arrow of diagram (6.4) is an acyclic cofibration, consider the following diagram in the category of cofibrant objects under X:

Now, let O'' denote the subset of $(Y')_0$ which is not in the image of X_0. Now we have the diagram in the category of cofibrant objects under $X \amalg O''$ with the same set in degree zero:

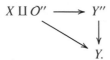

However, since we are now working in the fixed-object context, we know by Theorem 6.2.10 that $F_{O''}$ is the left adjoint of a Quillen equivalence, and therefore the map $F_{O''}Y' \rightarrow F_{O''}Y$ is a weak equivalence in $SC_{O''}$, and in particular a Dwyer–Kan equivalence when regarded as a map in SC. It follows that the map $FY' \rightarrow FY$ is a weak equivalence, from which we can conclude, using (6.4), that $FX \rightarrow FY$ is a weak equivalence. Thus F preserves acyclic cofibrations. $\square$

The following lemma again uses the perspective that a Segal category is a local diagram and a simplicial category is a strictly local diagram in $SeCat_f$; we omit the nerve notation for simplicity.

Lemma 6.6.3 [30, 8.5] *Let X be a cofibrant object in $SeCat_f$. Then the map $X \to FX$ is a Dwyer–Kan equivalence.*

Proof First consider a cofibrant object in $SeCat_f$ of the form $\amalg_i Q_{m_i,n_i}$. If Y is a fibrant object in $SeCat_f$, then we can use Proposition 2.4.10 and Remark 6.1.4 to obtain the weak equivalences

$$\text{Map}_{SeCat_f}\left(\coprod_i Q_{m_i,n_i}, Y\right) \simeq \prod_i \text{Map}_{SeCat_f}(Q_{m_i,n_i}, Y)$$

$$\simeq \prod_i \coprod_{v_0,\dots,v_n} \text{Map}_{SSets}(\Delta[m_i], \text{map}_Y(v_0, \dots, v_{n_i}))$$

$$\simeq \prod_i \coprod_{v_0,\dots,v_n} \text{Map}_{SSets}(\Delta[0], \text{map}_Y(v_0, \dots, v_{n_i}))$$

$$\simeq \text{Map}_{SeCat_f}\left(\coprod_i Q_{0,n_i}, Y\right)$$

$$\simeq \text{Map}_{SeCat_f}\left(\coprod_i \Delta[n_i]^t, Y\right).$$

Therefore, we have reduced to the case of objects of the form $\amalg_i \Delta[n_i]^t$, which are Segal categories. Indeed, they are simplicial nerves of ordinary categories and thus strictly local diagrams. It follows that the map

$$\coprod_i \Delta[n_i]^t \to F\left(\coprod_i \Delta[n_i]^t\right)$$

is a Dwyer–Kan equivalence.

Now suppose that X is any cofibrant object in $SeCat_f$. By Proposition 2.5.3, X can be written as $\text{colim}_{\Delta^{op}} X_j$, where each X_j has the form $\amalg_i \Delta[n_i]^t$. If Y is a fibrant object in $SeCat_f$ which is strictly local, we have

$$\text{Map}_{SeCat_f}(\text{colim}_{\Delta^{op}} X_j, Y) \cong \lim_{\Delta} \text{Map}_{SeCat_f}(X_j, Y)$$

$$\cong \lim_{\Delta} \text{Map}_{SeCat_f}(FX_j, Y)$$

$$\cong \text{Map}_{SeCat_f}(\text{colim}_{\Delta^{op}} FX_j, Y)$$

$$\cong \text{Map}_{SeCat_f}(F(\text{colim}_{\Delta^{op}}(FX_j)), Y).$$

We can now apply Proposition 2.8.11, which says that

$$F(\text{colim}(FX_j)) \simeq F(\text{colim} X_j).$$

Therefore we have

$$\text{Map}_{SeCat_f}(F(\text{colim}_{\Delta^{op}}(FX_j)), Y) \simeq \text{Map}_{SeCat_f}(FX, Y).$$

It follows that the map $X \to FX$ is a Dwyer–Kan equivalence. $\square$

We are now able to prove the main result of this section.

Theorem 6.6.4   [30, 8.6] *The Quillen pair*

$$F: SeCat_f \rightleftarrows SC: R$$

is a Quillen equivalence.

Proof   We first show that F reflects weak equivalences between cofibrant objects. Let $f: X \to Y$ be a map of cofibrant Segal precategories such that $Ff: FX \to FY$ is a weak equivalence of simplicial categories. (Since F preserves cofibrations, both FX and FY are again cofibrant.) Then consider the following diagram:

$$
\begin{array}{ccccc}
FX & \longrightarrow & L_f FX & \longleftarrow & L_f X \\
\simeq \downarrow & & \downarrow & & \downarrow \\
FY & \longrightarrow & L_f FY & \longleftarrow & L_f Y.
\end{array}
$$

By assumption, the leftmost vertical arrow is a Dwyer–Kan equivalence. The horizontal arrows of the left-hand square are also Dwyer–Kan equivalences by definition. Since X and Y are cofibrant, Lemma 6.6.3 shows that the horizontal arrows of the right-hand square are Dwyer–Kan equivalences. The commutativity of the whole diagram shows that the map $L_f FX \to L_f FY$ is a Dwyer–Kan equivalence and thus that the map $L_f X \to L_f Y$ is also. Therefore, F reflects weak equivalences between cofibrant objects.

Now, we show that, given any fibrant simplicial category Y, the map

$$F((RY)^c) \to Y$$

is a Dwyer–Kan equivalence. Consider a fibrant simplicial category Y and apply the functor R to obtain a Segal category which is levelwise fibrant and therefore fibrant in $SeCat_f$. Its cofibrant replacement is Dwyer–Kan equivalent to it in $SeCat_f$. Then, by the above argument, strictly localizing this object again yields a Dwyer–Kan equivalent simplicial category.   □

7
Quasi-Categories

In this chapter, we look at yet another model for $(\infty, 1)$-categories with composition only defined up to homotopy. While we still use simplicial techniques, this model has a substantially different flavor. Whereas Segal categories and complete Segal spaces are kinds of simplicial spaces, quasi-categories are instead simplicial sets.

7.1 Basic Definitions

We first recall the following definition, which was given previously as Definition 3.2.5.

Definition 7.1.1 A simplicial set K is a *quasi-category* if a lift exists in any diagram

for $0 < k < n$.

We saw there that the idea of a quasi-category is that it looks like the nerve of a category, but with nonunique composition. It this sense, it is similar in spirit to Segal categories and complete Segal spaces. However, one main difference is that a quasi-category is a simplicial set, rather than a simplicial space. In other words, we are encoding the same essential information but in a simpler object. This structural difference has advantages and disadvantages: one advantage is that a quasi-category is a simpler kind of object, while a disadvantage is that

157

the categorical properties are more compressed. A quasi-category is in some sense a more categorical, rather than topological, structure.

Since we have been looking at structures with weak composition through the lens of Segal conditions in the last two chapters, let us take the opportunity to connect the Segal condition to the inner horn-filling condition used in the definition of a quasi-category.

Proposition 7.1.2   *A simplicial set K is a quasi-category if and only if a lift exists in any diagram of the form*

for any $n \geq 2$.

Proof   Consider any diagram of the form

$$V[2,1] \longrightarrow K$$
$$\downarrow \quad \nearrow$$
$$\Delta[2] \qquad .$$

Since $V[2,1] = G(2)$, the existence of the dotted arrow lift is equivalent to surjectivity of the map

$$\varphi_2 \colon K_2 = \mathrm{Hom}(\Delta[2], K) \to \mathrm{Hom}(G(2), K) = K_1 \times_{K_0} K_1.$$

Now, for $n > 2$, consider the diagram

$$G(n) \longrightarrow K$$
$$\downarrow \quad {}^{f} \quad \nearrow$$
$$V[n,k] \quad {}_{g}$$
$$\downarrow$$
$$\Delta[n] \qquad .$$

First, suppose K is a quasi-category, and suppose that the Segal maps

$$\varphi_i \colon K_i \to \underbrace{K_1 \times_{K_0} \cdots \times_{K_0} K_1}_{i}$$

are surjective for all $i < n$. Since $V[n, i]$ can be obtained from $G(n)$ by taking pushouts along the Segal maps φ_i for $2 \leq i < n$, by induction the lift f exists. The existence of g follows from the fact that K is a quasi-category.

Conversely, suppose the maps φ_i are all surjective, and suppose the inner horn-filling conditions hold for dimensions less than n. Then the lift g exists by assumption, and the lift f can be defined as the composite of g with the horn inclusion. $\qquad\qquad\square$

Equivalently, this proposition says that K is a quasi-category if and only if the Segal maps $\varphi_n \colon K_n \to X_1 \times_{X_0} \cdots \times_{X_0} X_1$ are surjective for every $n \geq 2$.

We would like to prove that there is a model structure on the category of simplicial sets in which the fibrant objects are the quasi-categories. To begin, we set up to define appropriate weak equivalences.

As we want to consider quasi-categories as weak versions of simplicial categories, we need to consider simple maps, each of which includes a single point into a larger but weakly equivalent quasi-category. Just as for complete Segal spaces, we make use of the simplicial set E whose definition we recall.

Definition 7.1.3   Let E be the nerve of the groupoid with two objects and a single isomorphism between them.

Remark 7.1.4   The object E is denoted by E^1 by Dugger and Spivak [51, 52] and J by Joyal [73].

Definition 7.1.5   Suppose $f, g \colon A \to K$ are maps of simplicial sets with K a quasi-category. The maps f and g are *E-homotopic*, denoted by $f \sim g$, if there exists a map $H \colon A \times E \to K$ such that $Hi_0 = f$ and $Hi_1 = g$, where $i_0, i_1 \colon A \to A \times E$ are the natural inclusions.

One can check that E-homotopy defines an equivalence relation on the set of maps $A \to K$. Let $[A, K]_E$ denote the set of equivalence classes. Equivalently, we can regard $[A, K]_E$ as a coequalizer

$$\mathrm{Hom}(A \times E, K) \rightrightarrows \mathrm{Hom}(A, K) \to [A, K]_E.$$

Definition 7.1.6   A map $A \to B$ of simplicial sets is a *Joyal equivalence* if, for every quasi-category K, the induced map of sets

$$[B, K]_E \to [A, K]_E$$

is an isomorphism.

Joyal refers to these maps as *weak categorical equivalences*, whereas Lurie simply calls them *categorical equivalences* [88]. We are now able to describe our desired model structure.

Theorem 7.1.7   [51, 2.13], [88, 2.2.5.1] *There exists a cofibrantly generated model structure on the category of simplicial sets in which:*

1 the cofibrations are the monomorphisms,
2 the fibrant objects are the quasi-categories, and
3 the weak equivalences are the Joyal equivalences.

We denote this model structure by *QCat*. Note that it is left proper because every object is cofibrant, by Proposition 1.7.3.

The proof of Theorem 7.1.7 is given at the end of Section 7.3, after providing some preliminary results.

Now that we have two different model structures on the category of simplicial sets, both of which are used in this chapter, we make the following distinctions in terminology. In *QCat*, we use the adjective *Joyal*, and thus refer to *Joyal fibrations* and *Joyal equivalences*. To emphasize the difference, in the model structure *SSets* from Theorem 2.2.5, we use the adjective *Kan*, and so specify *Kan fibrations* and *Kan weak equivalences*. Since the classes of cofibrations agree in both model structures, we have no need for this distinction.

There are several proofs of this theorem, due to Joyal [73], Lurie [88], and Dugger and Spivak [51]. We closely follow the treatment of Dugger and Spivak.

We must first establish a number of preliminary results. The first feature we investigate in *QCat* is the interplay between two kinds of equivalences. We have already seen the definition of Joyal equivalence, but the notion of E-homotopy suggests the following definition.

Definition 7.1.8   A map $f: K \to L$ of simplicial sets is an *E-homotopy equivalence* if there is a map $g: L \to K$ such that fg and gf are E-homotopic to id_L and id_K, respectively.

Remark 7.1.9   [51, 2.10] The relationship between Joyal equivalence and E-homotopy equivalence is analogous to that of homotopy equivalence and weak homotopy equivalence in $\mathcal{T}op$. Specifically, any E-homotopy equivalence $f: X \to Y$ is a Joyal equivalence. If X and Y are quasi-categories, then the converse statement also holds. Observe that either of the two maps $\Delta[0] \to E$ is an E-homotopy equivalence.

7.2 Properties of Acyclic Cofibrations

The goal of this section is to prove the following result which shows that our desired acyclic cofibrations for *QCat* behave appropriately.

Theorem 7.2.1   *The class of cofibrations which are Joyal equivalences is closed under pushouts and transfinite composition.*

We prove this proposition by making an explicit comparison with the acyclic cofibrations in the Kan model structure on simplicial sets, which requires intermediate results which in turn require making several definitions. An important point is that the acyclic cofibrations in *QCat* are not quite so simple to characterize as those in *SSets*, and in particular they are not generated by the inner horn inclusions alone. However, looking at inner horn inclusions is a good place to start our investigation. The terminology given here is inspired by the fact that Kan acyclic cofibrations are often called *anodyne maps*.

Definition 7.2.2 A map of simplicial sets is *inner anodyne* if it can be constructed out of the maps $V[n, k] \to \Delta[n]$, where $n \geq 2$ and $0 < k < n$, by pushouts and transfinite compositions. A map of simplicial sets is an *inner fibration* if it has the right lifting property with respect to inner horn inclusions, and hence with respect to all inner anodyne maps.

By definition, any quasi-category is *inner fibrant*, in that it has the right lifting property with respect to all inner anodyne maps. Given maps of simplicial sets $f : A \to B$ and $g : C \to D$, consider the induced pushout-product map

$$A \times D \cup_{A \times C} B \times C \to B \times D.$$

We begin with two technical results concerning such maps. The first is proved in appendix B of [51].

Proposition 7.2.3 [51, 2.3] *Let K be a quasi-category. Then the map $K \to \Delta[0]$ has the right lifting property with respect to the maps*

$$A \times E \cup_{A \times \{0\}} B \times \{0\} \to B \times E$$

for every monomorphism $A \to B$.

Proposition 7.2.4 [51, 2.6] *If $V[n, k] \to \Delta[n]$ is an inner horn inclusion and $A \to B$ is any monomorphism, then the map*

$$V[n, k] \times B \cup_{V[n,k] \times A} \Delta[n] \times A \to \Delta[n] \times B$$

is inner anodyne.

The proof can be found in appendix A of [51]. An adjoint version can be used to prove the following result.

Proposition 7.2.5 [51, 2.7] *If $A \to B$ is a monomorphism and $K \to L$ is an inner fibration, then the pushout-product map*

$$K^B \to K^A \times_{L^A} L^B$$

is an inner fibration. In particular, if K is a quasi-category, then so is K^A for any simplicial set A.

To continue our investigation of Joyal acyclic cofibrations, we need to consider horn inclusions for which one of the outer edges can be thought of as invertible in some sense. To be more precise, we need to define the notion of equivalence within a quasi-category. The definition can be stated for any simplicial set.

Definition 7.2.6   Let K be a simplicial set. An *equivalence* in K is a 1-simplex $\Delta[1] \to K$ which can be extended to a map $E \to K$.

When K is a quasi-category, we expect these equivalences to resemble isomorphisms in a category. To get a sense of how equivalences in a quasi-category behave, we give the following equivalent characterizations. The proof can be found in appendix B of [51].

Proposition 7.2.7   [51, 2.2] *Let K be a quasi-category, and let $f\colon \Delta[1] \to K$ be a 1-simplex. The following statements are equivalent.*

1 *The 1-simplex f is an equivalence.*
2 *The 1-simplex f can be extended to a map $\mathrm{sk}_2(E) \to K$.*
3 *The 1-simplex f has a left inverse and a right inverse. That is, there exist 2-simplices in K of the form*

$$
\begin{array}{ccc}
a \xrightarrow{\;f\;} b & \qquad & b \xrightarrow{\;h\;} a \\
\text{id}_a \searrow \;\downarrow g & & \text{id}_b \searrow \;\downarrow f \\
a & & b.
\end{array}
$$

Our purpose in considering these equivalences within a quasi-category is to specify horns for which the image of certain edges is an equivalence. For the n-simplex $\Delta[n]$ and any $0 \le k < n$, we denote by $\Delta[n]_{k,k+1}$ the restriction of $\Delta[n]$ to the single 1-simplex given by $k \to k + 1$.

Definition 7.2.8   [51, A.3]   Let K be a simplicial set.

1 A map $h\colon V[n,k] \to K$ is a *special right horn* if $k = n$ and $h(\Delta[n]_{n-1,n})$ is an equivalence in K.
2 A map $h\colon V[n,k] \to K$ is a *special left horn* if $k = 0$ and $h(\Delta[n]_{0,1})$ is an equivalence in K.
3 A *special outer horn* is either a special left horn or a special right horn.
4 A horn h is *special* if it is either an inner horn or a special outer horn.

5 A map $f \colon K \to L$ is *special outer anodyne* if it is a composite of pushouts along special outer horns. It is *special anodyne* if it is the composite of pushouts along special horns.

Given any $n \geq 1$, consider the map

$$\Delta[0] \times \Delta[n] \cup_{\Delta[0] \times \partial \Delta[n]} E \times \partial \Delta[n] \to E \times \Delta[n]$$

induced from the inclusions $\Delta[0] \to E$ and $\partial \Delta[n] \to \Delta[n]$. Observe that the domain of this map can be simplified to

$$\Delta[n] \amalg_{\partial \Delta[n]} E \times \partial \Delta[n].$$

Lemma 7.2.9 [51, A.4] *For any $n \geq 1$, the map*

$$\Delta[n] \cup_{\partial \Delta[n]} E \times \partial \Delta[n] \to E \times \Delta[n]$$

is special anodyne.

Let us now consider some lifting properties involving special horns; for the first, we again refer the reader to appendix B of [51] for the proof.

Proposition 7.2.10 [51, B.11] *Let K and L be quasi-categories. A lift exists in any diagram*

$$
\begin{array}{ccc}
V[n,0] & \xrightarrow{\ p\ } & K \\
\downarrow & \nearrow & \downarrow \\
\Delta[n] & \xrightarrow{\ m\ } & L
\end{array}
$$

in which p is a special left horn and $K \to L$ is an inner fibration. Similarly, a lift exists when p is replaced by a special right horn $q \colon V[n,n] \to K$.

Proposition 7.2.11 [51, B.15] *Let $f \colon K \to L$ be an inner fibration between quasi-categories which has the right lifting property with respect to $\Delta[0] \to E$. Given any monomorphism $A \to B$, the map f has the right lifting property with respect to the map*

$$B \cup_A E \times A \to E \times B.$$

Equivalently, using adjointness, the map $X^E \to K \times_L L^E$ is a Kan acyclic fibration.

Proof By Lemma 7.2.9 and Proposition 7.2.10, we know that f has the right lifting property with respect to

$$\Delta[n] \amalg_{\partial \Delta[n]} E \times \partial \Delta[n] \to E \times \Delta[n]$$

for any $n > 0$. When $n = 0$, this map is just the inclusion $\Delta[0] \to E$, so this case holds by assumption. Since every monomorphism is generated by the boundary inclusions $\partial\Delta[n] \to \Delta[n]$, the result follows in the remaining cases by induction on simplices.   □

Thus, we can see that maps with the right lifting property with respect to $\Delta[0] \to E$ are also key.

Definition 7.2.12   A map $K \to L$ is a *special inner fibration* if it has the right lifting property with respect to the inclusion $\Delta[0] \to E$ and with respect to all inner horn inclusions.

Note that if K is a quasi-category then the unique map $K \to \Delta[0]$ is a special inner fibration, using the retraction $E \to \Delta[0]$.

We can now state the first of two tools that we will use to prove Theorem 7.2.1.

Proposition 7.2.13   [51, C.1] *Let K and L be quasi-categories. If $K \to L$ is a special inner fibration and $A \to B$ is a monomorphism, then the pullback-corner map $K^B \to K^A \times_{L^A} L^B$ is also a special inner fibration.*

Proof   We proved in Proposition 7.2.5 that the map $K^B \to K^A \times_{L^A} L^B$ is an inner fibration. Now we can use Proposition 7.2.11 to show that it also has the right lifting property with respect to the map $\Delta[0] \to E$, making it a special inner fibration.   □

The second result we need is the following lemma which gives two criteria for when Joyal equivalences are Kan acyclic fibrations.

Proposition 7.2.14   [51, C.3] *Let K and L be quasi-categories.*

1 *If a map $K \to L$ is a special inner fibration and a Joyal equivalence, then it is a Kan acyclic fibration.*
2 *If $A \to B$ is a monomorphism and a Joyal equivalence, then $K^B \to K^A$ is a Kan acyclic fibration.*

Proof   Given a quasi-category K, let $R(K)$ be the maximal Kan complex contained in K, and observe that $R(K)_0 = K_0$. If $\mathcal{Kan} \subseteq \mathcal{SSets}$ denotes the full subcategory of Kan complexes, then $R \colon \mathcal{SSets} \to \mathcal{Kan}$ is right adjoint to the inclusion $\mathcal{Kan} \to \mathcal{SSets}$. In particular, R preserves products.

If K and L are quasi-categories and $f \colon K \to L$ is a special inner fibration, we claim that the map $R(f) \colon R(K) \to R(K)$ is a Kan fibration. We first observe that $R(f)$ is an inner fibration using the diagram

$$
\begin{array}{ccccc}
V[n,k] & \longrightarrow & K & \longrightarrow & R(K) \\
\downarrow & \nearrow & \downarrow{\scriptstyle f} & & \downarrow{\scriptstyle R(f)} \\
\Delta[n] & \longrightarrow & L & \longrightarrow & R(L).
\end{array}
$$

Now observe that any outer horn is special in a Kan complex, so in particular in $R(K)$. Then we can apply Proposition 7.2.10 to see that $R(f)$ has the right lifting property with respect to outer horns. Thus, $R(f)$ is a Kan fibration.

Now, since f is a Joyal equivalence between quasi-categories, it is an E-homotopy equivalence by Remark 7.1.9. Using the fact that $R(K \times E) \cong R(K) \times R(E) \cong R(K) \times E$, we can see that $R(f)$ is also an E-homotopy equivalence. Restricting from E to $\Delta[1]$, it is therefore an ordinary homotopy equivalence. But a homotopy equivalence between Kan complexes is a Kan weak equivalence.

To prove (2), first recall by Proposition 7.2.5 that if K is a quasi-category, then so is K^C for any simplicial set C. Thus, the map $K^B \to K^A$ is a map of quasi-categories. By (1), it suffices to show that this map is both a special inner fibration and a Joyal equivalence. We can see that it is a special inner fibration by applying Proposition 7.2.13 to the maps $K \to \Delta[0]$ and $A \to B$.

Now we show that $K^B \to K^A$ is a Joyal equivalence. Since K^C is a quasi-category for any quasi-category K and simplicial set C, the map $[B, K^C]_E \to [A, K^C]_E$ is an isomorphism. Using adjointness, we obtain isomorphisms

$$
\begin{aligned}
[C, K^B]_E & \cong [C \times B, K]_E \\
& \cong [B, K^C]_E \\
& \cong [A, K^C]_E \\
& \cong [A \times C, K]_E \\
& \cong [C, K^A]_E.
\end{aligned}
$$

If we consider the cases $C = K^A$ and $C = K^B$, we can see that $K^B \to K^A$ is an E-homotopy equivalence. By Remark 7.1.9, it is a Joyal equivalence. $\qquad\square$

We are finally able to prove the main result of this section.

Proof of Theorem 7.2.1 First, recall that monomorphisms and maps defined by a left lifting property are closed under pushouts and transfinite compositions. Since Joyal acyclic cofibrations are monomorphisms, it suffices to prove that a monomorphism $f : A \to B$ is a Joyal equivalence if and only if it has the left lifting property with respect to special inner fibrations between quasi-categories.

First suppose that f is a monomorphism with the left lifting property with respect to special inner fibrations between quasi-categories. Then, for any quasi-category K, the map f has the left lifting property with respect to the map $K \to \Delta[0]$. Applying Proposition 7.2.13 to the monomorphism $\partial\Delta[1] \to E$ and the special inner fibration $K \to \Delta[0]$, we obtain that f has the right lifting property with respect to $K^E \to K^{\partial\Delta[1]}$. It follows that f is an E-homotopy equivalence and hence a Joyal equivalence.

Now suppose that f is a monomorphism and a Joyal equivalence. For any special inner fibration $K \to L$ between quasi-categories, consider the diagram

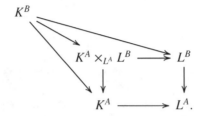

The map $K^B \to K^A \times_{L^A} L^B$ is a special inner fibration by Proposition 7.2.13. By Proposition 7.2.14(2), the maps $L^B \to L^A$ and $K^B \to K^A$ are Kan acyclic fibrations, so the pullback $K^A \times_{L^A} L^B \to K^A$ is also a Kan acyclic fibration. But then the maps $K^A \times_{L^A} L^B \to K^A$ and $K^B \to K^A$ are both Joyal equivalences, hence so is $K^B \to K^A \times_{L^A} L^B$. By Proposition 7.2.14(1), the map $K^B \to K^A \times_{L^A} L^B$ is an acyclic Kan fibration and therefore surjective. But then the map $f : A \to B$ has the left lifting property with respect to $K \to L$, as we needed to show. $\square$

7.3 The Model Structure

We are nearly ready to prove Theorem 7.1.7, establishing the model structure *QCat*. The following part of the proof is stated independently for future reference.

Proposition 7.3.1   *Every Kan acyclic fibration is a Joyal equivalence.*

Proof   Suppose $f : K \to L$ is a Kan acyclic fibration. Since every object of *SSets* is cofibrant, a lift exists in the diagram

If we call this lift s, then $fs = $ id, so we can conclude that $fsf = f$. Take H to be an E-homotopy between fsf and s. In the diagram

$$
\begin{array}{ccc}
K \amalg K & \xrightarrow{\text{id} \amalg sf} & K \\
\downarrow & \nearrow & \downarrow f \; \simeq \\
K \times E & \xrightarrow{\;\; H \;\;} & L
\end{array}
$$

the left-hand vertical map is a cofibration, and therefore a lift exists. Thus s provides an E-homotopy inverse to f, and f is hence a Joyal equivalence by Remark 7.1.9. □

We also give the following result about maps which are both cofibrations and Joyal equivalences; we go ahead and refer to them as Joyal acyclic cofibrations.

Proposition 7.3.2 [51, 2.11]

1 Let A be a simplicial set. If $C \to D$ is a Joyal acyclic cofibration, then so is $A \times C \to A \times D$.

2 If K is a quasi-category, then $K \to \Delta[0]$ has the right lifting property with respect to every Joyal acyclic cofibration.

3 Every inner horn inclusion $V[n, k] \to \Delta[n]$ is a Joyal equivalence.

Proof To prove (1), we need to show that the induced map $[A \times D, K]_E \to [A \times C, K]_E$ is an isomorphism for every quasi-category K. Using adjointness, we can equivalently consider the map $[D, K^A]_E \to [C, K^A]_E$. Since K^A is a quasi-category by Proposition 7.2.5, this second map is an isomorphism.

To prove part (2), we need to show that a dotted arrow lift exists in any diagram

where g is a Joyal acyclic cofibration and K is a quasi-category. Since g is a Joyal equivalence, we know that the map $[B, K]_E \to [A, K]_E$ is an isomorphism. Thus, associated to $f : A \to K$ there is a map $h : B \to K$ together with an E-homotopy from $h \circ g$ to f, producing a diagram

$$
\begin{array}{ccc}
A \times \{0\} & \xrightarrow{\;g\;} & B \times \{0\} \\
\downarrow & & \downarrow{\scriptstyle h} \\
A \times E & \xrightarrow{\;H\;} & K \\
\uparrow & \nearrow{\scriptstyle f} & \\
A \times \{1\} & &
\end{array}
$$

.

This homotopy produces a lift in the diagram

$$
\begin{array}{ccc}
(A \times E) \cup_{A \times \{0\}} (B \times \{0\}) & \longrightarrow & K \\
\downarrow & \nearrow & \\
B \times E & &
\end{array}
$$

using Proposition 7.2.3. Restricting this lift to $B \times \{1\}$ produces the desired lift in the original diagram.

Lastly, to prove (3), we need to prove that the map $[\Delta[n], K]_E \to [V[n,k], K]_E$ is an isomorphism for any quasi-category K. By definition of quasi-category, this map is surjective.

To show that it is injective, suppose $f, g \colon \Delta[n] \to K$ restrict to the same E-homotopy class in $[V[n,k], K]_E$. Then there is an E-homotopy from $f \circ i$ to $g \circ i$, where $i \colon V[n,k] \to \Delta[n]$ is the horn inclusion. We depict this E-homotopy via the diagram

$$
\begin{array}{ccc}
V[n,k] \times \{0\} & \xrightarrow{\;i\;} & \Delta[n] \times \{0\} \\
\downarrow & & \downarrow{\scriptstyle f} \\
V[n,k] \times E & \xrightarrow{\;H\;} & K \\
\uparrow & & \uparrow{\scriptstyle g} \\
V[n,k] \times \{1\} & \xrightarrow{\;i\;} & \Delta[n] \times \{1\}.
\end{array}
$$

We use this E-homotopy H to form the diagram

$$
\begin{array}{ccc}
(V[n,k] \times E) \cup (\Delta[n] \times \partial\Delta[1]) & \xrightarrow{\;H \times (f \amalg g)\;} & K \\
\downarrow & \nearrow & \\
\Delta[n] \times E & &
\end{array}
$$

where the vertical map is the pushout product of i and the inclusion $\partial\Delta[1] \to E$.

But the dotted arrow lift exists by Proposition 7.2.4, and this lift provides an
E-homotopy from f to g. □

While we do not need it for the model structure, the following corollary will
be used later.

Corollary 7.3.3 [51, 2.15] *Let* $f: A \to B$ *and* $g: C \to D$ *be Joyal cofibrations, and let* $h: K \to L$ *be a Joyal fibration. Then*

1 the map $A \times D \amalg_{A \times C} B \times C \to B \times D$ *is a Joyal cofibration which is acyclic if* f *or* g *is, and*
2 the map $K^B \to K^B \times_{L^A} K^A$ *is a Joyal fibration which is acyclic if* f *or* h *is.*

Proof To prove (1), first observe that this pushout-product map is a cofibration, since it is a monomorphism if both f and g are. We prove that this map
is a Joyal equivalence if f is; the argument for g is symmetric. Consider the
diagram

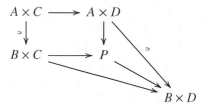

where P denotes the pushout of the square. The indicated maps are Joyal equivalences by Proposition 7.3.2(1). Since the left-hand vertical map is also a cofibration, it follows that the map $A \times D \to P$ is a Joyal acyclic cofibration. Thus
$P \to B \times D$ is also a Joyal equivalence by the two-out-of-three property.
The statement of (2) is adjoint to that of (1). □

Proof of Theorem 7.1.7 We use Theorem 2.7.8 to establish the model structure *QCat*. Thus, we must prove the following.

1 The Joyal equivalences are closed under retracts and satisfy the two-out-of-three property.
2 Every Kan acyclic fibration is a Joyal equivalence.
3 The class of cofibrations which are Joyal equivalences is closed under pushouts and transfinite composition.
4 The class of Joyal equivalences is an accessible class of maps.

Condition (1) follows by usual arguments applied to the definition of Joyal
equivalence. We established (2) in Proposition 7.3.1, and (3) in Theorem 7.2.1.

Thus, it remains to prove (4). Let

$$S = \{\Delta[0] \to E\} \cup \{V[n,k] \to \Delta[n] \mid n \geq 2, 0 < k < n\}.$$

Given a map $f\colon K \to L$ of simplicial sets, use the small object argument to factor it as

$$K \xrightarrow{i} P_f \xrightarrow{j} L,$$

where i is obtained by taking a transfinite composite of pushouts along maps in S and j has the right lifting property with respect to maps in S. Since the maps in S are Joyal acyclic cofibrations, and in particular Joyal equivalences, statement (3) implies that the map $i\colon K \to P_f$ is a Joyal equivalence.

If $c\colon K \to \Delta[0]$, we denote P_c instead by $P(K)$. Given $f\colon K \to L$ as above, consider the diagram

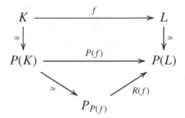

where the indicated maps are Joyal equivalences. Defining $R(f)$ to be the map so labeled in the diagram, observe that R defines a functor. Furthermore, notice that $R(f)$ is a special inner fibration between quasi-categories. Observe that f is a Joyal equivalence if and only if $R(f)$ is. But by Proposition 7.2.14(1) together with part (2) of the present proof, the map $R(f)$ is a Joyal equivalence if and only if it is a Kan acyclic fibration. So, the class of Joyal equivalences is given by $R^{-1}(\mathcal{K})$, where $\mathcal{K}$ is the class of Kan acyclic fibrations. But R is an accessible functor because it preserves large enough filtered colimits, and $\mathcal{K}$ is an accessible class by Lemma 2.7.7. So, by Proposition 2.7.6 the class of Joyal equivalences is also accessible, establishing statement (4).

Thus we have completed the proof of the existence of a model structure where the cofibrations are the monomorphisms and the weak equivalences are the Joyal equivalences.

To characterize the fibrant objects, note first that it follows from Proposition 7.3.2(3) that every fibrant object is a quasi-category. Conversely, let K be a quasi-category, and let $K \to K^f$ be a fibrant replacement in this model structure. By Proposition 7.3.2(2), there is a lift

But then K is a retract of the fibrant object K^f, and hence itself fibrant.   □

7.4 The Coherent Nerve and Rigidification Functors

Our next objective is to establish a Quillen equivalence between the model structure $QCat$ for quasi-categories and the model structure SC for simplicial categories. While it is not difficult to define an adaptation of the nerve functor which takes a simplicial category to a simplicial set, understanding its left adjoint is substantially more involved. The desired Quillen equivalence was established by Lurie [88] and in unpublished work of Joyal, but a more conceptual proof was given by Dugger and Spivak in [51] and [52], and we follow their approach here.

Let us recall the following definition, which we stated above in Definition 4.6.2. Recall from Section 3.5 that, given a category C, its free resolution defines a simplicial category F_*C.

Definition 7.4.1   [48] Let C be a simplicial category. Its *coherent nerve* is the simplicial set $\widetilde{N}(C)$ defined by

$$\widetilde{N}(C)_n = \operatorname{Hom}_{SC}(F_*[n], C).$$

One can check that the coherent nerve functor has a left adjoint, which we denote by $\mathfrak{C}$. Since quasi-categories can be thought of as structures like simplicial categories but with weak composition, and $\mathfrak{C}$ takes a quasi-category to a simplicial category, we regard $\mathfrak{C}$ as a rigidification functor. Much of the work in the next several sections is devoted to gaining a good understanding of this functor. First, we can make the following observation. Since $\widetilde{N}(C)$ is a simplicial set, we have an isomorphism

$$\widetilde{N}(C)_n = \operatorname{Hom}_{SSets}(\Delta[n], \widetilde{N}C).$$

Thus, since $(\mathfrak{C}, \widetilde{N})$ form an adjoint pair, we must have $\mathfrak{C}(\Delta[n]) = F_*[n]$. So, to begin our investigation of $\mathfrak{C}$, we look more closely at the simplicial category $F_*[n]$, and specifically at its mapping spaces.

For each $0 \le i, j \le n$ let $P_{i,j}$ denote the poset of all subsets of $\{i, i+1, \ldots, j\}$ containing both i and j. Taking the (ordinary) nerve of $P_{i,j}$, we see that

$$\text{nerve}(P_{i,j}) \cong \begin{cases} (\Delta[1])^{j-i-1} & \text{if } j > i \\ \Delta[0] & \text{if } j = i \\ \varnothing & \text{if } j < i. \end{cases}$$

These nerves naturally assemble as the mapping spaces of a simplicial category P with object set $\{0, \ldots, n\}$; composition is defined using the maps $P_{i,j} \times P_{j,k} \to P_{i,k}$ given by taking unions of sets. The proof of the following lemma can be found in [52, §A.7].

Lemma 7.4.2 [52, 2.5] *There is an isomorphism of simplicial categories* $F_*[n] \cong P$. *In particular, the simplicial category P is the result of applying* $\mathfrak{C}$ *to the simplicial set $\Delta[n]$.*

Now we consider $\mathfrak{C}(K)$ for any simplicial set K. By Lemma 2.2.2, K may be written as a colimit of simplices via the formula

$$K \cong \text{colim}_{\Delta[n] \to K} \Delta[n].$$

Then, since $\mathfrak{C}$ is a left adjoint functor and hence preserves colimits, we can define

$$\mathfrak{C}(K) \cong \text{colim}_{\Delta[n] \to K} \mathfrak{C}(\Delta[n]).$$

Remark 7.4.3   We have chosen one of several different approaches to defining $\mathfrak{C}$. In [88, §1.1.5], Lurie defines $\mathfrak{C}(\Delta[n])$ to be given by the simplicial category P, then extends to a functor on the category of simplicial sets and defines the coherent nerve functor to be its right adjoint. Joyal also starts with the coherent nerve functor, but defines it instead as an adjoint to a homotopy coherent diagram functor, so his method is also different as a consequence.

Now that we have defined the functor $\mathfrak{C}$, we state a few results which allow us to relate equivalences in a quasi-category K to homotopy equivalences in the corresponding simplicial category $\mathfrak{C}(K)$. The proof of the following lemma can be found in appendix B of [51].

Proposition 7.4.4 [51, 2.18] *Let K be a quasi-category, and let $f, g, h \in K_1$. Then $h = g \circ f$ in $\pi_0 \mathfrak{C}(K)$ if and only if there exists a 2-simplex $\sigma: \Delta[2] \to K$ such that $d_0\sigma = g$, $d_1\sigma = h$, and $d_2\sigma = f$.*

Using this result, we have the following consequence.

Corollary 7.4.5 [51, 2.19] *Let K be a quasi-category and $f \in K_1$. Then f is an equivalence if and only if its image in $\pi_0 \mathfrak{C}(K)$ is an isomorphism.*

We can use this corollary to prove the following result, which is key in establishing the Quillen equivalence between $\mathcal{Q}Cat$ and $\mathcal{SC}$.

Proposition 7.4.6 [51, 2.10] *If K is a quasi-category, then there is a natural bijection between $[\Delta[0], K]_E$ and the set of isomorphism classes of objects in $\pi_0 \mathfrak{C}(K)$.*

Proof We can produce a map between these two sets by considering them as the coequalizers in the diagram

$$
\begin{array}{ccccc}
\mathrm{Hom}(E, K) & \rightrightarrows & \mathrm{Hom}(\Delta[0], K) & \longrightarrow & [\Delta[0], K]_E \\
\downarrow & & \downarrow{\scriptstyle\cong} & & \downarrow \\
\mathrm{Hom}(\pi_0 \mathfrak{C}(E), \pi_0 \mathfrak{C}(K)) & \rightrightarrows & \mathrm{ob}(\pi_0 \mathfrak{C}(K)) & \longrightarrow & \mathrm{ob}(\pi_0 \mathfrak{C}(K))/ \cong .
\end{array}
$$

The fact that the bottom diagram is a coequalizer can be seen using the fact that the category $\pi_0 \mathfrak{C}(E)$ consists of two objects and a unique isomorphism between them. Since the middle vertical map is an isomorphism, we know that the right-hand vertical map is surjective. Thus, to complete the proof it remains to show that this map is injective.

Suppose a, b are isomorphic objects in $\pi_0 \mathfrak{C}(K)$. Then there exists a 1-simplex e in K which represents this isomorphism; by Corollary 7.4.5, e must be an equivalence. By Proposition 7.2.7, the 1-simplex e extends to a map $E \to K$, thereby identifying a and b in $[\Delta[0], K]_E$. $\qquad\square$

However, understanding the higher-order structure of a quasi-category K and its rigidification $\mathfrak{C}(K)$ is substantially more complicated. In particular, we need a better conceptual understanding of mapping spaces in $\mathfrak{C}(K)$, and for this purpose we introduce necklaces.

7.5 Necklaces and Their Rigidification

The objective of this section is to understand the mapping spaces of the simplicial categories in the image of $\mathfrak{C}$, using necklaces. In the definition of a necklace, we take wedge sums of simplices which are not given a specified basepoint. Thus, iterated wedge sums should be taken as "stringing together" simplicial sets in a chain, rather than as taking a bouquet of them at a common basepoint.

Definition 7.5.1   A *necklace* is a simplicial set of the form

$$\Delta[n_0] \vee \Delta[n_1] \vee \cdots \vee \Delta[n_k]$$

where each $n_i \geq 0$ and where in each wedge the terminal vertex of $\Delta[n_i]$ has
been identified with the initial vertex of $\Delta[n_{i+1}]$. We further assume either that
$k = 0$ or that each $n_i \geq 1$. Each $\Delta[n_i]$ is called a *bead* of the necklace. A *joint*
of the necklace is either an initial or a terminal vertex of some bead.

By our assumptions, $\Delta[0]$ can only be a bead of the necklace $\Delta[0]$ itself.
More generally, any n-simplex can be thought of as a necklace with one bead.
If S and T are two necklaces, then we can obtain a new necklace $S \vee T$ by
identifying the terminal vertex of S with the initial vertex of T.

We now give some notation for necklaces. Given a necklace T, let V_T be the
set of vertices of T, namely, the set T_0. Let $J_T \subseteq V_T$ be the set of joints of T.
Observe that we can consider both V_T and J_T as ordered sets, taking $a \leq b$ if
there is a directed path in T from a to b. We denote by α_T or simply α the initial
vertex of T, and we denote by ω_T or ω the terminal vertex of T. In particular,
α_T and ω_T are always joints of T.

Example 7.5.2   Consider the following necklace T:

The two triangles are meant to depict (filled in) 2-simplices. The vertex set is

$$V_T = \{v_1, v_2, v_3, v_4, v_5, v_6, v_7\}$$

and the joint set is

$$J_T = \{v_1, v_2, v_4, v_6, v_7\}.$$

The initial vertex α_T is v_1, and the final vertex ω_T is v_7.

It is helpful to think of necklaces as simplicial sets together with two specified 0-simplices, the initial and terminal vertices. We use the notation $SSets_{*,*}$
to denote the category $\partial\Delta[1] \downarrow SSets$ of such objects, and we define Nec to be
the full subcategory of $SSets_{*,*}$ whose objects are necklaces. A *map of necklaces* $f: S \to T$ must then be a map of simplicial sets such that $f(\alpha_S) = \alpha_T$
and $f(\omega_S) = \omega_T$.

A *spine* is a necklace in which every bead is a 1-simplex. Observe that a
spine is simply a simplicial set of the form $G(n)$ for some $n \geq 1$, where $G(n)$
is the simplicial set defined in Section 5.1. Any necklace T has an associated

spine $G[T]$, given by its ordered chain of 1-simplices. However, taking the spine is not functorial; as an example the unique map of necklaces $\Delta[1] \to \Delta[2]$ does not induce a map on spines, since the image of $\Delta[1]$ is not in the spine of $\Delta[2]$.

On the other hand, given a necklace T, we can associate to it the simplex on all the vertices of T; we denote this simplex by $\Delta[T]$. For any necklace T, we have inclusion maps $G[T] \hookrightarrow T \hookrightarrow \Delta[T]$. Unlike the spine construction, the assignment $T \mapsto \Delta[T]$ does define a functor.

Lemma 7.5.3 [51, 9.1] *For any necklace T, the maps $G[T] \hookrightarrow T \hookrightarrow \Delta[T]$ are both Joyal equivalences.*

Proof To begin, consider a necklace $T = \Delta[n] \vee \Delta[1]$, where $n \geq 1$. Then $\Delta[T] = \Delta[n + 1]$. We first want to prove that $\Delta[n] \vee \Delta[1] \to \Delta[n + 1]$ is a composite of maps obtained by taking pushouts along inner horn inclusions. To do so, define a filtration

$$\Delta[n] \vee \Delta[1] = \Delta[n + 1]^0 \subseteq \Delta[n + 1]^1 \subseteq \cdots \subseteq \Delta[n + 1]^{n-1} = \Delta[n + 1]$$

where, if the vertices of $\Delta[1]$ are labeled by n and $n+1$, $\Delta[n+1]^{i+1}$ is defined to be the union of $\Delta[n + 1]^i$ and all $(i + 2)$-simplices of $\Delta[n + 1]$ which contain the vertices n and $n + 1$. Observe that $\Delta[n + 1]^{n-2} = V[n + 1, n]$ and $\Delta[n + 1]^{n-1} = \Delta[n + 1]$ under this definition. Each inclusion $\Delta[n + 1]^i \to \Delta[n + 1]^{i+1}$ is a composite of pushouts along inner horn inclusions and therefore so is the map $\Delta[n] \vee \Delta[1] \to \Delta[n + 1]$. In particular, this map is inner anodyne.

Now, for any $m \geq 0$, the map $G[\Delta[m]] \hookrightarrow \Delta[m]$ can be factored by maps

$$G[\Delta[m]] = \Delta[1] \vee \cdots \vee \Delta[1] \to \Delta[2] \vee \Delta[1] \vee \Delta[1] \to \cdots$$
$$\to \Delta[m - 1] \vee \Delta[1] \to \Delta[m],$$

and so it is a Joyal equivalence. Given any necklace T, the map $G[T] \hookrightarrow \Delta[T]$ is of this form and hence a Joyal equivalence. The fact that $G[T] \hookrightarrow T$ is a Joyal equivalence can then be proven using pushouts along the maps $G[\Delta[r]] \to \Delta[r]$. The result then follows by the two-out-of-three property. $\square$

To understand necklaces more deeply, we briefly look at ordered simplicial sets, of which necklaces are examples.

Suppose that K is a simplicial set. Define a relation on K_0 by $k \leq_K k'$ if there is a directed path of 1-simplices from k to k' in K. In other words, there exists a spine T and a map $T \to K$ which sends α_T to k and ω_T to k'. If $k \leq_K k'$ but $k \neq k'$, we write $k <_K k'$. Observe that this relation is reflexive and transitive. In general, it is not antisymmetric, as there could be directed paths in both directions between distinct vertices x and y. For the moment we want to restrict ourselves to simplicial sets for which this relation is antisymmetric.

Definition 7.5.4 A simplicial set K is *ordered* if:

1 the relation $\leq_K$ on K_0 is antisymmetric, and
2 a simplex $k \in K_n$ is completely determined by its ordered sequence of vertices

$$k(0) \leq_K \cdots \leq_K k(n).$$

Any simplex $\Delta[n]$ is an ordered simplicial set, but, for example, $\Delta[1]/\partial\Delta[1]$ is not, since both the degenerate 1-simplex and the nondegenerate 1-simplex have the same single vertex. We can form a category of ordered simplicial sets, in which morphisms preserve the ordering on the vertices. We consider some properties of ordered simplicial sets and of the category they form.

Lemma 7.5.5 [52, 3.3]

1 *The category of ordered simplicial sets is closed under finite limits.*
2 *Every necklace is an ordered simplicial set.*
3 *Let $f: K \to L$ be a map between ordered simplicial sets. If f_0 is injective, then so is f.*
4 *Let K be an ordered simplicial set. The image of an n-simplex $\sigma: \Delta[n] \to K$ is of the form $\Delta[m] \hookrightarrow K$ for some $m \leq n$.*

Proof To prove (1), it suffices to prove that the category of ordered simplicial sets has a terminal object and admits pullbacks, by Proposition 1.1.22. The terminal object in the category of ordered simplicial sets is simply $\Delta[0]$ with its unique ordering. Consider a diagram of ordered simplicial sets $K \to M \leftarrow L$, and let A be its pullback in the category of simplicial sets. Let us define a natural ordering on A by $(k, \ell) \leq_A (k', \ell')$ precisely if both $k \leq_K k'$ and $\ell \leq_L \ell'$. Under this definition, if both $\leq_K$ and $\leq_L$ are antisymmetric, then $\leq_A$ must be also. Similarly, simplices in A are determined by their vertices because they are determined coordinate-wise by vertices in K and L.

Statement (2) follows from the fact that every necklace has a spine and is a wedge of simplices. Statement (3) follows from the fact that n-simplices are completely determined by their vertices.

Lastly, we prove (4). In general, the vertices $k(0), \ldots, k(n)$ of the n-simplex K need not be distinct. Let $d: \Delta[m] \to \Delta[n]$ be a face such that $k \circ d$ contains all vertices $k(j)$ without multiplicity; then the composite $k \circ d$ must be injective by (3). The original n-simplex k must be a degeneracy of $k \circ d$, from which it follows that $k \circ d: \Delta[m] \hookrightarrow K$ is the image of k. $\square$

Remark 7.5.6 We make a few observations about the morphisms in the category $\mathcal{N}ec$, using the fact that necklaces are ordered simplicial sets. If $S \to T$

is a map of necklaces, then every joint of T is the image of a joint of S. If, additionally, the map $S \to T$ is surjective, then, using the fact that S and T are ordered, every joint of S must map to a joint of T. In particular, each bead of T is mapped onto by a unique bead of S, while every bead of S maps onto either a bead or a joint of T.

Now that we have introduced necklaces, we would like to understand the effect of the rigidification functor $\mathfrak{C}$ on them. As we will see, they provide a good intermediate case from which we can understand this functor on more general simplicial sets. We now introduce some further notation.

Let T be a necklace and consider two vertices $a, b \in T_0$. Using the ordering on the vertices of T, define the set

$$V_T(a, b) = \{v \in T_0 \mid a \leq_T v \leq_T b\}.$$

Let $J_T(a, b)$ denote the union of $\{a, b\}$ with the set of joints between a and b. There is a unique subnecklace of T whose joints are given by the set $J_T(a, b)$ and whose vertices are given by $V_T(a, b)$. Let $B_0, \ldots, B_k$ denote the beads of this subnecklace. If a is not a joint of T, then B_0 is a proper face of a bead of T, and similarly for B_k if b is not a joint of T. In any case, there are canonical inclusions of each bead B_i into T.

Let us look at the relationship between $\mathfrak{C}(T)$ and each $\mathfrak{C}(B_i)$. Let j_i and j_{i+1} be the joints of the bead B_i. The inclusion maps $B_i \to T$ and the composition of mapping spaces in $\mathfrak{C}(T)$ induce a natural map

$$\mathrm{Map}_{\mathfrak{C}(B_0)}(a, j_1) \times \mathrm{Map}_{\mathfrak{C}(B_1)}(j_1, j_2) \times \cdots \times \mathrm{Map}_{\mathfrak{C}(B_k)}(j_k, b) \to \mathrm{Map}_{\mathfrak{C}(T)}(a, b).$$

Since each B_i is a simplex, Lemma 7.4.2 provides an explicit description of $\mathrm{Map}_{\mathfrak{C}(B_i)}(j_i, j_{i+1})$. We want to use this description to understand $\mathrm{Map}_{\mathfrak{C}(T)}(a, b)$.

Generalizing our description of $\mathfrak{C}(\Delta[n])$, let $C_T(a, b)$ denote the poset whose elements are the subsets of $V_T(a, b)$ which contain $J_T(a, b)$, ordered by inclusion. Given $a, b, c \in T_0$, we can use inclusion of subsets to induce a map

$$C_T(a, b) \times C_T(b, c) \to C_T(a, c).$$

Applying this construction for all triples of objects of T_0 and taking the nerve of each poset, we obtain a simplicial category C_T with object set T_0. For $a, b \in T_0$, an n-simplex in $\mathrm{Map}_{C_T}(a, b)$ is determined by a flag of sets $T^0 \subseteq \cdots \subseteq T^n$, where $J_T \subseteq T^0$ and $T^n \subseteq V_T$.

Now we prove that, just as for simplices, this simplicial category is precisely $\mathfrak{C}(T)$.

Proposition 7.5.7 [52, 3.8] *For any necklace T, there is a natural isomorphism of simplicial categories $\mathfrak{C}(T) \cong C_T$.*

Proof Let $T = B_1 \vee B_2 \vee \cdots \vee B_k$, where each B_i is a bead of T. Since $\mathfrak{C}$ is a left adjoint functor, it preserves colimits, so we can write

$$\mathfrak{C}(T) = \mathfrak{C}(B_1) \amalg_{\mathfrak{C}(\Delta[0])} \cdots \amalg_{\mathfrak{C}(\Delta[0])} \mathfrak{C}(B_k). \tag{7.1}$$

Note that $\mathfrak{C}(\Delta[0])$ is just the (simplicial) category with one object and no non-identity morphisms.

By Lemma 7.4.2, for each i we obtain an isomorphism of simplicial categories $\mathfrak{C}(B_i) \cong C_{B_i}$. Composing with the inclusion of the bead B_i into the necklace T, we then have simplicial functors $\mathfrak{C}(B_i) \to C_{B_i} \to C_T$ which assemble to define a functor $f\colon \mathfrak{C}(T) \to C_T$. This functor is an isomorphism on objects; it remains to prove that it also defines isomorphisms on all mapping spaces.

For any $a, b \in T_0$, consider the induced map of simplicial sets $\mathrm{Map}_{\mathfrak{C}(T)}(a, b) \to \mathrm{Map}_{C_T}(a, b)$. If $b <_T a$, then these mapping spaces are both empty; if $a = b$, then they each consist of a single point. So, let us assume that $a <_T b$. Let B_r be the bead containing a; if a is a joint, take B_r to be the bead of which a is the initial vertex. Similarly, let B_s be the bead containing b; if b is a joint, take B_s be the bead of which b is a terminal vertex. Let $j_r, \ldots, j_{s+1}$ denote the ordered elements of $J_T(a, b)$, where the joints of the bead B_i are j_i and j_{i+1}. In particular, $j_r = a$ and $j_{s+1} = b$.

Consider an n-simplex $x \in \mathrm{Map}_{C_T}(a, b)_n$, and decompose it uniquely as a composite of n-simplices $x_s \circ \cdots \circ x_r$, where $x_i \in \mathrm{Map}_{C_T}(j_i, j_{i+1})_n$. Since j_i and j_{i+1} are vertices within the same bead B_i of T, we may consider each x_i as an n-simplex in its respective $\mathrm{Map}_{\mathfrak{C}(B_i)}(j_i, j_{i+1})$. Taking the associated n-simplices in $\mathrm{Map}_{\mathfrak{C}(T)}(j_i, j_{i+1})$ for each i and composing them produces an n-simplex $\widetilde{x} \in \mathrm{Map}_{\mathfrak{C}(T)}(a, b)$.

Now, we define $g\colon \mathrm{Map}_{C_T}(a, b) \to \mathrm{Map}_{\mathfrak{C}(T)}(a, b)$ as the function sending each n-simplex x to its associated n-simplex $\widetilde{x}$. One can check that g is a well-defined map of simplicial sets, and that $f \circ g = \mathrm{id}$.

To see that f is an isomorphism, it suffices to show that g is surjective. From the expression (7.1) for $\mathfrak{C}(T)$ as a colimit of the categories $\mathfrak{C}(B_i)$, it follows that every map in $\mathrm{Map}_{\mathfrak{C}(T)}(a, b)$ can be written as a composite of maps from each $\mathfrak{C}(B_i)$. Thus, the map g is surjective. $\square$

In particular, we have the following description of the simplicial category $\mathfrak{C}(T)$ associated to a necklace T.

Corollary 7.5.8 [52, 3.10] *Let* $T = B_0 \vee \cdots \vee B_k$ *be a necklace, and let* $a, b \in T_0$ *be such that* $a <_T b$. *Let* $j_r, \ldots, j_{s+1}$ *be the ordered elements of* $J_T(a, b)$, *and let* B_i *be the bead containing* j_i *and* j_{i+1} *for each* $r \leq i \leq s$. *Then*

1 the map

$$\mathrm{Map}_{\mathfrak{C}(B_r)}(j_r, j_{r+1}) \times \cdots \times \mathrm{Map}_{\mathfrak{C}(B_s)}(j_s, j_{s+1}) \to \mathrm{Map}_{\mathfrak{C}(T)}(a, b)$$

is an isomorphism;

2 there is an isomorphism $\mathrm{Map}_{\mathfrak{C}(T)}(a, b) \cong \Delta[1]^N$, *where* $N = |V_T(a, b) \setminus J_T(a, b)|$; *and*

3 in particular, $\mathrm{Map}_{\mathfrak{C}(T)}(a, b)$ *is contractible if* $a \le b$ *and empty otherwise.*

7.6 Rigidification of Simplicial Sets

In the previous section, we used our understanding of $\mathfrak{C}$ applied to simplices to give a description of the result of applying $\mathfrak{C}$ to a necklace. Let us now use this information to apply $\mathfrak{C}$ to an arbitrary simplicial set K. Specifically, we analyze the mapping spaces of $\mathfrak{C}(K)$ via the mapping spaces of $\mathfrak{C}(T)$, where T is a necklace together with a map $T \to K$.

Let K be a simplicial set, and fix vertices $a, b \in K_0$. We denote by $K_{a,b}$ the simplicial set K regarded as an object of the category $SSets_{*,*} = \partial\Delta[1] \downarrow SSets$, where, if 0 and 1 denote the vertices of $\partial\Delta[1]$, the map $\partial\Delta[1] \to K$ is given by $0 \mapsto a$ and $1 \mapsto b$. For any necklace T and map $T \to K_{a,b}$ in the category $SSets_{*,*}$, so that $\alpha_T \mapsto a$ and $\omega_T \mapsto b$, there is an induced map of simplicial sets

$$\mathrm{Map}_{\mathfrak{C}(T)}(\alpha, \omega) \to \mathrm{Map}_{\mathfrak{C}(K)}(a, b).$$

For any fixed K, we would like to use maps of this form to define a simplicial category E_K which is isomorphic to $\mathfrak{C}(K)$. Let $Nec \downarrow K_{a,b}$ denote the category whose objects are pairs $(T, T \to K_{a,b})$ and whose morphisms are maps of necklaces $T \to U$ given by commutative triangles over K. Taking the colimit over this category, we obtain a map

$$\mathrm{Map}_{E_K}(a, b) := \mathrm{colim}_{T \to K} \mathrm{Map}_{\mathfrak{C}(T)}(\alpha, \omega) \to \mathrm{Map}_{\mathfrak{C}(K)}(a, b). \qquad (7.2)$$

We can define composition

$$\mathrm{Map}_{E_K}(a, b) \times \mathrm{Map}_{E_K}(b, c) \to \mathrm{Map}_{E_K}(a, c) \qquad (7.3)$$

as follows. Given necklaces T and U, two maps $T \to K_{a,b}$ and $U \to K_{b,c}$ assemble to a map $T \vee U \to K_{a,c}$. The composite

$$\mathrm{Map}_{\mathfrak{C}(T)}(\alpha_T, \omega_T) \longrightarrow \mathrm{Map}_{\mathfrak{C}(T \vee U)}(\alpha_T, \omega_T) \times \mathrm{Map}_{\mathfrak{C}(T \vee U)}(\alpha_U, \omega_U)$$

$$\downarrow$$

$$\mathrm{Map}_{\mathfrak{C}(T \vee U)}(\alpha_T, \omega_U)$$

induces the composition map (7.3), where the first map in the composite can be understood using Proposition 7.5.7 and the fact that $\omega_T = \alpha_U$.

One can check that these constructions define a simplicial category E_K with object set K_0, and that the construction of E_K is functorial in K. Furthermore, the maps (7.2) assemble to a map of simplicial categories $E_K \to \mathfrak{C}(K)$.

Proposition 7.6.1   [52, 4.3] *For every simplicial set K, the map $E_K \to \mathfrak{C}(K)$ is an isomorphism of simplicial categories.*

Proof   We first consider the special case when K is itself a necklace. Then the identity map $K \to K$ is a terminal object in $\mathcal{N}ec \downarrow K_{a,b}$, from which it follows that the map of simplicial sets $\mathrm{Map}_{E_K}(a,b) \to \mathrm{Map}_{\mathfrak{C}(K)}(a,b)$ is an isomorphism for any $a, b \in K_0$.

Now suppose K is an arbitrary simplicial set. Consider the commutative diagram of simplicial categories

$$
\begin{array}{ccc}
\mathrm{colim}_{\Delta[k]\to K}\, E_{\Delta[k]} & \xrightarrow{\;t\;} & E_K \\
\downarrow & & \downarrow \\
\mathrm{colim}_{\Delta[k]\to K}\, \mathfrak{C}(\Delta[k]) & \xrightarrow{\;=\;} & \mathfrak{C}(K).
\end{array}
$$

The bottom horizontal map is an equality by the definition of $\mathfrak{C}$. The left-hand vertical map defines an isomorphism on mapping spaces, since each $\Delta[k]$ is a necklace. It follows that t is injective on mapping spaces. Therefore, to prove that, for any $a, b \in K_0$, the map $\mathrm{Map}_{E_K}(a,b) \to \mathrm{Map}_{\mathfrak{C}(K)}(a,b)$ is an isomorphism of simplicial sets, it suffices to prove that t is surjective on mapping spaces.

Any n-simplex $x \in \mathrm{Map}_{E_K}(a,b)_n$ is represented by a necklace T, a map $f: T \to K_{a,b}$, and an element $\widetilde{x} \in \mathrm{Map}_{\mathfrak{C}(T)}(\alpha, \omega)$. In the commutative diagram

$$
\begin{array}{ccc}
\mathrm{colim}_{\Delta[k]\to T}\, \mathrm{Map}_{\mathfrak{C}(\Delta[k])}(\alpha,\omega) & \longrightarrow & \mathrm{Map}_{\mathfrak{C}(T)}(\alpha,\omega) \\
\uparrow & & \uparrow \\
\mathrm{colim}_{\Delta[k]\to T}\, \mathrm{Map}_{E_{\Delta[k]}}(\alpha,\omega) & \longrightarrow & \mathrm{Map}_{E_T}(\alpha,\omega) \\
f\downarrow & & \downarrow E_f \\
\mathrm{colim}_{\Delta[k]\to K}\, \mathrm{Map}_{E_{\Delta[k]}}(a,b) & \xrightarrow{\;t\;} & \mathrm{Map}_{E_K}(a,b)
\end{array}
$$

the n-simplex in $\mathrm{Map}_{E_T}(\alpha,\omega)$ represented by $(T, \mathrm{id}_T, \widetilde{x})$ is sent to x via E_f. Notice, however, that the top horizontal map and both vertical maps in the upper square are isomorphisms, from which it follows that the middle horizontal map is also an isomorphism. In particular, it is surjective, from which it follows that x is in the image of t. □

We can use this result to obtain the following explicit description of the simplicial set $\mathrm{Map}_{\mathfrak{C}(K)}(a, b)$.

Corollary 7.6.2 [52, 4.4] *For any simplicial set K and $a, b \in K_0$, an n-simplex in the mapping space $\mathrm{Map}_{\mathfrak{C}(K)}(a, b)$ consists of an equivalence class of triples $(T, T \to K, \vec{T})$ where:*

1 T is a necklace;
2 $T \to K$ is a map of simplicial sets which sends α_T to a and ω_T to b; and
3 $\vec{T}$ is a flag of sets $T^0 \subseteq \cdots \subseteq T^n$ such that T^0 contains the joints of T and T^n is contained in the set of vertices of T.

The equivalence relation is generated by identifying $(T, T \to K, \vec{T})$ and $(U, U \to K, \vec{U})$ if there exists a map of necklaces $f\colon T \to U$ over K with $\vec{U} = f_*(\vec{T})$. The ith face map omits the set T^i in the flag, and the ith degeneracy map repeats the set T^i.

Proof Recall from (7.2) that we defined the simplicial set

$$\mathrm{Map}_{E_K}(a, b) = \mathrm{colim}_{T \to K}(\mathrm{Map}_{\mathfrak{C}(T)}(\alpha, \omega)).$$

Using this colimit and Proposition 7.6.1, we can write $\mathrm{Map}_{\mathfrak{C}(T)}(a, b)$ as the coequalizer of the diagram

$$\coprod_{U \to T \to K} \mathrm{Map}_{\mathfrak{C}(T)}(\alpha, \omega) \rightrightarrows \coprod_{T \to K} \mathrm{Map}_{\mathfrak{C}(K)}(\alpha, \omega).$$

By Proposition 7.5.7, simplices of $\mathrm{Map}_{\mathfrak{C}(T)}(\alpha, \omega)$ are given by flags of subsets of V_T containing J_T. In particular, the simplices of $\coprod_{T \to K} \mathrm{Map}_{\mathfrak{C}(T)}(\alpha, \omega)$ are given precisely by triples $(T, T \to K, \vec{T})$. The relation on these triples described in the statement of this corollary exactly corresponds to the relation which defines the coequalizer above. □

We now formalize this extra structure on a necklace as the following definition.

Definition 7.6.3 A *flagged necklace* is a pair $(T, \vec{T})$, where T is a necklace and

$$\vec{T} = T^0 \subseteq T^1 \subseteq \cdots \subseteq T^n$$

is a flag of subsets of V_T which all contain J_T; we say that such a flag has *length* n. A *map of flagged necklaces* $(T, \vec{T}) \to (U, \vec{U})$ is a map of necklaces $f\colon T \to U$ such that $f(T^i) \subseteq U^i$ for each i.

Observe in particular that maps of flagged necklaces are only defined between necklaces whose associated flags have the same length.

Definition 7.6.4   A flagged necklace $(T, \vec{T})$, where $\vec{T} = (T^0 \subseteq \cdots \subseteq T^n)$, is
flanked if $T^0 = J_T$ and $T^n = V_T$.

Note that if $(T, \vec{T})$ and $(U, \vec{U})$ are both flanked, then the image of every
morphism $(T, \vec{T}) \to (U, \vec{U})$ is a subnecklace of U having the same vertices and
joints as U, so such a morphism must be surjective.

For the following result, we use the equivalence relation of Corollary 7.6.2.

Lemma 7.6.5   [52, 4.5] *Any triple $(T, T \to K, \vec{T})$ is equivalent to a triple for
which the flag is flanked. Two flanked triples are equivalent if and only if they
can be connected by a zigzag of morphisms of flagged necklaces in which every
triple of the zigzag is flanked.*

Proof Let $(T, T^0 \subseteq \cdots \subseteq T^n)$ be a flagged necklace. Consider the unique
subnecklace $T' \subseteq T$ whose set of joints is T^0 and whose vertex set is T^n.
Then the flagged necklace $(T', T^0 \subseteq \cdots \subseteq T^n)$ is flanked. This assignment is
functorial, since if $f : (T, \vec{T}) \to (U, \vec{U})$ is a map of flagged necklaces, then T'
must map into U', producing a map $(T', \vec{T}) \to (U', \vec{U})$. Then we can use the
inclusion map $T' \to T$ to verify that the flagged necklaces $(T, T \to K, \vec{T})$ and
$(T', T' \to T \to K, \vec{T})$ are equivalent.

Now suppose we have equivalent flanked flagged necklaces $(U, U \to K, \vec{U})$
and $(V, V \to K, \vec{V})$. We know that there is a zigzag of maps of flagged necklaces
between them, but the intermediate flagged necklaces need not be flanked.
However, applying the above construction produces a zigzag of flanked flagged
necklaces.   □

Definition 7.6.6   Let T be a necklace and K a simplicial set. A map $T \to K$
is *totally nondegenerate* if the image of each bead of T is a nondegenerate
simplex of K.

At times we simply say that the necklace T is totally nondegenerate, leaving
the map $T \to K$ implicit. Observe that a totally nondegenerate map need not
be injective. For example, letting $K = \Delta[1]/\partial\Delta[1]$, the map $\Delta[1] \to K$ picking
out the nondegenerate 1-simplex of K is totally nondegenerate.

Now recall that each n-simplex of a simplicial set K corresponds to a map
$\Delta[n] \to K$ in $SSets$, and each map $\sigma : \Delta[k] \to \Delta[n]$ is induced by a map $[k] \to$
$[n]$ in Δ. Under this correspondence, surjections $\Delta[k] \to \Delta[n]$ correspond to
composites of codegeneracy maps. Thus, if $\Delta[n] \to K$ is degenerate, there
is a nondegenerate simplex $\Delta[k] \to K$ and a unique surjection $\Delta[n] \to \Delta[k]$
making the appropriate triangle commute. Given any map from a necklace

$T \to K$, applying this factorization to each bead of T produces a necklace $\overline{T}$, a map $\overline{T} \to K$ which is totally nondegenerate, and a surjection of necklaces $T \to \overline{T}$ making the resulting triangle commute. Furthermore, the necklace $\overline{T}$ is unique up to isomorphism. We use these ideas in the proof of the following proposition.

Proposition 7.6.7 [52, 4.7] *Let S be a simplicial set and $a, b \in K_0$.*

1 Suppose that T and U are necklaces, $t: T \to K_{a,b}$ is a totally nondegenerate map, and $u: U \to K_{a,b}$ is any map. Then there is at most one surjection $f: U \to T$ such that $u = tf$.

2 In any diagram

where T, U, and V are flagged necklaces, the map t is totally nondegenerate, and the maps f and g are surjective maps which are surjective on each flag, a unique dotted arrow lift of flagged necklaces exists.

Proof Let us first prove (1). Observe that if $T = \Delta[0]$ (in which case we must also have $a = b$), then there is exactly one such map $U \to T$. Now consider more general T, and suppose there exist two surjections $f, f': U \to T$ such that $tf = tf' = u$. Recall from Remark 7.5.6 the behavior of necklace maps on individual beads. If we assume $f \neq f'$, let B be the first bead on which f and f' differ. Let j denote the initial vertex of B, and let C be the bead of T whose initial vertex is $f(j) = f'(j)$.

If f collapses B to a point, then so must u. If f' maps B onto C, then the map $C \to S$ factors through the point $u(B)$, contradicting that $T \to K$ is totally nondegenerate. Therefore, f cannot collapse B to a point and hence must map B onto C; similarly, f' cannot collapse B' to a point. Then the simplex $B \to U \to K$ can be identified with a degeneracy of the nondegenerate simplex $C \to K$. Since degeneracies are unique, we conclude that f and f' must agree on B, and hence $f = f'$.

Now we prove (2). Since the map f is surjective, if $V \to T$ exists then it must be surjective; by (1), it follows that it must be unique. Thus, it suffices to prove that this lift exists.

Using the same reasoning as in (1), if B is a bead in U which maps to a joint in V, then it must also map to a joint in T. Define a necklace U' by collapsing every bead of U that maps to a single point in V; this necklace fits into a

diagram

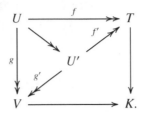

Making such a replacement if necessary, we assume that U and V have the same number of beads.

Let $B_1, \ldots, B_m$ denote the beads of U, and let $C_1, \ldots, C_m$ denote the beads of V. Assume inductively that the lift $V \to T$ can be defined on the beads $C_1, \ldots, C_{i-1}$. We know f maps B_i surjectively onto a bead D of T. Thus we have a commutative diagram

$$
\begin{array}{ccc}
B_i & \xrightarrow{\ f\ } & D \\
{\scriptstyle g}\downarrow & & \downarrow{\scriptstyle t} \\
C_i & \xrightarrow{\ v\ } & K
\end{array}
$$

where f and g are surjective maps between simplices. Therefore, they represent composites of codegeneracy maps s_f and s_g in Δ. Since the simplex t of K is nondegenerate by assumption, by Lemma 2.1.5, there must be a codegeneracy map s_h for which $v = s_h t$. Since s_h corresponds to a surjective map of simplices $C_i \to D$ making the above square commute, we extend the lift to C_i via this map.

It remains to show that the lift $\ell \colon V \to T$ is a map which respects flags. Suppose the flagged necklaces T, U, and V have flags of length n. For any $0 \le i \le n$, we need to show that $\ell(V^i) \subseteq T^i$. But using the fact that f and g are surjective on flags and the fact that $\ell \circ g = f$, we have

$$
\ell(V^i) = \ell(g(U^i)) = f(U^i) = T^i,
$$

completing the proof. □

Now, we use this result to prove that every flagged necklace is equivalent, not only to a flanked flagged necklace, but to one which is totally nondegenerate.

Corollary 7.6.8 [52, 4.8] *Let K be a simplicial set and $a, b \in K_0$. Every flagged necklace $(T, T \to K_{a,b}, \vec{T})$ is equivalent to a unique flagged necklace $(U, U \to K_{a,b}, \vec{U})$ which is flanked and totally nondegenerate.*

Proof Given a flagged necklace $(T, T \to K_{a,b}, \vec{T})$, we know that it is equivalent to the flanked totally nondegenerate flagged necklace $(\overline{T'}, \overline{T'} \to K_{a,b}, \vec{T})$ as described above. Thus, we need only prove uniqueness.

Suppose $(U, U \to K_{a,b}, \vec{U})$ and $(V, V \to K_{a,b}, \vec{V})$ are flanked totally nondegenerate flagged necklaces which are equivalent in $\mathrm{Map}_{\mathfrak{C}(K)}(a, b)_n$. By Lemma 7.6.5 there is a zigzag

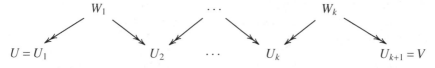

of maps between flanked necklaces over K. Using Proposition 7.6.7(2), we inductively construct surjections of flanked necklaces $U_i \to U$ over K, using diagrams of the form

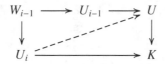

from which we obtain a surjection $V \to U$ over K. Inducting across the diagram in the reverse direction, we obtain a surjection $U \to V$ over K. By Proposition 7.6.7(1), these surjections must be unique and inverse to one another. □

Now, we bring these results together as follows.

Proposition 7.6.9 [52, 4.10] *Let K be a simplicial set and $a, b \in K_0$. Consider the category $FNec_n$ of flagged necklaces over $K_{a,b}$ of length n. For every $n \geq 0$, the nerve of $FNec_n$ is homotopy discrete in SSets.*

Proof Let $FFNec_n$ denote the full subcategory of $FNec_n$ consisting of flanked flagged necklaces. In the proof of Lemma 7.6.5, we defined a functor

$$\varphi \colon FNec_n \to FFNec_n.$$

Since, for any necklace T, the resulting necklace $\varphi(T)$ is a subnecklace of T, the inclusion of flanked necklaces defines a natural transformation from φ to the identity. It follows that the inclusion functor $FFNec_n \hookrightarrow FNec_n$ induces a weak equivalence after taking the nerve. Thus, to complete the proof it suffices to show that the nerve of $FFNec_n$ is homotopy discrete.

By Corollary 7.6.8, every component of $FFNec_n$ contains a unique necklace which is both flanked and totally nondegenerate. Furthermore, every necklace in the same component as such a necklace admits a unique map to it. Thus, each component has a final object, and therefore its nerve is contractible. Since every component is homotopy contractible, the nerve of $FFNec_n$ is homotopy discrete.   □

We conclude this section by considering the special case when K is an ordered simplicial set. In this situation, we can understand $\mathrm{Map}_{\mathfrak{C}(K)}(a, b)$ more explicitly because there are fewer necklaces which map into K than in the case of a general simplicial set.

Lemma 7.6.10   [52, 4.12] *Let K be an ordered simplicial set and let $a, b \in K_0$. Then every n-simplex in $\mathrm{Map}_{\mathfrak{C}(K)}(a, b)$ is represented by a unique flanked necklace $(T, T \to K, \vec{T})$ of length n for which the map $T \to K$ is injective.*

Proof   By Corollary 7.6.8, we know that every n-simplex in $\mathrm{Map}_{\mathfrak{C}(K)}(a, b)$ is represented by a unique flagged necklace $(T, T \to K, \vec{T})$ which is both flanked and totally nondegenerate. Using Lemma 7.5.5(4) and the fact that K is ordered, we can conclude that any totally nondegenerate map $T \to K$ is injective.   □

Example 7.6.11   [52, 4.14] Consider the ordered simplicial set

$$K = \Delta[2] \amalg_{\Delta[1]} \Delta[2]$$

which we can depict by the following picture:

We want to understand the mapping space $\mathrm{Map}_{\mathfrak{C}(K)}(0, 3)$.

Since K is ordered, by Lemma 7.6.10, it suffices to consider flanked necklaces that map injectively to $K_{0,3}$. There are only four such necklaces. First, there is the necklace $T = \Delta[1] \vee \Delta[1]$, which maps to K in two different ways f, g, i.e., by $0 \to 1 \to 3$ and by $0 \to 2 \to 3$. Then each of the necklaces $U = \Delta[1] \vee \Delta[1] \vee \Delta[1]$, $V = \Delta[1] \vee \Delta[2]$, and $W = \Delta[2] \vee \Delta[1]$ maps uniquely onto $K_{0,3}$; these three maps are all surjective on vertices.

The mapping space $\mathrm{Map}_{\mathfrak{C}(K)}(0, 3)$ has three 0-simplices:

$$(T, \{0, 1, 3\}), \quad (T, \{0, 2, 3\}), \quad (U, \{0, 1, 2, 3\}).$$

There are two nondegenerate 1-simplices:

$$(V, \{0, 1, 3\} \subseteq \{0, 1, 2, 3\}), \quad (W, \{0, 2, 3\} \subseteq \{0, 1, 2, 3\}).$$

These 1-simplices connect the three 0-simplices, resulting in two 1-simplices with a common final vertex. There are no higher nondegenerate simplices. Thus $\mathrm{Map}_{\mathfrak{C}(K)}(0, 3)$ can be depicted as

7.7 Properties of the Rigidification Functor

Now that we have a description of mapping spaces of $\mathfrak{C}(K)$, for K any simplicial set, in terms of necklaces, we would like to investigate a few closely related constructions and prove that they are equivalent. We then use these constructions to establish important properties of the functor $\mathfrak{C}$.

The first variant of $\mathfrak{C}$ is very similar to the original, but we use a homotopy colimit, rather than a colimit, of maps from necklaces to define each mapping space. Specifically, for any simplicial set K and any $a, b \in K_0$, define

$$\mathrm{Map}_{\mathfrak{C}^{\mathrm{hoc}}(K)}(a, b) = \mathrm{hocolim}_T \, \mathrm{Map}_{\mathfrak{C}(T)}(\alpha, \omega),$$

where the homotopy colimit is taken over all objects in $\mathcal{N}ec \downarrow K_{a,b}$. These mapping spaces assemble to a simplicial category $\mathfrak{C}^{\mathrm{hoc}}(K)$.

Let us look more closely at these mapping spaces, using the description of homotopy colimits from Remark 2.3.2. For a given $a, b \in K_0$, the simplicial set $\mathrm{Map}_{\mathfrak{C}^{\mathrm{hoc}}(K)}(a, b)$ is the diagonal of the bisimplicial set given by

$$[k] \mapsto \coprod_{T_0 \to \cdots \to T_n \to K_{a,b}} \mathrm{Map}_{\mathfrak{C}(T_0)}(a, b). \tag{7.4}$$

Looking at both simplicial directions at once, we can describe this functor instead as given by

$$([k], [\ell]) \mapsto \coprod_{T_0 \to \cdots \to T_n \to K_{a,b}} \mathrm{Map}_{\mathfrak{C}(T_0)}(a, b)_\ell.$$

If we fix ℓ and let k vary, then the resulting simplicial set is the nerve of the category $F\mathcal{N}ec_\ell$ of flagged necklaces of length ℓ mapping to $K_{a,b}$. Thus, we have defined the functor $\mathfrak{C}^{\mathrm{hoc}}$.

Recall from Proposition 2.3.4 that there is a canonical natural transformation from the homotopy colimit to the colimit of that diagram. Hence, there is a morphism $\mathrm{Map}_{\mathfrak{C}^{\mathrm{hoc}}(K)}(a, b) \to \mathrm{Map}_{\mathfrak{C}(K)}(a, b)$, using Proposition 7.6.1. As this construction is natural in $a, b \in K_0$, there is a natural transformation $\mathfrak{C} \to \mathfrak{C}^{\mathrm{hoc}}$.

The next variant, $\mathfrak{C}^{\mathrm{nec}}$, is defined more explictly, using categories of necklaces. For any $a, b \in K_0$, recall that the objects of the category $\mathcal{N}ec \downarrow K_{a,b}$ are the pairs $(T, T \to K_{a,b})$, where T is a necklace. Given $a, b, c \in K_0$, define the functor

$$(Nec \downarrow K_{a,b}) \times (Nec \downarrow K_{b,c}) \to (Nec \downarrow K_{a,c}) \tag{7.5}$$

by sending $((T_1, T_1 \to K_{a,b}), (T_2, T_2 \to K_{b,c}))$ to $(T_1 \vee T_2, T_1 \vee T_2 \to K_{a,c})$.

If we take the nerve of each category $Nec \downarrow K_{a,b}$, as the vertices a and b vary, we obtain a simplicial category $\mathfrak{C}^{nec}(K)$ in which composition is induced by the functors just described. Specifically, we have

$$\mathrm{Map}_{\mathfrak{C}^{nec}(K)}(a, b) = \mathrm{nerve}(Nec \downarrow K_{a,b}).$$

However, observe that the simplicial set $\mathrm{Map}_{\mathfrak{C}^{nec}(K)}(a, b)$ can also be described as the homotopy colimit of the constant functor $Nec \downarrow K_{a,b} \to SSets$ which sends all objects to $\Delta[0]$, via Proposition 2.3.5. The map

$$\mathrm{Map}_{\mathfrak{C}^{hoc}(K)}(a, b) \to \mathrm{Map}_{\mathfrak{C}^{nec}(K)}(a, b)$$

is the induced map on homotopy colimits.

Theorem 7.7.1   [52, 5.3] *For every simplicial set K, the maps*

$$\mathfrak{C}(K) \leftarrow \mathfrak{C}^{hoc}(K) \to \mathfrak{C}^{nec}(K)$$

are Dwyer–Kan equivalences.

Proof   Using Corollary 7.5.8, the spaces $\mathrm{Map}_{\mathfrak{C}(T)}(\alpha, \omega)$ are all contractible simplicial sets, from which it follows by Proposition 2.3.6 that the induced map on homotopy colimits $\mathrm{Map}_{\mathfrak{C}^{hoc}(K)}(a, b) \to \mathrm{Map}_{\mathfrak{C}^{nec}(K)}(a, b)$ is a weak equivalence. Since the object sets of these simplicial categories agree, we thus obtain a Dwyer–Kan equivalence $\mathfrak{C}^{hoc}(K) \to \mathfrak{C}^{nec}(K)$.

It remains to show that for each $a, b \in K_0$, the natural map $\mathrm{Map}_{\mathfrak{C}^{hoc}(K)}(a, b) \to \mathrm{Map}_{\mathfrak{C}(K)}(a, b)$ is a weak equivalence of simplicial sets.

Recall that $\mathrm{Map}_{\mathfrak{C}^{hoc}(K)}(a, b)$ is the diagonal of the bisimplicial set (7.4), and that the ℓth row is the nerve of the category $FNec_\ell$ of flagged necklaces of length ℓ mapping to K. Also, we can deduce from Corollary 7.6.2 that the simplicial set $\mathrm{Map}_{\mathfrak{C}(K)}(a, b)$ has ℓ-simplices given by $\pi_0(\mathrm{nerve}(FNec_\ell))$.

But by Proposition 7.6.9, the nerve of $FNec_\ell$ is homotopy discrete for all ℓ, so $\mathrm{nerve}(FNec_\ell) \simeq \pi_0(FNec_\ell)$. It follows that $\mathrm{Map}_{\mathfrak{C}(K)}(a, b) \to \mathrm{Map}_{\mathfrak{C}^{hoc}(K)}(a, b)$ is also a weak equivalence in $SSets$. $\square$

Indeed, we can define a broader class of functors from the category of simplicial sets to the category of small simplicial categories satisfying similar properties to the three functors that we just compared.

Definition 7.7.2 [52, 5.5] A subcategory $\mathcal{G}$ of $SSets_{*,*}$ is a *category of gadgets* if:

1 $\mathcal{G}$ contains $\mathcal{N}ec$;
2 for every object G of $\mathcal{G}$ and every necklace T, all maps $T \to G$ are in $\mathcal{G}$; and
3 for any object G of $\mathcal{G}$, the simplicial set $\mathrm{Map}_{\mathfrak{C}(G)}(\alpha, \omega)$ is contractible.

The category $\mathcal{G}$ is *closed under wedge product* if:

4 for any objects G and H of $\mathcal{G}$, $G \vee H$ is also in $\mathcal{G}$.

Let $\mathcal{G}$ be a category of gadgets. Given a simplicial set K, and $a, b \in K_0$, define $\mathrm{Map}_{\mathfrak{C}^{\mathcal{G}}(K)}(a, b) = \mathrm{nerve}(\mathcal{G} \downarrow K_{a,b})$; if $\mathcal{G}$ is closed under wedge product, then a composition law can be defined for these mapping spaces, from which we obtain a functor $SSets \to SC$.

Proposition 7.7.3 [52, 5.6] *Let $\mathcal{G}$ be a category of gadgets closed under wedge product. For any simplicial set K and any $a, b \in K_0$, the simplicial functor $\mathfrak{C}^{\mathrm{nec}}(K) \to \mathfrak{C}^{\mathcal{G}}(K)$ induced by the inclusion $\mathcal{N}ec \to \mathcal{G}$ is a Dwyer–Kan equivalence.*

Proof Let $j \colon \mathcal{N}ec \downarrow K_{a,b} \to \mathcal{G} \downarrow K_{a,b}$ be the functor induced by the inclusion functor $\mathcal{N}ec \to \mathcal{G}$. Applying the nerve functor results in the map

$$\mathrm{Map}_{\mathfrak{C}^{\mathrm{nec}}(K)}(a, b) \to \mathrm{Map}_{\mathfrak{C}^{\mathcal{G}}(K)}(a, b),$$

which we need to prove is a weak equivalence of simplicial sets. To do so, we apply Quillen's Theorem A (Theorem 2.3.9).

Given an object $(G, G \to K)$ in $\mathcal{G} \downarrow K_{a,b}$, we need to show that the nerve of the overcategory $j \downarrow (G, G \to K)$ is contractible. However, this category is precisely the category $\mathcal{N}ec \downarrow G_{\alpha,\omega}$, whose nerve is $\mathrm{Map}_{\mathfrak{C}^{\mathrm{nec}}(G)}(\alpha, \omega)$. By Theorem 7.7.1, this nerve is weakly equivalent to $\mathrm{Map}_{\mathfrak{C}(G)}(\alpha, \omega)$, which is contractible by part (3) of Definition 7.7.2.

Since both categories have the same objects, and since we have proved that their mapping spaces are weakly equivalent, we have established that the simplicial functor $\mathfrak{C}^{\mathrm{nec}}(K) \to \mathfrak{C}^{\mathcal{G}}(K)$ is a Dwyer–Kan equivalence. $\square$

Consider the following example. Let $T_1, \dots, T_n$ be necklaces, which are ordered simplicial sets by Lemma 7.5.5(2). Using Lemma 7.5.5(1), the product $T_1 \times \cdots \times T_n$ is again an ordered simplicial set. Define $\mathcal{P}$ to be the full subcategory of $SSets_{*,*}$ whose objects are products of necklaces equipped with a specified map $f \colon \partial\Delta[1] \to T_1 \times \cdots \times T_n$ which satisfies $\alpha = f(0) \le f(1) = \omega$. Observe that the map f is part of the data, and is not canonical for a given product of necklaces.

Proposition 7.7.4 [52, 6.1] *The category $\mathcal{P}$ is a category of gadgets which is closed under wedge product.*

The proof is technical, so we refer the reader to Dugger and Spivak [52, A.6]. We want to use the category $\mathcal{P}$ to prove that $\mathfrak{C}$ preserves products.

For any simplicial sets K and L, both $\mathfrak{C}(K \times L)$ and $\mathfrak{C}(K) \times \mathfrak{C}(L)$ are simplicial categories with object set $K_0 \times L_0$; we would like to know that the natural maps $\mathfrak{C}(K \times L) \to \mathfrak{C}(K)$ and $\mathfrak{C}(K \times L) \to \mathfrak{C}(L)$ induced by projections induce a Dwyer–Kan equivalence between $\mathfrak{C}(K \times L)$ and $\mathfrak{C}(K) \times \mathfrak{C}(L)$.

Proposition 7.7.5 [52, 6.2] *Let K and L be simplicial sets. For any $k_1, k_2 \in K$ and $\ell_1, \ell_2 \in L$, the map*

$$\mathrm{Map}_{\mathfrak{C}(K \times L)}((k_1, \ell_1), (k_2, \ell_2)) \to \mathrm{Map}_{\mathfrak{C}(K)}(k_1, k_2) \times \mathrm{Map}_{\mathfrak{C}(L)}(\ell_1, \ell_2)$$

induced by $\mathfrak{C}(K \times L) \to \mathfrak{C}(K)$ and $\mathfrak{C}(K \times L) \to \mathfrak{C}(L)$ is a weak equivalence of simplicial sets. Consequently, $\mathfrak{C}(K \times L) \to \mathfrak{C}(K) \times \mathfrak{C}(L)$ is a Dwyer–Kan equivalence.

Proof By Theorem 7.7.1 and Proposition 7.7.3, it suffices to prove the analogous statement when we apply the functor $\mathfrak{C}^{\mathcal{P}}$ rather than $\mathfrak{C}$.

Observe that the category $\mathcal{P}$ is closed under finite products. Consider the functors

$$\varphi \colon G \downarrow (K \times L)_{(k_1, \ell_1), (k_2, \ell_2)} \rightleftarrows (G \downarrow K_{k_1, k_2}) \times (G \downarrow L_{\ell_1, \ell_2}) \colon \theta$$

given by

$$\varphi \colon (G, G \to K \times L) \mapsto ((G, G \to K \times L \to K), (G, G \to K \times L \to L))$$

and

$$\theta \colon ((G, G \to K), (H, H \to L)) \mapsto (G \times H, G \times H \to K \times L).$$

Using diagonal maps, we can define a natural transformation $\mathrm{id} \to \theta\varphi$, and, using projections, we can define a natural transformation $\varphi\theta \to \mathrm{id}$. It follows that applying the nerve functor to the functors θ and φ produces a homotopy equivalence. □

Recall that $E(n)$ denotes the nerve of the groupoid with object set $\{0, 1, \ldots, n\}$ and a single isomorphism between any pair of objects, and let $\{x\}$ denote the simplicial category with a single object and only an identity morphism.

Lemma 7.7.6 [52, 6.3] *For every $n \geq 0$, the simplicial functor $\mathfrak{C}(E(n)) \to \{x\}$ is a Dwyer–Kan equivalence.*

Proof If C is a simplicial category, then $C \to \{x\}$ is a Dwyer–Kan equivalence if and only if, for every $a, b \in \mathrm{ob}(C)$, the mapping space $\mathrm{Map}_C(a, b)$ is contractible. Thus, it suffices to prove that the mapping spaces of $\mathfrak{C}(E(n))$ are all contractible. By Theorem 7.7.1, we can equivalently prove that the mapping spaces of $\mathfrak{C}^{\mathrm{nec}}(E(n))$ are contractible.

Each mapping space $\mathrm{Map}_{\mathfrak{C}^{\mathrm{nec}}(E(n))}(i, j)$, where $i, j \in \{0, \ldots, n\}$, is defined to be the nerve of the category $\mathcal{N}ec \downarrow E(n)_{i,j}$. If T is a necklace, then any map $T \to E(n)$ extends uniquely over $\Delta[T]$, since maps into $E(n)$ are determined by their values on 0-simplices and $T \hookrightarrow \Delta[T]$ is an isomorphism on 0-simplices.

Consider the functors $f, g \colon \mathcal{N}ec \downarrow E(n)_{i,j} \to \mathcal{N}ec \downarrow E(n)_{i,j}$ given by

$$f \colon (T, T \xrightarrow{x} E(n)) \mapsto (\Delta[T], \Delta[T] \xrightarrow{\bar{x}} E(n))$$

and

$$g \colon (T, T \xrightarrow{x} E(n)) \mapsto (\Delta[1], \Delta[1] \xrightarrow{z} E(n)),$$

where $\bar{x}$ is the unique extension of x to $\Delta[T]$, and z is the unique 1-simplex of $E(n)$ connecting i to j. Observe that g is the constant functor and that there are natural transformations $\mathrm{id} \to f \leftarrow g$. Since g factors through the terminal category $\{x\}$, after taking nerves the identity map is null-homotopic. Hence, the nerve of $\mathcal{N}ec \downarrow E(n)_{i,j}$ is contractible. $\qquad\square$

Lemma 7.7.7 [52, 6.4] *The functor* $\mathfrak{C} \colon S\!Sets \to S\!C$ *takes monomorphisms to cofibrations.*

Proof Because monomorphisms of simplicial sets are all obtained from boundary inclusions by composites and pushouts, it suffices to show that, for every $n \geq 0$, the map

$$\mathfrak{C}(\partial \Delta[n]) \to \mathfrak{C}(\Delta[n])$$

is a cofibration in $S\!C$.

Label the vertices of $\Delta[n]$ and $\partial \Delta[n]$ by $0, \ldots, n$, and let T be a necklace; for any $0 \leq i, j \leq n$, consider the simplicial sets $\Delta[n]_{i,j}$ and $\partial \Delta[n]_{i,j}$. If $i > 0$ or $j < n$, then every map $T \to \Delta[n]_{i,j}$ factors through $\partial \Delta[n]_{i,j}$, since its image must have dimension less than n. It follows in this case that there is an isomorphism of categories

$$\mathcal{N}ec \downarrow \Delta[n]_{i,j} \cong \mathcal{N}ec \downarrow \partial \Delta[n]_{i,j}.$$

Therefore we can use Proposition 7.6.1 to obtain an isomorphism of simplicial sets

$$\mathrm{Hom}_{\mathfrak{C}(\partial \Delta[n])}(i, j) \to \mathrm{Hom}_{\mathfrak{C}(\Delta[n])}(i, j).$$

It remains to consider the case when $i = 0$ and $j = n$. By Lemma 7.4.2, the simplicial set $\mathrm{Map}_{\mathfrak{C}(\Delta[n])}(0, n)$ can be identified with the cube $(\Delta[1])^{n-1}$, and the simplicial set $\mathrm{Map}_{\mathfrak{C}(\partial\Delta[n])}(0, n)$ is precisely the boundary of this cube. Further, the inclusion of mapping spaces is precisely given by this boundary inclusion, which we denote by i.

Recall the functor $U\colon SSets \to SC$ which takes a simplicial set K to a simplicial category UK with objects x and y and $\mathrm{Map}_{UK}(x, y) = K$. There is a map

$$U(\partial(\Delta[n]^{n-1})) \to \mathfrak{C}(\partial\Delta[n])$$

which takes x to 0 and y to n. Taking the pushout of $U(i)$ along this map results in the simplicial category $\mathfrak{C}(\Delta[n])$. Since U takes monomorphisms to cofibrations (by definition of generating cofibration in SC), the functor $U(i)$ is a cofibration. Since the functor $\mathfrak{C}(\partial\Delta[n]) \to \mathfrak{C}(\Delta[n])$ is obtained as a pushout along a cofibration, it must also be a cofibration.   □

The previous lemma can be used to prove that the adjoint functor $\widetilde{N}$ preserves fibrant objects.

Lemma 7.7.8   [52, 6.5] *If $\mathcal{D}$ is a fibrant simplicial category, then $\widetilde{N}\mathcal{D}$ is a quasi-category.*

Proof   We need to show that a lift exists in any diagram of the form

$$
\begin{array}{ccc}
V[n, k] & \longrightarrow & \widetilde{N}\mathcal{D} \\
\downarrow & \nearrow & \\
\Delta[n] & &
\end{array}
$$

where $n \geq 1$ and $0 < k < n$. The existence of such a lift is equivalent, by adjointness, to the existence of a lift

$$
\begin{array}{ccc}
\mathfrak{C}(V[n, k]) & \longrightarrow & \mathcal{D} \\
\downarrow & \nearrow & \\
\mathfrak{C}(\Delta[n]) & &
\end{array}
$$

namely, that each map $\mathfrak{C}(V[n, k]) \to \mathfrak{C}(\Delta[n])$ is an acyclic cofibration in SC. We have already proved it is a cofibration in Lemma 7.7.7, so it remains only to prove that it is a Dwyer–Kan equivalence. Similarly to the argument in the proof of Lemma 7.7.7, one can check that

$$\mathrm{Map}_{\mathfrak{C}(V[n,k])}(i, j) \to \mathrm{Map}_{\mathfrak{C}(\Delta[n])}(i, j)$$

is an isomorphism unless $i = 0$ and $j = n$. We can apply an argument similar to the one in Example 7.6.11 to identify $\mathrm{Map}_{\mathfrak{C}(V[n,k])}(0, n)$ with the space obtained by removing one face from the boundary of $(\Delta[1])^{n-1}$; this latter space is weakly equivalent to $(\Delta[1])^{n-1}$. $\qquad\square$

Proposition 7.7.9 [52, 6.6] *If $K \to L$ is a Joyal equivalence of simplicial sets, then $\mathfrak{C}(K) \to \mathfrak{C}(L)$ is a Dwyer–Kan equivalence of simplicial categories.*

Proof For any simplicial set K, the map $\mathfrak{C}(K \times E(n)) \to \mathfrak{C}(K)$ induced by projection is a weak equivalence in SC, since it is the composite of the maps

$$\mathfrak{C}(K \times E(n)) \xrightarrow{\simeq} \mathfrak{C}(K) \times \mathfrak{C}(E(n)) \xrightarrow{\simeq} \mathfrak{C}(K)$$

which are weak equivalences by Proposition 7.7.5 and Lemma 7.7.6. Recall that $E(1) = E$. Since the map $K \amalg K \to K \times E$ is a cofibration of simplicial sets, the functor

$$\mathfrak{C}(K) \amalg \mathfrak{C}(K) = \mathfrak{C}(K \amalg K) \to \mathfrak{C}(K \times E)$$

is a cofibration in SC by Lemma 7.7.7. It follows that $\mathfrak{C}(K \times E)$ is a cylinder object for $\mathfrak{C}(K)$ in SC.

If $\mathcal{D}$ is a fibrant simplicial category, the set of homotopy classes of maps $[\mathfrak{C}(K), \mathcal{D}]$ may be obtained as the coequalizer of the diagram

$$\mathrm{Hom}_{SC}(\mathfrak{C}(K \times E), \mathcal{D}) \rightrightarrows \mathrm{Hom}_{SC}(\mathfrak{C}(K), \mathcal{D}).$$

Using the adjunction $(\mathfrak{C}, \widetilde{N})$, we can equivalently consider the coequalizer of the diagram

$$\mathrm{Hom}_{SSets}(K \times E, \widetilde{N}\mathcal{D}) \rightrightarrows \mathrm{Hom}_{SSets}(K, \widetilde{N}\mathcal{D}).$$

Denoting the latter coequalizer by $[K, N\mathcal{D}]_E$, as in the definition of Joyal equivalence, we obtain a bijection

$$[\mathfrak{C}(K), \mathcal{D}] \cong [K, N\mathcal{D}]_E. \tag{7.6}$$

Now suppose that $K \to L$ is a Joyal equivalence. Then $\mathfrak{C}(K) \to \mathfrak{C}(L)$ is a simplicial functor between cofibrant simplicial categories. To prove it is a Dwyer–Kan equivalence, it is sufficient to prove that the induced map on homotopy classes $[\mathfrak{C}(L), \mathcal{D}] \to [\mathfrak{C}(K), \mathcal{D}]$ is a bijection for every fibrant object $\mathcal{D}$ in SC. Since $\widetilde{N}\mathcal{D}$ is a quasi-category by Lemma 7.7.8 and $K \to L$ is a Joyal equivalence, it follows that there is a bijection $[L, N\mathcal{D}]_E \to [K, N\mathcal{D}]_E$. The result follows after applying the bijection (7.6). $\qquad\square$

7.8 The Equivalence With Simplicial Categories

With all the tools we have built up to understand the functors $\widetilde{N}$ and $\mathfrak{C}$, we would like to show that they define a Quillen equivalence of model categories between SC and $QCat$. The next step in this process is to understand models for mapping spaces in a quasi-category. We take an abbreviated approach, as there are many different ways to define these mapping spaces; we refer the reader to Dugger and Spivak [51] and Lurie [88] for more details. Here, we make use of cosimplicial resolutions, for which we need the join construction on simplicial sets.

Definition 7.8.1   For any simplicial sets K and L, the *join* $K \star L$ is a simplicial set with

$$(K \star L)_n = \coprod_{-1 \le i \le n} K_i \times L_{n-i-1}$$

where $K_{-1} = L_{-1} = \Delta[0]$.

Note that $K \star \varnothing = \varnothing \star K = K$, so in particular there are natural inclusions $K \hookrightarrow K \star L$ and $L \hookrightarrow K \star L$. Note as well that $K \star \Delta[0]$ and $\Delta[0] \star K$ are cones on K, and that $\Delta[n] \star \Delta[r] \cong \Delta[n + r + 1]$.

Let $C(K) = (K \star \Delta[0])/K$. Let $C^\bullet$ denote the cosimplicial space $[n] \mapsto C(\Delta[n])$. Note that there are canonical maps $\partial\Delta[1] \to C^\bullet \to \Delta[1]$, where $\partial\Delta[1]$ and $\Delta[1]$ are regarded constant cosimplicial objects.

Proposition 7.8.2   [51, 4.5]

1 *The cosimplicial object $C^\bullet$ is cofibrant in the Reedy model structure on the category of cosimplicial objects in $QCat_{*,*}$.*
2 *For each $n \ge 0$, the map $C^n \to \Delta[1]$ is a Joyal weak equivalence.*
3 *Consequently, $C^\bullet$ is a cosimplicial resolution of $\Delta[1]$ in the category of cosimplicial objects in $QCat_{*,*}$.*

Observe that, in this proposition, we use $QCat_{*,*}$ rather than $SSets_{*,*}$ to specify that the model structure in question is the Joyal model structure. Let $QCat_{*,*}^\Delta$ denote the category of cosimplicial objects in $QCat_{*,*}$.

Just as we did for complete Segal spaces and Segal categories, we want to define mapping spaces between objects in a quasi-category. Taking the objects of a quasi-category K to be the set K_0, let us use the cosimplicial objects above to define the mapping space from one object to another. In fact, the definition can be applied to any simplicial set, not just a quasi-category.

Definition 7.8.3   Let K be a simplicial set and $a, b \in K_0$. The *mapping space* from a to b in K is

$$\mathrm{map}_K(a, b) = \mathrm{Map}_{QCat_{*,*}^\Delta}(C^\bullet, K_{a,b}),$$

where $K_{a,b}$ is treated as a constant cosimplicial object.

Remark 7.8.4   At times, we would also like to take mapping spaces in simplicial sets which may not be quasi-categories. Since general simplicial sets are not fibrant in $QCat$, the model we have just used may not be as well-behaved. So, if K is a simplicial set and $a, b \in K_0$, we define

$$\text{map}_K(a, b) = \text{Map}^h_{QCat_{*,*}}(\Delta[1], K_{a,b})$$

and simply leave the specific model for the homotopy mapping space ambiguous.

To see how this definition works in practice, let us first look at the mapping space from the initial to the terminal object in a necklace.

Proposition 7.8.5   [51, 4.10] *Let T be a necklace. Then $\text{map}_T(\alpha, \omega)$ is weakly equivalent to $\Delta[0]$.*

Proof   By Lemma 7.5.3, the map $T \to \Delta[T]$ is a Joyal equivalence. Also, $\Delta[T]$ is fibrant in the Joyal model structure, since it is the nerve of a category. We may therefore model

$$\text{map}_T(\alpha, \omega) = \text{Map}^h_{QCat_{*,*}}(\Delta[1], T_{\alpha,\omega})$$

by

$$\text{Map}_{QCat^\Delta_{*,*}}(C^\bullet, \Delta[T]_{\alpha,\omega}),$$

where $\Delta[T]_{\alpha,\omega}$ is regarded as a constant cosimplicial space. In the category $QCat_{*,*}$, for each n there is a unique map $C^n \to \Delta[T]$, factoring through $\Delta[1]$. Therefore, $\text{Map}_{QCat^\Delta_{*,*}}(C^\bullet, \Delta[T]_{\alpha,\omega}) = \Delta[0]$.   □

Let $\mathcal{Y}$ denote the full subcategory of $QCat_{*,*}$ whose objects are simplicial sets Y such that both $\text{map}_Y(\alpha, \omega)$ and $\text{Map}_{\mathfrak{C}(Y)}(\alpha, \omega)$ are weakly contractible. Note that $\mathcal{Y}$ contains the category $\mathcal{N}ec$ by Proposition 7.8.5 and Corollary 7.5.8, and therefore $\mathcal{Y}$ is a category of gadgets in the sense of Definition 7.7.2.

Let $\mathcal{Y}_q$ denote the full subcategory of $\mathcal{Y}$ whose objects are quasi-categories. Let $\mathfrak{C}^{\mathcal{Y}}$ and $\mathfrak{C}^{\mathcal{Y}_q}$ be the corresponding functors $SSets \to SC$, as defined just after Definition 7.7.2.

For a simplicial set K, recall from Definition 2.2.1 the simplex category of K, denoted by $\Delta \downarrow K$.

Proposition 7.8.6   [51, 5.2] *If K is a quasi-category and $a, b \in K_0$, then there* *is a natural commutative diagram of weak equivalences of simplicial sets:*

$$\mathrm{Map}_{\mathfrak{C}^{\mathrm{nec}}(K)}(a,b) \xrightarrow{\ \simeq\ } \mathrm{Map}_{\mathfrak{C}^y(K)}(a,b) \xleftarrow{\ \simeq\ } \mathrm{Map}_{\mathfrak{C}^{y_q}(K)}(a,b)$$

$$\uparrow{\scriptstyle\simeq} \qquad\qquad\qquad\qquad\qquad \uparrow{\scriptstyle\simeq}$$

$$\mathrm{nerve}(\Delta \downarrow \mathrm{map}_K(a,b)) \xleftarrow{\ \simeq\ } \mathrm{nerve}(\Delta \downarrow \mathrm{map}_K(a,b).)$$

We omit the proof of this result, since it relies on technicalities about cosimplicial resolutions and precise models for homotopy mapping spaces which we do not want to address here. Its importance lies in the following consequence.

Corollary 7.8.7 [51, 5.3] *For any quasi-category K and $a, b \in K_0$, there is a natural zigzag of weak equivalences between $\mathrm{Map}_{\mathfrak{C}(K)}(a,b)$ and $\mathrm{map}_K(a,b)$.*

Proof There is a natural zigzag of weak equivalences between $\mathrm{Map}_{\mathfrak{C}(K)}(a,b)$ and $\mathrm{Map}_{\mathfrak{C}^{\mathrm{nec}}(K)}(a,b)$ by Theorem 7.7.1. Proposition 7.8.6 gives a zigzag of weak equivalences between $\mathrm{Map}_{\mathfrak{C}^{\mathrm{nec}}(K)}(a,b)$ and nerve $(\Delta \downarrow \mathrm{map}_K(a,b))$. Finally, using Proposition 2.2.7, we complete the zigzag to $\mathrm{Map}_K(a,b)$.   □

Proposition 7.8.8 [51, 5.9] *Let $\mathcal{D}$ be a fibrant simplicial category. Then the counit map $\mathfrak{C}(\widetilde{N}\mathcal{D}) \to \mathcal{D}$ is a Dwyer–Kan equivalence.*

Proof Since $\mathfrak{C}(\widetilde{N}\mathcal{D})$ is a simplicial category with the same object set as $\mathcal{D}$, it suffices to show that, for every $a, b \in \mathrm{ob}(\mathcal{D})$, the map

$$\mathrm{Map}_{\mathfrak{C}(\widetilde{N}\mathcal{D})}(a,b) \to \mathrm{Map}_{\mathcal{D}}(a,b)$$

is a weak equivalence of simplicial sets.

As above, define $C^n = (\Delta[n] \star \Delta[0])/\Delta[n]$. Observe that $\mathfrak{C}(C^n)$ is a simplicial category with two objects, which we denote by 0 and 1, and let Q^n denote the mapping space $\mathrm{Map}_{\mathfrak{C}(C^n)}(0,1)$. By Propositions 7.7.9 and 7.8.2(2), the map

$$Q^n \to \mathrm{Map}_{\mathfrak{C}(\Delta[1])}(0,1) = \Delta[0]$$

is a weak equivalence, and hence Q^n is weakly contractible.

Consider the following commutative diagram of simplicial sets:

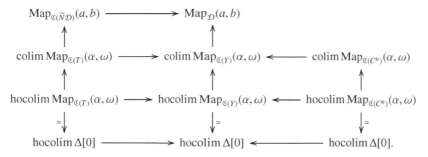

The (homotopy) colimits in the left-hand column are indexed by the category $\mathcal{N}ec \downarrow \widetilde{N}\mathcal{D}_{a,b}$. Those in the middle column are indexed by $\mathcal{Y} \downarrow \widetilde{N}\mathcal{D}_{a,b}$, where $\mathcal{Y}$ is the category of gadgets described above. Finally, the (homotopy) colimits in the right-hand column are indexed by the simplex category $\Delta \downarrow \mathrm{map}_{\widetilde{N}\mathcal{D}}(a,b)$. Except for the topmost one, the horizontal maps are induced by maps on the indexing categories. The top vertical map in the middle column is obtained by taking the adjoint map $\mathfrak{C}(Y) \to \mathcal{D}$ of the map $Y \to \widetilde{N}\mathcal{D}$, then taking the induced map $\mathrm{Map}_{\mathfrak{C}(Y)}(\alpha,\omega) \to \mathrm{Map}_{\mathcal{D}}(a,b)$. The indicated maps are weak equivalences since the mapping spaces in $\mathfrak{C}(T)$, $\mathfrak{C}(Y)$, and $\mathfrak{C}(C^n)$ are all contractible. The bottom horizontal row can be rewritten as

$$\mathrm{Map}_{\mathfrak{C}^{\mathrm{nec}}(\widetilde{N}\mathcal{D})}(a,b) \to \mathrm{Map}_{\mathfrak{C}^{\mathcal{Y}}(\widetilde{N}\mathcal{D})}(a,b) \leftarrow \mathrm{nerve}(\Delta \downarrow \mathrm{map}_{\widetilde{N}\mathcal{D}}(a,b)),$$

and these maps are weak equivalences by Proposition 7.8.6. It follows that the horizontal maps in the third row must all be weak equivalences.

Now consider the composite

$$\mathrm{hocolim}_{[n],\mathfrak{C}(C^n)\to\mathcal{D}} \mathrm{Map}_{\mathfrak{C}(C^n)}(\alpha,\omega) \to \mathrm{colim}_{[n],\mathfrak{C}(C^n)\to\mathcal{D}} \mathrm{Map}_{\mathfrak{C}(C^n)}(\alpha,\omega)$$
$$\to \mathrm{Map}_{\mathcal{D}}(a,b).$$

But a map $\mathfrak{C}(C^n) \to \mathcal{D}$ over (a,b) is precisely given by a map $\mathrm{Map}_{\mathfrak{C}(C^n)}(\alpha,\omega) \to \mathrm{Map}_{\mathcal{D}}(a,b)$. So, the above maps may instead be written as

$$\mathrm{hocolim}_{[n],Q^n\to\mathrm{Map}_{\mathcal{D}}(a,b)} Q^n \to \mathrm{colim}_{[n],Q^n\to\mathrm{Map}_{\mathcal{D}}(a,b)} Q^n \to \mathrm{Map}_{\mathcal{D}}(a,b).$$

By Lemma 7.8.9 below, using that $\mathrm{Map}_{\mathcal{D}}(a,b)$ is a Kan complex, this composite is a weak equivalence of simplicial sets. Referring back to the large diagram, we can conclude that the map

$$\mathrm{hocolim}_{Y\to\widetilde{N}\mathcal{D}} \mathrm{Map}_{\mathfrak{C}(Y)}(\alpha,\omega) \to \mathrm{Map}_{\mathcal{D}}(a,b)$$

is a Kan weak equivalence of simplicial sets.

Finally, by Theorem 7.7.1, the map

$$\text{hocolim}_{T \to \widetilde{N}\mathcal{D}} \, \text{Map}_{\mathfrak{C}(T)}(\alpha, \omega) \to \text{Map}_{\mathfrak{C}(\widetilde{N}\mathcal{D})}(a, b)$$

is a Kan weak equivalence, since its domain is exactly $\text{Map}_{\mathfrak{C}^{\text{hoc}}(\widetilde{N}\mathcal{D})}(a, b)$. It follows that

$$\text{Map}_{\mathfrak{C}(\widetilde{N}\mathcal{D})}(a, b) \to \text{Map}_{\mathcal{D}}(a, b)$$

is a weak equivalence, as desired.   □

Lemma 7.8.9   [51, 5.10] *For every Kan complex X, the composite*

$$\text{hocolim}_{[n], C^n \to X} \, C^n \to \text{colim}_{[n], C^n \to X} \, C^n \to X$$

is a weak equivalence.

We state one more technical result concerning homotopy mapping spaces.

Proposition 7.8.10   [51, 6.8] *Suppose that $f : K \to L$ is a map of quasi-categories and that, for all $a, b \in K_0$, the induced map*

$$\text{map}_K(a, b) \to \text{map}_L(fa, fb)$$

is a Kan weak equivalence. Then, for any $g, h : \Delta[1] \to K$, taken as a single map $\partial\Delta[1] \times \Delta[1] \to K$, the induced map

$$\text{Map}^h(\Delta[1] \times \Delta[1], K) \to \text{Map}^h(\Delta[1] \times \Delta[1], L),$$

of homotopy mapping spaces, taken in the category of simplicial sets under $\partial\Delta[1] \times \Delta[n]$, is also a Kan weak equivalence.

Now, given a quasi-category K and two objects $a, b \in K_0$, we can think of $\text{Map}^h_{QCat_{*,*}}(\Delta[1], K_{a,b})$ as the space of maps from a to b within the quasi-category K. In other words, we can regard a quasi-category as a weak version of a simplicial category in a more concrete way. Furthermore, we can now define Dwyer–Kan equivalences between quasi-categories, in analogy with the definition for simplicial categories.

Definition 7.8.11   [51, 7.1] A map $f : K \to L$ of simplicial sets is a *Dwyer–Kan equivalence* if:

1 for every $a, b \in X_0$, the induced map

$$\text{map}_K(a, b) \to \text{map}_L(fa, fb)$$

is a Kan weak equivalence of simplicial sets, and

2 the induced map

$$\mathrm{Hom}_{\mathrm{Ho}(QCat)}(\Delta[0], K) \to \mathrm{Hom}_{\mathrm{Ho}(QCat)}(\Delta[0], L)$$

is an isomorphism of sets.

In fact, in the presence of condition (2), we can replace condition (1) in this definition with a number of equivalent conditions which we now state.

Proposition 7.8.12 [51, 7.2] *Let* $f\colon K \to L$ *be a Joyal fibration between quasi-categories, and assume that f satisfies condition (1) of Definition 7.8.11. Then the following are equivalent:*

1 *the map f has the right lifting property with respect to the maps* $\varnothing \to \Delta[0]$ *and* $\partial\Delta[1] \to E$;
2 *the induced map* $[\Delta[0], K]_E \to [\Delta[0], L]_E$ *is an isomorphism of sets;*
3 *the map f has the right lifting property with respect to the map* $\varnothing \to \Delta[0]$, *i.e., f is surjective; and*
4 *the map f satisfies condition (2) of Definition 7.8.11.*

Proof Since E is a cylinder object for $\Delta[0]$ in *QCat*, statements (2) and (4) are equivalent, using the construction of the homotopy category of a model category.

Next, we use the map of coequalizer diagrams defining $[-, -]_F$, to see that (1) implies (2), as follows. Consider the diagram

$$
\begin{array}{ccccc}
\mathrm{Hom}(E, K) & \rightrightarrows & \mathrm{Hom}(\Delta[0], K) & \longrightarrow & [\Delta[0], K]_E \\
\downarrow & & \downarrow & & \downarrow \\
\mathrm{Hom}(E, L) & \rightrightarrows & \mathrm{Hom}(\Delta[0], L) & \longrightarrow & [\Delta[0], L]_E
\end{array}
$$

induced by the map $f\colon K \to L$. Since we have assumed that f has the right lifting property with respect to the maps $\varnothing \to \Delta[0]$, we can conclude that the map $[\Delta[0], K]_E \to [\Delta[0], L]_E$ is surjective. Now suppose that $g, h\colon \Delta[0] \to K$ are maps such that their images agree in $[\Delta[0], L]_E$. By definition of E-homotopy, there exists a map $\varphi\colon E \to L$ such that the diagram

$$
\begin{array}{ccc}
\partial\Delta[1] & \xrightarrow{g \amalg h} & K \\
\downarrow & \nearrow & \downarrow f \\
E & \xrightarrow{\varphi} & L
\end{array}
$$

commutes; the lift exists since we assumed that f has the right lifting property with respect to the map $\partial\Delta[1] \to E$. But the existence of this lift shows that g and h are E-homotopic, and hence the map $[\Delta[0], K]_E \to [\Delta[0], L]_E$ is injective.

We next prove that (2) implies (3). So, assume condition (2) holds, and consider a map $a\colon \Delta[0] \to L$. Since the map $[\Delta[0], K]_E \to [\Delta[0], L]_E$ is surjective, there are maps $b\colon \Delta[0] \to K$ and $h\colon E \to L$ such that $h(0) = f(b)$ and $h(1) = a$. Since $K \to L$ is a Joyal fibration between quasi-categories, it has the right lifting property with respect to the map $\Delta[0] \to E$ whose image is the vertex 0. Then the lift in the diagram

$$
\begin{array}{ccc}
\Delta[0] & \xrightarrow{\ b\ } & K \\
\downarrow & \nearrow & \downarrow{\scriptstyle f} \\
E & \xrightarrow{\ h\ } & L
\end{array}
$$

sends the vertex 1 to a 0-simplex which is sent to a via f. Thus f has the right lifting property with respect to the map $\varnothing \to \Delta[0]$, which is precisely the statement of (3).

Finally, we want to prove that (3) implies (1). Let $a, b\colon \Delta[0] \to K$, and consider a diagram

$$
\begin{array}{ccc}
\partial\Delta[1] & \xrightarrow{\ a \amalg b\ } & K \\
\downarrow & {\scriptstyle \beta}\nearrow & \downarrow{\scriptstyle f} \\
E & \xrightarrow{\ \gamma\ } & L.
\end{array}
$$

Since we have assumed that $K \to L$ has the right lifting property with respect to $\Delta[0] \to E$, a lift β exists such that $\beta(0) = a$. Let $a' = \beta(1)$; then $f(a') = \gamma(1)$.

Let F be the fiber of f over $\gamma(1)$, and let $\partial\Delta[1] \to F$ be given by $0 \mapsto a'$ and $1 \mapsto b$. Then we obtain a pullback square

$$
\begin{array}{ccc}
\mathrm{map}_F(a, b) & \longrightarrow & \mathrm{map}_K(a', b) \\
\downarrow & & \downarrow{\scriptstyle \varphi} \\
\mathrm{map}_{\Delta[0]}(0, 0) & \longrightarrow & \mathrm{map}_L(\gamma(1), \gamma(1)),
\end{array}
$$

where φ is a Kan acyclic fibration by our assumptions on f. Since the pullback of φ is hence also a Kan acyclic fibration, the simplicial set $\mathrm{map}_F(a', b)$ is weakly contractible. By Corollary 7.8.7, it follows that $\mathrm{Map}_{\mathfrak{C}(F)}(a', b)$ is weakly contractible. We can use analogous arguments to show that $\mathrm{Map}_{\mathfrak{C}(F)}(a', a')$, $\mathrm{Map}_{\mathfrak{C}(F)}(b, b)$, and $\mathrm{Map}_{\mathfrak{C}(F)}(b, a')$ are also weakly contractible. Thus, a' and b are isomorphic in $\pi_0\mathfrak{C}(F)$. Hence, by Corollary 7.4.5 and Proposition 7.2.7, there is a map $E \to F$ connecting a' and b.

Let δ denote the composite $E \to F \to K$, and let h be the composite $E(2) \to$

$E(1) = E \xrightarrow{\gamma} Y$, where the map $E(2) \to E(1)$ is defined by $0 \mapsto 0$, $1 \mapsto 1$, and $2 \mapsto 1$. We then have a commutative square

$$
\begin{array}{ccc}
E \vee E & \xrightarrow{\beta \vee \gamma} & K \\
\downarrow & & \downarrow f \\
E(2) & \xrightarrow{h} & L.
\end{array}
$$

The left vertical map is a Joyal acyclic cofibration, since it is an inclusion and both the domain and the codomain are weakly contractible in *QCat*, and so there is a lift $E(2) \to K$. Precomposing this lift with the inclusion $E(1) \to E(2)$ defined by $0 \mapsto 0$ and $1 \mapsto 2$, we obtain a lift of the original map γ. □

The key result we need to prove in order to establish the Quillen equivalence between *SC* and *QCat* is that Dwyer–Kan equivalences in *QCat* agree exactly with Joyal equivalences. Once that result is proved, the Quillen equivalence is not so difficult to obtain, since the weak equivalences in both model categories have the same kind of description. However, we need a number of intermediate results before we can show that the two kinds of equivalences in *QCat* coincide.

Proposition 7.8.13 [51, 7.3] *Let K, L, K', and L' be quasi-categories.*

1 If $f: K \to L$ is a Joyal fibration and a Dwyer–Kan equivalence, and $g: L' \to L$ is any map, then the pullback $K \times_L L' \to L'$ is a Joyal fibration and a Dwyer–Kan equivalence.

2 Let

$$
\begin{array}{ccc}
K & \longrightarrow & L \\
& \searrow & \downarrow \\
& & L'
\end{array}
$$

be a diagram in which all maps are Joyal fibrations. If two of the three maps are Dwyer–Kan equivalences, then so is the third.

3 Consider a diagram

$$
\begin{array}{ccccc}
K & \longrightarrow & L & \longleftarrow & M \\
\downarrow & & \downarrow & & \downarrow \\
K' & \longrightarrow & L' & \longleftarrow & M'
\end{array}
$$

in which all the maps are Joyal fibrations and the vertical maps are Dwyer–Kan equivalences. Let P denote the pullback of the top row and P' the pullback of the bottom row. Assume that the induced map $P \to P'$ and the maps

$K \to K' \times_{L'} L$ and $M \to M' \times_{M'} M$ are all Joyal fibrations. Then the map $P \to P'$ is a Dwyer–Kan equivalence.

Proof   Let us first prove (1). Since f is a Joyal fibration, so is the pullback map $K \times_L L' \to L'$, and furthermore its domain $K \times_L L'$ is a quasi-category. Since f has the right lifting property with respect to the map $\varnothing \to \Delta[0]$, the pullback map does also, from which we obtain condition (2) of Definition 7.8.11. To check condition (1), let $(a_1, a_2), (b_1, b_2) \in (K \times_L L')_0$. In the pullback square

$$
\begin{array}{ccc}
\mathrm{map}_{X \times_Y Y'}((x_1, y_1), (x_2, y_2)) & \longrightarrow & \mathrm{map}_X(x_1, x_2) \\
\downarrow & & \downarrow \\
\mathrm{map}_{Y'}(y_1, y_2) & \longrightarrow & \mathrm{map}_Y(z_1, z_2),
\end{array}
$$

where z_i denotes the common image of x_i and y_i in Y, the horizontal maps are Kan fibrations. By assumption, the bottom horizontal map is a Kan equivalence, hence so is the top horizontal map.

For (2), it suffices to observe that the two-out-of-three property holds for isomorphisms of sets and Kan weak equivalences of simplicial sets.

Finally, we prove (3). By part (1), the maps $K' \times_{L'} L \to K'$ and $M' \times_{L'} L \to M'$ are Joyal fibrations and Dwyer–Kan equivalences. Therefore, the maps $K \to K' \times_{L'} L$ and $M \to M' \times_{L'} L$, which are Joyal fibrations, are also Dwyer–Kan equivalences by (2). In particular, they have the right lifting property with respect to the map $\varnothing \to \Delta[0]$.

Let $a = (a_1, a_3) \in P'_0$, and a_2 the image of a_1 and a_3 in Y'_0. Since $L \to L'$ is a Joyal fibration and a Dwyer–Kan equivalence, the 0-simplex a_2 can be lifted to some $b_2 \in L_0$. It follows that $(a_1, b_2) \in (K' \times_{L'} L)_0$, which can be lifted to some $b_1 \in K_0$. Similarly, $(a_3, b_2) \in (M' \times_{L'} L)_0$, so it has a lift to some $b_3 \in M_0$. The pair $(b_1, b_3) \in P_0$ is a lift of a, and hence $P \to P'$ has the right lifting property with respect to $\varnothing \to \Delta[0]$, establishing condition (2) of Definition 7.8.11.

To see that $f \colon P \to P'$ also satisfies condition (1) of Definition 7.8.11, let $a, b \in P_0$. Then $a = (x_1, z_1)$ and $b = (x_2, z_2)$; let y_i denote the image of both x_i and z_i in L. In the diagram

$$
\begin{array}{ccccc}
\mathrm{map}_K(x_1, x_2) & \longrightarrow & \mathrm{map}_L(y_1, y_2) & \longleftarrow & \mathrm{map}_M(z_1, z_2) \\
\simeq \downarrow & & \downarrow \simeq & & \downarrow \simeq \\
\mathrm{map}_{K'}(fx_1, fx_2) & \longrightarrow & \mathrm{map}_{L'}(fy_1, fy_2) & \longleftarrow & \mathrm{map}_{M'}(fz_1, fz_2)
\end{array}
$$

the vertical maps are Kan acyclic fibrations. But the induced map on pullbacks must also be a Kan equivalence, as we needed to show. $\square$

Lemma 7.8.14   [51, 7.5] *Let K and L be quasi-categories and $f \colon K \to L$ a*

Joyal fibration and Dwyer–Kan equivalence. Then for every $n \geq 0$ the following maps are also Joyal fibrations and Dwyer–Kan equivalences:

1 $K^{\Delta[n]} \to L^{\Delta[n]}$,

2 $K^{V[n,k]} \to L^{V[n,k]}$ *for every* $0 < k < n$,

3 $K^{\partial\Delta[n]} \to L^{\partial\Delta[n]}$, *and*

4 $K^{\Delta[n]} \to L^{\Delta[n]} \times_{L^{\partial\Delta[n]}} K^{\partial\Delta[n]}$.

Proof To begin, observe that all the maps are Joyal fibrations between quasi-categories by Corollary 7.3.3, using cofibrations with empty domain for the first three statements. We first prove that $K^{\Delta[1]} \to L^{\Delta[1]}$ is a Dwyer–Kan equivalence. Condition (1) in Definition 7.8.11 is verified by Proposition 7.8.10. Using Proposition 7.8.12, it suffices to prove that the map $K^{\Delta[1]} \to L^{\Delta[1]}$ is surjective, or, equivalently, that the original map $K \to L$ has the right lifting property with respect to the map $\varnothing \to \Delta[1]$.

Using Lemma 7.8.12, we know that $K \to L$ is surjective, and hence that, given a map $g\colon \Delta[1] \to L$, we may lift $g(0)$ and $g(1)$ to points a and b in K_0. Since the map

$$\mathrm{map}_K(a, b) \to \mathrm{map}_L(\gamma(0), \gamma(1))$$

is a Kan acyclic fibration, and g represents a 0-simplex in the target, we can lift g to a 0-simplex in the domain.

Now consider the simplicial set $G(n)$; we want to prove by induction that $K^{G(n)} \to L^{G(n)}$ is a Dwyer–Kan equivalence for all $n \geq 0$. The case when $n = 0$ recovers the map $K \to L$, and we already established the case $n = 1$ since $G(1) = \Delta[1]$. So, assume that $K^{G(n)} \to L^{G(n)}$ is a Dwyer–Kan equivalence for some $n \geq 1$. Since $G(n + 1)$ can be written as the pushout of the diagram $G(n) \leftarrow \Delta[0] \to \Delta[1]$, we can obtain $K^{G(n+1)}$ as the pullback of the induced diagram $K^{G(n)} \to K \leftarrow K^{\Delta[1]}$. Since the diagram

$$
\begin{array}{ccccc}
K^{G(n)} & \longrightarrow & K & \longleftarrow & K^{\Delta[1]} \\
\downarrow & & \downarrow & & \downarrow \\
L^{G(n)} & \longrightarrow & L & \longleftarrow & L^{\Delta[1]}
\end{array}
$$

satisfies all the hypotheses of Proposition 7.8.13(3), the induced map on pullbacks, which is precisely $K^{G(n+1)} \to L^{G(n+1)}$, is a Dwyer–Kan equivalence.

To complete the proof of (1), observe that the inclusion $G(n) \to \Delta[n]$ is a Joyal acyclic cofibration, so $K^{\Delta[n]} \to K^{G(n)}$ is a Joyal acyclic fibration; in particular, it is a Dwyer–Kan equivalence. Applying Proposition 7.8.13(2) to the diagram

$$
\begin{array}{ccc}
K^{\Delta[n]} & \longrightarrow & K^{G(n)} \\
\downarrow & & \downarrow \\
L^{\Delta[n]} & \longrightarrow & L^{G(n)}
\end{array}
$$

shows that $K^{\Delta[n]} \to L^{\Delta[n]}$ is a Dwyer–Kan equivalence.

Now we turn to (2). For any $0 < k < n$, the inclusion $V[n,k] \to \Delta[n]$ is a Joyal acyclic cofibration, so the map $K^{\Delta[n]} \to K^{V[n,k]}$ is a Joyal acyclic fibration. Using part (1), it follows that $K^{V[n,k]} \to L^{V[n,k]}$ is a Dwyer–Kan equivalence.

We prove part (3) by induction; the cases $n = 0$ and $n = 1$ follow by hypothesis. For $n \geq 2$, note that $\partial\Delta[n]$ is the pushout of the diagram

$$
\Delta[n-1] \leftarrow \partial\Delta[n-1] \to V[n, n-1].
$$

Then the map $K^{\partial\Delta[n]} \to L^{\partial\Delta[n]}$ is the induced map on pushouts of the diagram

$$
\begin{array}{ccc}
K^{V[n,n-1]} & \longrightarrow & K^{\partial\Delta[n-1]} & \longleftarrow & K^{\Delta[n-1]} \\
\downarrow & & \downarrow & & \downarrow \\
L^{V[n,n-1]} & \longrightarrow & L^{\partial\Delta[n-1]} & \longleftarrow & L^{\Delta[n-1]}.
\end{array}
$$

We have already proved that the left and right vertical maps are Dwyer–Kan equivalences, and the middle vertical map is a Dwyer–Kan equivalence by our inductive hypothesis. Since this diagram satisfies the hypotheses of Proposition 7.8.13(3), the map $K^{\partial\Delta[n]} \to L^{\partial\Delta[n]}$ is a Dwyer–Kan equivalence.

Finally, to prove (4), let $n \geq 0$ and consider the diagram

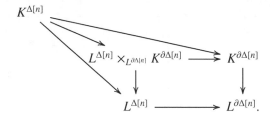

We proved in part (3) that the right vertical map is a Dwyer–Kan equivalence, hence so is the pullback $L^{\Delta[n]} \times_{L^{\partial\Delta[n]}} K^{\partial\Delta[n]} \to L^{\Delta[n]}$ by Proposition 7.8.13(1). We have also proved in (1) that $K^{\Delta[n]} \to L^{\Delta[n]}$ is a Dwyer–Kan equivalence, so the same is true for $K^{\Delta[n]} \to L^{\Delta[n]} \times_{L^{\partial\Delta[n]}} K^{\partial\Delta[n]}$ by Proposition 7.8.13(2). □

Proposition 7.8.15 [51, 7.6] *If $K \to L$ is a Joyal fibration and Dwyer–Kan*

equivalence between quasi-categories, then $K \to L$ is a Kan acyclic fibration, and in particular a Joyal equivalence.

Proof By Lemma 7.8.14(4), for every $n \geq 0$, the map

$$K^{\Delta[n]} \to K^{\partial\Delta[n]} \times_{L^{\partial\Delta[n]}} L^{\Delta[n]}$$

is a Joyal fibration and Dwyer–Kan equivalence. In particular, it has the right lifting property with respect to the map $\varnothing \to \Delta[0]$. Equivalently, any square

$$
\begin{array}{ccc}
\partial\Delta[n] & \longrightarrow & K \\
\downarrow & \nearrow & \downarrow \\
\Delta[n] & \longrightarrow & L
\end{array}
$$

has a lift, so $K \to L$ is a Kan acyclic fibration. It is hence a Joyal equivalence by Proposition 7.3.1. $\qquad\square$

Theorem 7.8.16 [51, 8.1] *For a map $g\colon K \to L$ of simplicial sets, the following are equivalent.*

1 The map g is a Joyal equivalence.
2 The functor $\mathfrak{C}(g)\colon \mathfrak{C}(K) \to \mathfrak{C}(L)$ is a Dwyer–Kan equivalence of simplicial categories.
3 The map g is a Dwyer–Kan equivalence of simplicial sets.

Proof The implication (1) $\Rightarrow$ (2) has already been proved in Proposition 7.7.9.

To establish the equivalence of (2) and (3), we claim that it suffices to consider the case where K and L are quasi-categories. To prove this claim, consider any diagram

$$
\begin{array}{ccc}
K & \longrightarrow & K' \\
g\downarrow & & \downarrow g' \\
L & \longrightarrow & L'
\end{array}
$$

in which the horizontal maps are Joyal equivalences. Using the definition of Dwyer–Kan equivalence of simplicial sets, one can check that g is a Dwyer–Kan equivalence if and only if g' is. Using the implication (1) $\Rightarrow$ (2), one can also see that $\mathfrak{C}(g)$ is a Dwyer–Kan equivalence of simplicial categories if and only if $\mathfrak{C}(g')$ is. Thus, if K and L are not quasi-categories, we can apply fibrant replacement in *QCat* and consider the induced map instead.

Now recall from Proposition 7.4.6 that if K is a quasi-category, then the set $[\Delta[0], K]_{\mathcal{L}}$ is in bijection with the set of isomorphism classes in the category

$\pi_0 \mathfrak{C}(K)$. Furthermore, by Corollary 7.8.7, for any $a, b \in K_0$, the simplicial set $\mathrm{map}_K(a, b)$ is connected to $\mathrm{Map}_{\mathfrak{C}(K)}(a, b)$ by a natural zigzag of weak equivalences. The equivalence of (2) and (3) now follows from the definitions.

Lastly, we prove that (2) implies (1). Let $L \to L^f$ be a fibrant replacement in *QCat*. Factor the composite map $K \to L \to L^f$ as a Joyal acyclic cofibration followed by a Joyal fibration to produce a square

$$
\begin{array}{ccc}
K & \xrightarrow{\ g\ } & L \\
{\scriptstyle \simeq}\downarrow & & \downarrow{\scriptstyle \simeq} \\
K^f & \xrightarrow{\ g^f\ } & L^f
\end{array}
$$

in which the vertical maps are Joyal equivalences. They are therefore Dwyer–Kan equivalences of simplicial categories after applying the functor $\mathfrak{C}$. It follows that $\mathfrak{C}(g^f)$ is also a weak equivalence, and therefore g^f is a Dwyer–Kan equivalence by the fact that (2) implies (3). Then by Proposition 7.8.15, the map g^f is a Joyal equivalence, and hence the same is true for g by the two-out-of-three property. $\qquad\square$

Corollary 7.8.17 [51, 8.2] *The adjoint functors* $\mathfrak{C}\colon QCat \leftrightarrows SC\colon \widetilde{N}$ *define a Quillen equivalence.*

Proof The functor $\mathfrak{C}$ preserves cofibrations and acyclic cofibrations by Lemma 7.7.7 and Proposition 7.7.9. Thus, the functors $(\mathfrak{C}, \widetilde{N})$ define a Quillen pair.

To prove that it is a Quillen equivalence, we first need to know that, for any fibrant simplicial category $\mathcal{D}$, the map $\mathfrak{C}(\widetilde{N}\mathcal{D}) \to \mathcal{D}$ is a weak equivalence in SC. This fact has been proved in Proposition 7.8.8.

It remains to prove that, for any simplicial set K and any fibrant replacement $\mathfrak{C}(K) \to \mathcal{D}$ in SC, the induced map $K \to \widetilde{N}(\mathfrak{C}(K)) \to \widetilde{N}\mathcal{D}$ is a Joyal equivalence. By Theorem 7.8.16, it suffices to prove that $\mathfrak{C}(K) \to \mathfrak{C}\widetilde{N}\mathfrak{C}(K) \to \mathfrak{C}(\widetilde{N}\mathcal{D})$ is a weak equivalence in SC. Consider the diagram

$$
\begin{array}{ccc}
\mathfrak{C}(K) & \longrightarrow \ \mathfrak{C}\widetilde{N}\mathfrak{C}(K) \ \longrightarrow & \mathfrak{C}(\widetilde{N}\mathcal{D}) \\
& {\scriptstyle =}\searrow \quad\ \ \downarrow & \quad \downarrow{\scriptstyle \simeq} \\
& \mathfrak{C}(K) \ \xrightarrow{\ \simeq\ } & \mathcal{D}.
\end{array}
$$

It follows that the composite of the top vertical maps is a weak equivalence, completing the proof. $\qquad\square$

7.9 The Equivalence With Complete Segal Spaces

In this section, we prove that the model categories $QCat$ and CSS are Quillen equivalent. This result is due to Joyal and Tierney [74], and our treatment here draws from their proof, as well as from the analogous proof in the dendroidal setting of Cisinski and Moerdijk [45]. We should add that Joyal and Tierney give two such Quillen equivalences, using two different adjoint pairs of functors, and they also establish similar Quillen equivalences between quasi-categories and Segal categories. We do not give these Quillen equivalences here, however.

Of particular importance in this section is the fact that a complete Segal space, or more generally, any simplicial space, can be regarded as a simplicial object in the category of simplicial sets in two different ways. Typically, given a simplicial space W, we regard it as a functor $W \colon \Delta^{op} \to SSets$ and denote the simplicial set in degree n by W_n. For increased clarity, in this section we denote this simplicial set by $W_{n,*}$. If we instead take the other simplicial direction, we denote the simplicial set in degree n by $X_{*,n}$. Whereas we have $X_{n,*} = \mathrm{Map}(\Delta[n]^t, X)$, looking at the other orientation we can replace $\Delta[n]^t$ by the constant simplicial set $\Delta[n]$ to see that $W_{*,n} = \mathrm{Map}(\Delta[n], W)$. In other words, the roles of the two "constant" simplicial spaces associated to a simplicial set are reversed when we change perspective in this way. Joyal distinguishes between the two simplicial directions by calling one the "vertical" direction and the other the "horizontal" direction, but there does not seem to be consistency in which direction people visualize as which, so we avoid any of this kind of terminology.

First, we consider the following adjoint pair between the categories of simplicial spaces and simplicial sets. Consider the inclusion $SSets \to SSets^{\Delta^{op}}$ given by taking a simplicial set K to the simplicial space K^t. This functor has a right adjoint which takes a simplicial space W to the simplicial set $W_{*,0}$. Our goal in this section is to prove that this adjunction defines a Quillen equivalence $QCat \rightleftarrows CSS$.

Let us first recall some facts about the model structure CSS, and in particular how we can characterize its fibrant objects by lifting properties. Since the fibrant objects of CSS are precisely the complete Segal spaces, they satisfy three conditions: Reedy fibrancy, the Segal condition, and the completeness condition.

If W is Reedy fibrant, then the map $W \to \Delta[0]$ has the right lifting property with respect to the Reedy acyclic cofibrations, and in particular the generating set

$$\{\partial\Delta[n] \times \Lambda[m]^t \cup \Delta[n] \times V[m,k]^t \to \Delta[n] \times \Delta[m]^t\}$$

where $n \geq 0$, $m \geq 1$, and $0 \leq k \leq m$.

If W is additionally a Segal space, then we know that, for every $n \geq 2$, the map

$$\mathrm{Map}(\Delta[n]^t, W) \rightarrow \mathrm{Map}(G(n)^t, W)$$

is a weak equivalence of simplicial sets; it is furthermore a fibration since $G(n)^t \rightarrow \Delta[n]^t$ is a cofibration of simplicial spaces, by Proposition 2.4.6. Thus, a lift exists in any diagram of the form

$$\begin{array}{ccc} \partial\Delta[m] & \longrightarrow & \mathrm{Map}(\Delta[n]^t, W) \\ \downarrow & \nearrow & \downarrow \\ \Delta[m] & \longrightarrow & \mathrm{Map}(G(n)^t, W) \end{array}$$

where $m \geq 0$. Using adjointness of products and mapping spaces, such a lift exists precisely when the map $W \rightarrow \Delta[0]$ has the right lifting property with respect to the maps

$$\{\partial\Delta[m] \times \Delta[n]^t \cup \Delta[m] \times G(n)^t \rightarrow \Delta[m] \times \Delta[n]^t\}$$

for $m \geq 0$ and $n \geq 2$.

Using a similar argument, if W is a complete Segal space, then the map $W \rightarrow \Delta[0]$ has the right lifting property with respect to the maps in the set

$$\{\partial\Delta[m] \times E^t \cup \Delta[m] \times \Delta[0]^t \rightarrow \Delta[m] \times E^t\}$$

where $m \geq 0$.

We can assemble this information into the following statement.

Proposition 7.9.1 *A simplicial space W is a complete Segal space if and only if the map $W \rightarrow \Delta[0]$ has the right lifting property with respect to the following maps:*

1 $\partial\Delta[n] \times \Delta[m]^t \cup \Delta[n] \times V[m,k]^t \rightarrow \Delta[n] \times \Delta[m]^t$ for $n \geq 0$, $m \geq 1$, and $0 \leq k \leq m$;
2 $\partial\Delta[m] \times \Delta[n]^t \cup \Delta[m] \times G(n)^t \rightarrow \Delta[m] \times \Delta[n]^t$ for $m \geq 0$ and $n \geq 2$; and
3 $\partial\Delta[m] \times E^t \cup \Delta[m] \times \Delta[0]^t \rightarrow \Delta[m] \times E^t$ for $m \geq 0$.

Now, we would like to define an equivalent model structure on the same category, but in which simplicial spaces are viewed with the opposite orientation, as described above, and for which we use the Joyal model structure on simplicial sets. Specifically, we want to start with a Reedy model structure in which we take the weak equivalences to be the maps $W \rightarrow Z$ such that $W_{*,n} \rightarrow Z_{*,n}$ is a Joyal equivalence of simplicial sets for all $n \geq 0$. Cofibrations are still the monomorphisms. We denote this model structure by $QCat^{\Delta^{op}}$. Let us describe the fibrant objects of this model structure in terms of lifting properties, as we did above.

If W is fibrant in $QCat^{\Delta^{op}}$, then for every $n \geq 0$ the map $W_{*,n} \to \mathrm{cosk}^t_{n-1}(W)_{*,n}$ is a fibration of simplicial sets. Here, we use the "transpose" notation to indicate that we are taking coskeleta in the opposite simplicial direction from the usual one. Further, note that if W is fibrant, then this map is a fibration between quasi-categories. Thus, it has the right lifting property with respect to inner horns $V[m,k] \to \Delta[m]$ for $m \geq 2$ and the inclusion map $\Delta[0] \to E$. However, having the right lifting property with respect to inner horns is equivalent to having the right lifting property with respect to the maps $G(m) \to \Delta[m]$ for all $m \geq 2$, by Proposition 7.1.2.

Now, a map $G(m) \to W_{*,n}$ corresponds to a map $G(m) \to \mathrm{Map}(\Delta[n], W)$, which in turn corresponds, by adjointness, to a map

$$\Delta[n] \times G(m)^t \to W.$$

Similarly, a map $\Delta[m] \to \mathrm{cosk}^t_{n-1}(W)_{*,n}$ is given by $\Delta[m] \to \mathrm{Map}(\partial\Delta[n], W)$, which corresponds to a map

$$\partial\Delta[n] \times \Delta[m]^t \to W.$$

Assembling such maps together, we obtain that a lift in a diagram

corresponds to a lift in the diagram

By a similar construction, we see that a lift in a diagram

is equivalent to a lift in the corresponding diagram

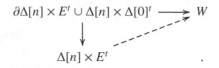

Thus, we see that fibrant objects W in $QCat^{\Delta^{op}}$ satisfy exactly the lifting properties which correspond to the Segal and completeness conditions.

Now, we would like a localization of this model structure so that the fibrant objects are precisely the complete Segal spaces. While we have seen that the fibrant objects are already local with respect to the Segal and completeness conditions, there is no reason for them to be fibrant in the usual Reedy model structure on simplicial spaces. We claim that the following condition is the one that we need.

Definition 7.9.2 An object W of $QCat^{\Delta^{op}}$ is *locally constant* if the maps $W_{*,0} \to W_{*,m}$ induced from the maps $[m] \to [0]$ in Δ are Joyal equivalences for all $n \geq 1$.

Proposition 7.9.3 *Let W be a fibrant object of $QCat^{\Delta^{op}}$. Then W is locally constant if and only if $W \to \Delta[0]$ has the right lifting property with respect to the maps*

$$V[m,k] \times \Delta[n]^t \cup \Delta[m] \times \partial\Delta[n]^t \to \Delta[m] \times \Delta[n]^t$$

for all $m \geq 1$, $0 \leq k \leq m$, and $n \geq 0$.

Proof The simplicial space W has the desired lifting property if and only if each map $\mathrm{Map}(\Delta[m], W) \to \mathrm{Map}(V[m,k], W)$ is an acyclic fibration, i.e., has the right lifting property with respect to the maps $\partial\Delta[n] \to \Delta[n]$. Let us first consider the case where $m = 1$. In this case, $V[1,k] \cong \Delta[0]$ for both values of k, so on mapping spaces we get precisely that the map $W_{*,1} \to W_{*,0}$ is an acyclic fibration.

We can prove the result for higher values of m using induction; for simplicity, we present the argument for $m = 2$. Observe that for any possible value of k, the simplicial set $V[2,k]$ can be written as a pushout $\Delta[1] \amalg_{\Delta[0]} \Delta[1]$ (where the precise gluing depends on the value of k). Then we have

$$\begin{aligned}
\mathrm{Map}(V[2,k], W) &\simeq \mathrm{Map}(\Delta[1] \amalg_{\Delta[0]} \Delta[1], W) \\
&\simeq \mathrm{Map}(\Delta[1], W) \times_{\mathrm{Map}(\Delta[0],W)} \mathrm{Map}(\Delta[1], W_0) \\
&\simeq W_{*,1} \times_{W_{*,0}} W_{*,1} \\
&\simeq W_{*,1}
\end{aligned}$$

where the last equivalence follows from the inductive hypothesis that $W_{*,1} \to W_{*,0}$ is a weak equivalence. Thus, $\mathrm{Map}(\Delta[2], W) \to \mathrm{Map}(V[2,k], W)$ is a weak equivalence if and only if $W_{*,2} \to W_{*,1}$ is a weak equivalence. We can thus conclude that the desired lifting property holds if and only if W is locally constant. □

Now, if we localize $QCat^{\Delta^{op}}$ with respect to the maps $\Delta[n] \to \Delta[0]$, we obtain a model structure on simplicial spaces in which the fibrant objects are locally constant. The arguments above establish the following result.

Proposition 7.9.4 *The locally constant model structure on simplicial spaces is precisely the model structure CSS.*

Proof In both model structures, the cofibrations are exactly the monomorphisms of simplicial spaces. By Proposition 1.4.10, it suffices to prove that they have the same fibrant objects. But we have already shown that the fibrant objects in both model structures are characterized by the same lifting conditions. □

However, thinking of this model structure as the locally constant one allows us to prove the following result.

Theorem 7.9.5 [74, 4.11] *The inclusion functor $i\colon QCat \to QCat^{\Delta^{op}}$, taking a simplicial set K to the simplicial space Z with $Z_{*,n} = K$ for all n, has a right adjoint, the evaluation map ev_0 taking a simplicial space W to the simplicial set $W_{*,0}$. This adjoint pair defines a Quillen equivalence*

$$i\colon QCat \rightleftarrows CSS\colon \mathrm{ev}_0.$$

Proof The fact that evaluation at zero is right adjoint to the inclusion is not hard to check. To prove that the adjoint pair is a Quillen pair, first observe that the inclusion functor preserves cofibrations, since they are precisely the monomorphisms in each category. In fact, this functor preserves weak equivalences by definition of the locally constant model structure, and in particular it preserves acyclic cofibrations.

To prove this Quillen pair is a Quillen equivalence, let K be a simplicial set. We need to prove that the map $K \to \mathrm{ev}_0 Z^f$ is a Joyal equivalence. A fibrant replacement Z^f of Z is a locally constant simplicial space with a quasi-category in each simplicial degree. Thus, $\mathrm{ev}_0 Z^f$ is simply a fibrant replacement for K in $QCat$.

Finally, let W be a fibrant object of the locally constant model structure. Then $i(\mathrm{ev}_0 W)^c = i(\mathrm{ev}_0 W) = W_{*,0}$, regarded as a constant simplicial space.

However, since W is locally constant, it follows that the map $W_{*,0} \to W$ is a weak equivalence in the locally constant model structure, as we needed to show.   □

8

Relative Categories

The final model that we consider takes us back to one of our motivations for the subject. From the perspective of homotopy theory, an $(\infty, 1)$-category should model a category with weak equivalences. So, we want to have a model structure on the category of small categories with weak equivalences.

As it is less cumbersome, in this chapter we adopt the terminology of Barwick and Kan [11] and refer to categories with weak equivalences as *relative categories*. The results of this chapter are taken from their paper [11]. The main idea is that a pair of adjoint functors is used to define a model structure on the category of small relative categories in such way that it is Quillen equivalent to the complete Segal space model structure.

8.1 Basic Definitions

Although we have seen relative categories already, let us state a formal definition.

Definition 8.1.1 A *relative category* (or *category with weak equivalences*) is a pair $(C, \mathcal{W})$ consisting of

1. a category C, called the *underlying category*, and
2. a subcategory $\mathcal{W}$ of C containing all the objects of C, whose morphisms are called *weak equivalences*.

A *relative functor* $F \colon (C, \mathcal{W}) \to (\mathcal{D}, \mathcal{V})$ is given by a functor $F \colon C \to \mathcal{D}$ such that the image of $\mathcal{W}$ is contained in $\mathcal{V}$. A *relative inclusion* is a relative functor such that the underlying functor $F \colon C \to \mathcal{D}$ is the inclusion of a subcategory and $\mathcal{W} = \mathcal{V} \cap C$.

We denote by $\mathcal{R}el\mathcal{C}at$ the category of small relative categories and relative

213

functors between them. When referring to a relative category $(C, \mathcal{W})$, we frequently simply write C, leaving the subcategory $\mathcal{W}$ of weak equivalences implicit.

Given a category C, we can consider the two extreme cases of relative category structures, one in which all morphisms are weak equivalences, and the other in which only the identity morphisms are weak equivalences.

Definition 8.1.2   A relative category $(C, \mathcal{W})$ is *maximal* if $\mathcal{W} = C$, and it is *minimal* if $\mathcal{W}$ has no nonidentity morphisms.

We denote the maximal relative category structure on a category C by $C_{\max}$, and the minimal relative category structure by $C_{\min}$. Of particular importance here are the maximal and minimal relative category structures on the category $[n] = \{0 \to \cdots \to n\}$.

We have the following nice property of the category of small relative categories.

Proposition 8.1.3   [11, 7.1] *The category* $\mathcal{R}elCat$ *is cartesian closed.*

Proof   We first need to show that, for any relative category $\mathcal{A}$, the functor $- \times \mathcal{A} \colon \mathcal{R}elCat \to \mathcal{R}elCat$ has a right adjoint $(-)^{\mathcal{A}}$. Given a relative category $\mathcal{B}$, define $\mathcal{B}^{\mathcal{A}}$ to be the relative category whose objects are the relative functors $\mathcal{A} \to \mathcal{B}$, whose morphisms are the relative functors $\mathcal{A} \times [1]_{\min} \to \mathcal{B}$, and whose weak equivalences are the relative functors $\mathcal{A} \times [1]_{\max} \to \mathcal{B}$. One can check that this construction is functorial and defines the desired adjoint functor.

Secondly, for relative categories $\mathcal{A}$, $\mathcal{B}$, and C, we need to prove that there is an isomorphism $C^{\mathcal{A} \times \mathcal{B}} \cong (C^{\mathcal{B}})^{\mathcal{A}}$. Define a functor which on objects takes a relative functor $f \colon \mathcal{A} \times \mathcal{B} \to C$ to the relative functor $\mathcal{A} \to C^{\mathcal{B}}$ which takes an object a of $\mathcal{A}$ to the functor $b \mapsto f(a, b)$ and behaves similarly on morphisms and weak equivalences of $\mathcal{A}$. On morphisms, a relative functor $\mathcal{A} \times \mathcal{B} \times [1]_{\min} \to C$ is assigned to the relative functor $\mathcal{A} \times [1]_{\min} \to C^{\mathcal{B}}$ adjoint to

$$(\mathcal{A} \times \mathcal{B}) \times [1]_{\min} \cong (\mathcal{A} \times [1]_{\min}) \times \mathcal{B} \to C.$$

The definition on weak equivalences is similar. Again, one can check that this construction gives the required isomorphism.   □

We can restrict ourselves to relative categories whose underlying categories are posets.

Definition 8.1.4   [11, 4.1] A *relative poset* is a relative category $(\mathcal{P}, \mathcal{W})$ such that $\mathcal{P}$, and hence also $\mathcal{W}$, is a poset.

Let $\mathcal{RelPos}$ be the full subcategory of $\mathcal{RelCat}$ whose objects are relative posets.

We now turn to a definition of homotopy between relative functors.

Definition 8.1.5 [11, 3.3]

1 Given two relative functors $f, g\colon C \to \mathcal{D}$, a *strict homotopy* from f to g is given by a natural weak equivalence $H\colon C \times [1]_{\max} \to \mathcal{D}$ in $\mathcal{RelCat}$ such that $H(C, 0) = f(C)$ and $H(C, 1) = g(C)$ for every object C of C, and likewise for morphisms in C.
2 A relative functor $f\colon C \to \mathcal{D}$ is a *homotopy equivalence* if there is a relative functor $g\colon \mathcal{D} \to C$ such that gf is strictly homotopic to the identity on C and fg is strictly homotopic to the identity on $\mathcal{D}$.

This homotopy relation is compatible with the cartesian closure of $\mathcal{RelCat}$, in the folllowing sense.

Proposition 8.1.6 [11, 7.2] *If two relative functors* $f, g\colon \mathcal{A} \to \mathcal{B}$ *are strictly homotopic, then for any relative category* C, *the induced maps* $f^*, g^*\colon C^{\mathcal{B}} \to C^{\mathcal{A}}$ *are strictly homotopic. In particular, if* $h\colon \mathcal{A} \to \mathcal{B}$ *is a strict homotopy equivalence, then so is* $h^*\colon C^{\mathcal{B}} \to C^{\mathcal{A}}$ *for any relative category* C.

Proof Suppose $H\colon \mathcal{A} \times [1]_{\max} \to \mathcal{B}$ is a strict homotopy. Consider the composite

$$C^{\mathcal{B}} \to C^{\mathcal{A} \times [1]_{\max}} \cong (C^{\mathcal{A}})^{[1]_{\max}}.$$

Its adjoint $C^{\mathcal{B}} \times [1]_{\max} \to \mathcal{Z}^{\mathcal{A}}$ is the desired strict homotopy. □

We make use of strong homotopies of relative functors in the following definition, which is key in the definition of cofibration between relative categories.

Definition 8.1.7 Let $i\colon C \to \mathcal{D}$ be a relative inclusion. A *strong deformation retraction* of $\mathcal{D}$ onto C is given by

1 a relative functor $r\colon \mathcal{D} \to C$ such that $ri = \mathrm{id}_C$, and
2 a strict homotopy S from r to id_C.

Before we can define the key maps of interest in this section, we need a categorical definition.

Definition 8.1.8 Let $\mathcal{D}$ be a category and C a subcategory of $\mathcal{D}$.

1 The subcategory C is a *sieve* in $\mathcal{D}$ if any morphism $d \to c$ in $\mathcal{D}$, where c is an object of C, is a morphism in C.
2 Dually, C is a *cosieve* in $\mathcal{D}$ if any morphism $c \to d$ in $\mathcal{D}$, where c is an object of C, is a morphism in C.

Remark 8.1.9 Equivalently, C is a sieve in $\mathcal{D}$ if there exists a functor $\alpha: \mathcal{D} \rightarrow [1]_{max}$ such that $\alpha^{-1}(0) = C$. The function α is called a *characteristic relative functor*. Dually, C is a cosieve in $\mathcal{D}$ if there is a functor $\beta: \mathcal{D} \rightarrow [1]_{max}$ such that $\beta^{-1}(1) = C$.

The following kinds of relative functors are key in developing the model structure.

Definition 8.1.10 [11, 3.5] A relative inclusion $(C, \mathcal{W}) \rightarrow (\mathcal{D}, \mathcal{V})$ is a *Dwyer inclusion* if:

1 the category C is a sieve in $\mathcal{D}$, and
2 the category C is a strong deformation retract of the smallest cosieve ZC of $\mathcal{D}$ containing C.

A relative functor $(C, \mathcal{W}) \rightarrow (\mathcal{D}, \mathcal{V})$ is a *Dwyer map* if it admits a (unique) factorization $(C, \mathcal{W}) \rightarrow (C', \mathcal{W}') \rightarrow (\mathcal{D}, \mathcal{V})$ in $\mathcal{R}elCat$ where $(C, \mathcal{W}) \rightarrow (C', \mathcal{W}')$ is an isomorphism and $(C', \mathcal{W}') \rightarrow (\mathcal{D}, \mathcal{V})$ is a Dwyer inclusion.

Proposition 8.1.11 [11, 9.1] *Dwyer maps are closed under retracts.*

Proof Let $\mathcal{A} \rightarrow \mathcal{B}$ be a Dwyer map of relative categories; replacing $\mathcal{A}$ if necessary, assume it is a Dwyer inclusion. Consider a diagram

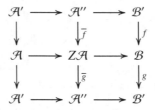

in which $gf = \mathrm{id}_{\mathcal{B}'}$ and the horizontal maps are relative inclusions, and suppose that (r, S) is a strong deformation retraction of $Z\mathcal{A}$ onto $\mathcal{A}$. One can check that $\mathcal{A}'$ is a sieve on $\mathcal{B}'$ and that $\mathcal{A}'' = Z\mathcal{A}'$. It remains to define a strong deformation retraction of $\mathcal{A}''$ onto $\mathcal{A}'$. Let $r' = \overline{g} r \overline{f}$ and $S' = \overline{g} S \overline{f} \colon \overline{g} f \overline{f} \rightarrow \overline{g} f = \mathrm{id}_{\mathcal{A}''}$. Then (r', S') is precisely the strong deformation retraction that we need. □

Proposition 8.1.12 [11, 9.3] *Dwyer maps are closed under transfinite composition.*

Proof As in the proof of the previous proposition, assume that all Dwyer maps in question are relative inclusions. Suppose $\mathcal{A}_0 \rightarrow \mathcal{A}_1$ and $\mathcal{A}_1 \rightarrow \mathcal{A}_2$ are Dwyer maps, with (r_{01}, S_{01}) a strong deformation retraction of $Z(\mathcal{A}_0, \mathcal{A}_1)$

onto $\mathcal{A}_0$ and (r_{12}, S_{12}) a strong deformation retraction of $Z(\mathcal{A}_1, \mathcal{A}_2)$ onto $\mathcal{A}_1$. We want to show that the composite $\mathcal{A}_0 \to \mathcal{A}_2$ is again a Dwyer map. It is not hard to check that $\mathcal{A}_0$ is a sieve in $\mathcal{A}_2$.

It remains to define a strong deformation retraction (r_{02}, S_{02}) of $Z(\mathcal{A}_0, \mathcal{A}_2)$ onto $\mathcal{A}_2$. Let (r'_{12}, S'_{12}) be the restriction of (r_{12}, S_{12}) to $Z(\mathcal{A}_0, \mathcal{A}_1) \subseteq Z(\mathcal{A}_1, \mathcal{A}_2)$. Then define $r_{02} = r_{01} r'_{12}$ and $S_{02} = S'_{12} S_{01}$ to be the desired retraction.

A similar argument can be made for transfinite compositions. □

Proposition 8.1.13   [11, 9.2] *Let*

$$
\begin{array}{ccc}
\mathcal{A} & \xrightarrow{\ s\ } & C \\
{\scriptstyle i}\downarrow & & \downarrow{\scriptstyle j} \\
\mathcal{B} & \xrightarrow{\ t\ } & \mathcal{D}
\end{array}
$$

be a pushout diagram of relative categories in which $i\colon \mathcal{A} \to C$ *is a Dwyer map.*

1 *The relative functor* $j\colon C \to \mathcal{D}$ *is a Dwyer map, and* $ZC \cong Z\mathcal{A} \amalg_{\mathcal{A}} C$.
2 *Let* $X\mathcal{A}$ *denote the full relative subcategory of* $\mathcal{B}$ *spanned by objects not in the image of* $\mathcal{A}$. *The map* $t\colon \mathcal{B} \to \mathcal{D}$ *restricts to isomorphisms* $X\mathcal{A} \cong XC$ *and* $X\mathcal{A} \cap Z\mathcal{A} \cong XC \cap ZC$.
3 *If* $\mathcal{A}$, $\mathcal{B}$, *and* C *are relative posets, then so is* $\mathcal{D}$.

Proof   To prove (1), we first assume as usual that i is a relative inclusion. Let us first prove that C is a sieve in $\mathcal{D}$. Consider the characteristic relative functor $\mathcal{B} \to [1]_{\max}$ and the relative functor $C \to [1]_{\max}$ which takes all of C to the object 0. There is an induced relative functor $\mathcal{D} \to [1]_{\max}$ for which the preimage of 0 is C, showing that C is a sieve in $\mathcal{D}$. An analogous argument can be used to show that $Z\mathcal{A} \amalg_{\mathcal{A}} C$ is a cosieve in $\mathcal{D}$.

Since i is a Dwyer map, there is a strong deformation retraction (r, S) of $Z\mathcal{A}$ onto $\mathcal{A}$. Define a strong deformation retraction (r', S') of $Z\mathcal{A} \amalg_{\mathcal{A}} C$ onto C by

$$ r' = r \amalg_{\mathcal{A}} C \colon Z\mathcal{A} \amalg_{\mathcal{A}} C \to \mathcal{A} \amalg_{\mathcal{A}} C = C $$

and

$$ S' = S \amalg_{\mathcal{A}} C \colon r \amalg_{\mathcal{A}} C \to \mathrm{id}_{Z\mathcal{A}} \amalg_{\mathcal{A}} C = \mathrm{id}_{Z\mathcal{A} \amalg_{\mathcal{A}} C}. $$

Thus we have shown that C deformation retracts onto $Z\mathcal{A} \amalg_{\mathcal{A}} C$. Since we have that $Z\mathcal{A} \amalg_{\mathcal{A}} C$ is a cosieve in $\mathcal{D}$, we can conclude that $ZC \cong Z\mathcal{A} \amalg_{\mathcal{A}} C$.

To prove (2), consider the relative inclusion

$$ [0]_{\max} = \mathcal{A} \amalg_{\mathcal{A}} [0]_{\max} \to \mathcal{B} \amalg_{\mathcal{A}} [0]_{\max}. $$

Observe that $Z[0]_{\max} = Z\mathcal{A} \amalg_{\mathcal{A}} [0]_{\max}$ can be obtained from $X\mathcal{A} \cap Z\mathcal{A}$ by adjoining a single object 0 and a single weak equivalence $0 \to B$ for every object B of $X\mathcal{A} \cap Z\mathcal{A}$. We can build $\mathcal{B} \amalg_{\mathcal{A}} [0]_{\max}$ similarly from $X\mathcal{A}$, and hence we can build $\mathcal{D} \amalg_C [0]_{\max}$ from $XC \cap ZC$ analogously. Furthermore, the relative functor $\mathcal{B} \to \mathcal{D}$ induces an isomorphism

$$\mathcal{B} \amalg_{\mathcal{A}} [0]_{\max} \cong \mathcal{D} \amalg_C [0]_{\max}.$$

The result follows from the construction of these two relative categories, as just described.

Finally, we prove (3) by showing that there is at most one morphism between any two objects of $\mathcal{D}$. Suppose D and E are two objects of $\mathcal{D}$. If they are both in C, then there is at most one map between them, since C is a relative poset. We know from (2) that $XC \cong X\mathcal{A}$, and since $X\mathcal{A} \subseteq \mathcal{A}$, we know XC must be a relative poset. Therefore if D and E are both in XC, there is at most one morphism between them.

Now suppose that D is in C and E is in XC. Since C is a sieve in $\mathcal{D}$, we conclude there can be no map $E \to D$ in $\mathcal{D}$. If there is a map $g \colon D \to E$, then E is in ZC, and $g = (S'E)(r'g)$, where (r', S') is as in (1). Since $r'g \colon D \to r'E$ is in the relative poset C, the map g must be unique. $\square$

8.2 Subdivision Functors

We now turn to defining the subdivision of a relative poset. This construction is critical in developing the model structure on the category of small relative categories.

Definition 8.2.1 [11, 4.2] The *terminal subdivision* of a relative poset $\mathcal{P}$ is the relative poset $\xi_t \mathcal{P}$ with

1 objects the monomorphisms $[n]_{\min} \to \mathcal{P}$ in $\mathcal{RelPos}$, for any $n \geq 0$,
2 morphisms

$$(x_1 \colon [n_1]_{\min} \to \mathcal{P}) \to (x_2 \colon [n_2]_{\min} \to \mathcal{P})$$

given by commutative diagrams

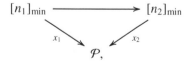

and

3 weak equivalences the commutative triangles as above for which the induced map $x_1(n_1) \to x_2(n_2)$ is a weak equivalence in $\mathcal{P}$.

The terminal subdivision of a relative poset is equipped with a *terminal projection functor* $\pi_t \colon \xi_t \mathcal{P} \to \mathcal{P}$ which sends an object $x \colon [n]_{\min} \to \mathcal{P}$ of $\xi_t \mathcal{P}$ to the object $x(n)$ of $\mathcal{P}$ and a commutative triangle as above to the morphism $x_1(n_1) \to x_2(n_2)$ in $\mathcal{P}$. In particular, a map in $\xi_t \mathcal{P}$ is a weak equivalence if and only if its image under π_t is a weak equivalence in $\mathcal{P}$.

We have the analogous definition of the initial subdivision; note the reversal of the direction of the horizontal arrow in the diagram specifying a morphism.

Definition 8.2.2 [11, 4.2] The *initial subdivision* of a relative poset $\mathcal{P}$ is the relative poset $\xi_i \mathcal{P}$ with

1 objects the monomorphisms $[n]_{\min} \to \mathcal{P}$ in $\mathcal{R}el\mathcal{P}os$, for any $n \geq 0$,
2 morphisms

$$(x_1 \colon [n_1]_{\min} \to \mathcal{P}) \to (x_2 \colon [n_2]_{\min} \to \mathcal{P})$$

given by commutative diagrams

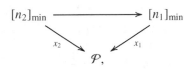

and
3 weak equivalences the commutative triangles as above for which the induced map $x_2(0) \to x_1(0)$ is a weak equivalence in $\mathcal{P}$.

The initial subdivision of a relative poset is equipped with an *initial projection functor* $\pi_i \colon \xi_i \mathcal{P} \to \mathcal{P}$ which sends an object $x \colon [n]_{\min} \to \mathcal{P}$ of $\xi_i \mathcal{P}$ to the object $x(0)$ of $\mathcal{P}$ and a commutative triangle as above to $x_2(0) \to x_1(0)$ in $\mathcal{P}$. In particular, a map in $\xi_i \mathcal{P}$ is a weak equivalence if and only if its image under π_i is a weak equivalence in $\mathcal{P}$.

We can extend these constructions to define functors ξ_t and ξ_i from the category $\mathcal{R}el\mathcal{P}os$ to itself. Given a functor $f \colon \mathcal{P} \to Q$ of relative posets, we take a monomorphism $[n]_{\min} \to \mathcal{P}$ to the unique monomorphism $[m]_{\min} \to Q$ for which $[n]_{\min} \to [m]_{\min}$ is an epimorphism and the diagram

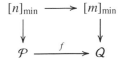

commutes.

We often want to apply both subdivision functors to a relative poset.

Definition 8.2.3 Let $\mathcal{P}$ be a relative poset. Define the *two-fold subdivision* of $\mathcal{P}$ to be $\xi\mathcal{P} = \xi_t\xi_i\mathcal{P}$.

Note that we have made a choice here of which subdivision to take first. We could equally have chosen to perform the subdivisions in the opposite order; Barwick and Kan refer to this construction as the *conjugate* two-fold subdivision. We do not make use of this construction here, however.

Proposition 8.2.4   [11, 7.3]

1 *The subdivision functors ξ_t, ξ_i, and ξ preserve homotopies between maps whose domains are finite relative posets, and therefore they preserve homotopy equivalences between finite relative posets.*
2 *For any $m, n \geq 0$, all the maps in the diagram*

$$\xi([n]_{\min} \times [m]_{\max}) \xrightarrow{\pi_i\xi_i} \xi_i([n]_{\min} \times [m]_{\max}) \xrightarrow{\pi_i} [n]_{\min} \times [m]_{\max}$$
$$\downarrow \qquad\qquad\qquad \downarrow \qquad\qquad\qquad \downarrow$$
$$\xi[n]_{\min} \xrightarrow{\pi_i\xi_i} \xi_i[n]_{\min} \xrightarrow{\pi_i} [n]_{\min},$$

in which the vertical maps are induced by the projection $[n]_{\min} \times [m]_{\max} \rightarrow [n]_{\min}$, are homotopy equivalences.

Proof   We leave the proof of (1) as an exercise. To prove (2), first observe that the map $[n]_{\min} \times [m]_{\max} \rightarrow [n]_{\min}$ is a homotopy equivalence, from which we can conclude that the other two vertical maps are homotopy equivalences by (1). It remains to show that all the horizontal maps are homotopy equivalences. Consider the diagram

$$\xi_t\xi_i[n]_{\min} \xrightarrow{\pi_i\xi_i} \xi_i[n]_{\min}$$
$$\xi_i\pi_i \downarrow \qquad\qquad\qquad \downarrow \pi_i$$
$$\xi_t[n]_{\min} \xrightarrow{\pi_t} [n]_{\min}.$$

The map π_t has a homotopy inverse, given by sending any object i of $[n]_{\min}$ to the object $(0, \ldots, i)$ of $\xi_t[n]_{\min}$. Similarly, the map π_i has a homotopy inverse given by sending any object i of $[n]_{\min}$ to the object $(n - i, \ldots, n)$ of $\xi_i[n]_{\min}$. Another application of (1) shows that $\pi_t\pi_i$ is a homotopy equivalence, from which it follows from the two-out-of-three property that $\pi_t\xi_i$ is as well. Thus we have proved that the bottom horizontal maps in the large diagram are homotopy equivalences; yet another application of the two-out-of-three property shows that the top horizontal maps must also be homotopy equivalences. $\square$

Proposition 8.2.5 [11, 9.4] *Suppose that* $\mathcal{P} \to \mathcal{Q}$ *is a relative inclusion of relative posets such that* $\mathcal{P}$ *is a sieve or a cosieve in* $\mathcal{Q}$. *Then the induced inclusion* $\xi_t\mathcal{P} \to \xi_t\mathcal{Q}$ *is a Dwyer map.*

Proof We prove the case in which $\mathcal{P}$ is a cosieve, the case of a sieve being similar. First, we need to prove that $\xi_t\mathcal{P}$ is a sieve in $\xi_t\mathcal{Q}$. Consider a morphism $(q_1, \ldots, q_n) \to (p_1, \ldots, p_m)$ in $\xi_t\mathcal{Q}$ with $(p_1, \ldots, p_m)$ in $\xi_t\mathcal{P}$. By definition of $\xi_t\mathcal{Q}$, such a morphism is given by a map $\delta \colon [n] \to [m]$ in Δ over $\mathcal{Q}$, so that $q_i = p_{\delta(i)}$ for all $0 \le i \le n$. But then each q_i must be in $\mathcal{P}$, and hence $(q_1, \ldots, q_n)$ is in $\xi_t\mathcal{P}$.

It remains to find a strong deformation retraction of $Z\xi_t\mathcal{P}$ onto $\xi_t\mathcal{P}$. Let $(q_1, \ldots, q_n)$ be an object of $\xi_t\mathcal{Q}$. Since $\mathcal{P}$ is a cosieve in $\mathcal{Q}$, either $q_i \notin \mathcal{P}$ for all $0 \le i \le n$, or there is some j such that $q_i \in \mathcal{P}$ for all $i \ge j$. In the latter case, we have that $(q_j, \ldots, q_n)$ is an object of $\xi_t\mathcal{P}$, and $(q_0, \ldots, q_n)$ is an object of $Z\xi_t\mathcal{P}$. So define the required strong deformation retraction (r, S) by $r(q_0, \ldots, q_n) = (q_j, \ldots, q_n)$ and $S \colon (q_j, \ldots, q_n) \mapsto (q_0, \ldots, q_n)$. $\square$

8.3 The Model Structure and Equivalence With Complete Segal Spaces

Finally, we define a pair of adjoint functors between the category of small relative categories and the category of simplicial spaces which induces a model structure on $\mathcal{R}el\mathcal{C}at$ which is Quillen equivalent to the model category $\mathcal{CSS}$.

Given a relative category $(C, \mathcal{W})$, we also have the alternative notation we C for the subcategory $\mathcal{W}$ of weak equivalences. Given a relative category C, the category $C^{[n]}$ of functors $[n] \to C$ can be given the structure of a relative category in which the weak equivalences are the natural transformations of functors whose component maps are weak equivalences in C. We use this notation in the following generalization of the classifying diagram to the setting of relative categories.

Definition 8.3.1 Let C be a relative category. The *classification diagram* of C is the simplicial space $N\mathcal{D}$ defined by

$$(NC)_n = \text{nerve(we } C^{[n]}).$$

This construction defines a functor $N \colon \mathcal{R}el\mathcal{C}at \to \mathcal{SSets}^{\Delta^{op}}$.

Alternatively, the set $(NC)_{n,m}$ can be defined to be the set of functors

$$[n]_{\min} \times [m]_{\max} \to C.$$

Proposition 8.3.2 [11, 5.3] *The classification diagram functor N has a left adjoint K, given by*

$$K(\Delta[n]^t \times \Delta[m]) = [n]_{\min} \times [m]_{\max}.$$

Define the functor $\xi = \xi_t \xi_i$, where ξ_t denotes terminal subdivision and ξ_i denotes initial subdivision.

Proposition 8.3.3 [11, 5.3] *The functor $K_\xi \colon SSets^{\Delta^{op}} \to RelCat$ defined by*

$$K_\xi(\Delta[n]^t \times \Delta[m]) = \xi([n]_{\min} \times [m]_{\max})$$

has a right adjoint which we denote by N_ξ.

Observe that the natural transformation $\pi \colon \xi \to \mathrm{id}$ induces a natural transformation $\pi^* \colon N \to N_\xi$.

Lemma 8.3.4 [11, 5.4] *The natural transformation π^* is a Reedy weak equivalence.*

Proof We need to show that, for any relative category $\mathcal{A}$ and any $n \geq 0$, the map

$$\pi_n^* \colon (N\mathcal{A})_n \to (N_\xi\mathcal{A})_n$$

is a weak equivalence of simplicial sets. By definition, we have

$$(N\mathcal{A})_{n,m} = \mathrm{Hom}_{RelCat}([n]_{\min} \times [m]_{\max}, \mathcal{A})$$

and

$$(N_\xi\mathcal{A})_{n,m} = \mathrm{Hom}_{RelCat}(\xi([n]_{\min} \times [m]_{\max}), \mathcal{A}).$$

Define a simplicial space $F_n\mathcal{A}$ by

$$(F_n\mathcal{A})_{m,p} = \mathrm{Hom}_{RelCat}([n]_{\min} \times [m]_{\max}, \mathcal{A}^{[p]_{\max}}).$$

If we define $\overline{F}_n\mathcal{A}$ by

$$(\overline{F}_n\mathcal{A})_{m,p} = \mathrm{Hom}_{RelCat}([n]_{\min} \times [m]_{\max}, \mathcal{A}^{[0]_{\max}}),$$

observe that there is map

$$\overline{F}_n\mathcal{A} \to F_n\mathcal{A}$$

induced by the inclusion $[0]_{\max} \to [p]_{\max}$. By Proposition 8.1.6, if we restrict to level p we get a weak equivalence of simplicial sets

$$(\overline{F}_n\mathcal{A})_{*,p} \to (F_n\mathcal{A})_{*,p}.$$

Similarly, define a map $\overline{G}_n\mathcal{A} \to G_n\mathcal{A}$, where

$$(G_n\mathcal{A})_{m,p} = \mathrm{Hom}_{RelCat}(\xi([n]_{\min} \times [m]_{\max}), \mathcal{A}^{[p]_{\max}})$$

and

$$(\overline{G}_n\mathcal{A})_{m,p} = \mathrm{Hom}_{RelCat}(\xi([n]_{\min} \times [m]_{\max}), \mathcal{A}^{[0]_{\max}}),$$

and observe that the induced map

$$(\overline{G}_n\mathcal{A})_{*,p} \to (G_n\mathcal{A})_{*,p}$$

is a weak equivalence of simplicial sets.

Now, using adjointness and the map $\pi \colon \xi([n]_{\min} \times [m]_{\max}) \to [n]_{\min} \times [m]_{\max}$, we obtain a map

$$(F_n\mathcal{A})_{m,p} \cong \mathrm{Hom}_{RelCat}([p]_{\max}, \mathcal{A}^{[n]_{\min} \times [m]_{\max}})$$
$$\to \mathrm{Hom}_{RelCat}([p]_{\max}, \mathcal{A}^{\xi([n]_{\min} \times [m]_{\max})}) \cong (G_n\mathcal{A})_{m,p}.$$

We can then apply Proposition 8.2.4(2) to see that, if we fix m and let p vary, this map gives a weak equivalence of simplicial sets. We can complete the proof by observing that there are isomorphisms

$$(N\mathcal{A})_n \cong \mathrm{diag}\, \overline{F}_n\mathcal{A}$$

and

$$(N_\xi\mathcal{A})_n \cong \mathrm{diag}\, \overline{G}_n\mathcal{A}. \qquad \square$$

Lemma 8.3.5 *The functor N_ξ takes Dwyer maps to Reedy cofibrations.*

Proof Let $\mathcal{A} \to \mathcal{B}$ be a Dwyer map. For simplicity, assume that it is a Dwyer inclusion and so in particular a relative inclusion of relative posets. It suffices to prove that, for any $m, n \geq 0$, the map

$$N_\xi(\mathcal{A})_{n,m} \to N_\xi(\mathcal{B})_{n,m}$$

is a monomorphism of sets. Applying the definition of N_ξ, we consider the map

$$\mathrm{Hom}_{RelCat}(\xi([n]_{\min} \times [m]_{\max}), \mathcal{A}) \to \mathrm{Hom}_{RelCat}(\xi([n]_{\min} \times [m]_{\max}), \mathcal{B}).$$

Since the original map $\mathcal{A} \to \mathcal{B}$ is a relative inclusion, one can check that this induced map is indeed a monomorphism. $\qquad \square$

We recall the following definitions for simplicial spaces.

Definition 8.3.6 Two maps $W \to Z$ of simplicial spaces are *homotopic* if there is a homotopy $W \times \Delta[1]^t \to Z$ restricting to the maps in question on the endpoints of $\Delta[1]^t$, or if the maps can be connected by a zigzag thereof. A map $f: W \to Z$ is a *homotopy equivalence* if there exists a map $g: Z \to W$ such that gf is homotopic to id_W and fg is homotopic to id_Z.

The following result can be checked from the above definitions.

Proposition 8.3.7 [11, 7.4] *Any homotopy equivalence of simplicial spaces is a Reedy weak equivalence.*

Now we show that this definition of homotopy for simplicial spaces is compatible with the analogous notion for relative functors, via the functor N_ξ.

Proposition 8.3.8 [11, 7.5] *The functor N_ξ takes homotopic maps in $\mathcal{R}el\mathcal{C}at$ to homotopic maps in $\mathcal{S}Sets^{\Delta^{op}}$, and in particular takes homotopy equivalences to homotopy equivalences.*

Proof Suppose $H: \mathcal{A} \times [1]_{\max} \to \mathcal{B}$ is a homotopy in $\mathcal{R}el\mathcal{C}at$. Applying the functor N_ξ gives a map

$$N_\xi(\mathcal{A} \times [1]_{\max}) \to N_\xi\mathcal{B}.$$

Since N_ξ is a right adjoint and hence preserves products, there is an isomorphism

$$N_\xi\mathcal{A} \times N_\xi[1]_{\max} \cong N_\xi(\mathcal{A} \times [1]_{\max}).$$

Finally, there is a natural map $\Delta[1]^t \to N_\xi[1]_{\max}$ which induces a map

$$N_\xi\mathcal{A} \times \Delta[1]_{\max} \to N_\xi \times N_\xi[1]_{\max}.$$

Taking the composite of these maps gives the desired homotopy of simplicial spaces

$$N_\xi\mathcal{A} \times \Delta[1]^t \to N_\xi\mathcal{B}. \qquad \square$$

For the next result, we introduce some notation. Thinking of a simplicial space as a functor $\Delta^{op} \times \Delta^{op} \to \mathcal{S}ets$, let us denote by $\Delta[n, m]$ the representable object associated to the object $([n], [m])$ of $\Delta \times \Delta$. Note that $\Delta[n, m] = \Delta[n]^t \times \Delta[m]$. We can take its boundary $\partial\Delta[n, m]$ to be the largest subsimplicial space not containing the nondegenerate (n, m)-bisimplex; alternatively,

$$\partial\Delta[n, m] = \partial\Delta[n] \times \Delta[m]^t \cup \Delta[n] \times \partial\Delta[m]^t.$$

Proposition 8.3.9 [11, 9.5] *The inclusion* $\partial\Delta[n,m] \to \Delta[n,m]$ *induces a Dwyer map of relative posets*

$$K_\xi \partial\Delta[n,m] \to K_\xi \Delta[n,m].$$

Proof Recall that $K_\xi \Delta[n,m] = \xi([n]_{\min} \times [m]_{\max})$. Let us analogously define the functor $K_{\xi_i} : SSets^{\Delta^{op}} \to RelCat$ to be

$$K_{\xi_i} \Delta[n,m] = \xi_i([n]_{\min} \times [m]_{\max})$$

on representable objects. It can be extended to all objects, using the fact that it is a left adjoint functor and therefore preserves colimits, but here we only apply it to representable objects.

Observe in particular that

$$\xi_t K_{\xi_i} \Delta[n,m] = \xi([n]_{\min} \times [m]_{\max}) = K_\xi \Delta[n,m].$$

Let us first prove that the relative functor $K_{\xi_i} \partial\Delta[n,m] \to K_\xi \Delta[n,m]$ satisfies the hypotheses of Proposition 8.2.5, from which we can conclude that

$$\xi_t K_{\xi_i} \partial\Delta[n,m] \to \xi_t K_{\xi_i} \Delta[n,m] = K_\xi \Delta[n,m]$$

is a Dwyer map.

To do so, let $\mathcal{P}$ be the poset of relative subcategories of $[n]_{\min} \times [m]_{\max}$ of the form $[a]_{\min} \times [b]_{\max}$, where $[a]_{\min} \subseteq [n]_{\min}$ and $[b]_{\max} \subseteq [m]_{\max}$ are relative subcategories, and define the morphisms of $\mathcal{P}$ to be relative inclusions. If $[a_1]_{\min} \times [b_1]_{\max}$ and $[a_2]_{\min} \times [b_2]_{\max}$ are objects of $\mathcal{P}$ such that $[a_1]_{\min} \cap [a_2]_{\min} \neq \varnothing$ and $[b_1]_{\max} \cap [b_2]_{\max} \neq \varnothing$, then

$$([a_1]_{\min} \times [b_1]_{\max}) \cap ([a_2]_{\min} \times [b_2]_{\max})$$
$$= ([a_1]_{\min} \cap [a_2]_{\min}) \times ([b_1]_{\max} \cap [b_2]_{\max}).$$

In other words, the objects of $\mathcal{P}$ are closed under intersection. One can check that the same equality holds after applying the functors ξ_i and ξ. We can use this fact to see that

$$K_{\xi_i} \partial\Delta[n,m] = \bigcup \xi_i([a]_{\min} \times [b]_{\max}),$$

where the union is taken over objects of $\mathcal{P}$ such that $a \neq n$ or $b \neq m$.

Furthermore, given a morphism $f : [a_1]_{\min} \times [b_1]_{\max} \to [a_2]_{\min} \times [b_2]_{\max}$, the map $\xi_i f$ is a relative inclusion and $\xi_i([a_1]_{\min} \times [b_1]_{\max})$ is a cosieve in $\xi_i([a_2]_{\min} \times [b_2]_{\max})$. It follows that $K_{\xi_i} \partial\Delta[n,m]$ is a cosieve in $K_{\xi_i} \Delta[n,m]$.

Thus, by Proposition 8.2.5, we can conclude that the map

$$\xi_t K_{\xi_i} \partial\Delta[n,m] \to K_\xi \Delta[n,m]$$

is a Dwyer map. It remains to show that

$$\xi_t K_{\xi_i} \partial\Delta[n,m] = K_\xi \Delta[n,m].$$

One can check, from the definitions and the decomposition given above, that there is a natural inclusion

$$K_\xi \partial \Delta[n, m] \to \xi_t K_{\xi_i} \partial \Delta[n, m].$$

We need only show that this map is surjective. Suppose that $h\colon x \to y$ is a morphism in $\xi_t K_{\xi_i} \partial \Delta[n, m]$. If we regard the object y as a function $[n] \to K_{\xi_i} \partial \Delta[n, m]$, the object $y(0)$ lies in some $\xi_i([a]_{\min} \times [b]_{\max})$; by the sieve property proved above, we can conclude that y is an object of $\xi([a]_{\min} \times [b]_{\max})$. Now note that, in analogy to this property, the image of a map f of $\mathcal{P}$ under ξ is a relative inclusion, but now the domain of such a map is a sieve, rather than a cosieve, in the codomain. We can use this result to see that the map h must be in $K_\xi \partial \Delta[n, m]$, which we wished to show. $\square$

Proposition 8.3.10   [11, 9.6] *Every monomorphism* $W \to Z$ *of simplicial spaces induces a Dwyer–Kan equivalence* $K_\xi W \to K_\xi Z$ *of relative posets.*

Proof   Assume that $W \to Z$ is an inclusion of simplicial spaces, and let Z^n denote the simplicial space whose nondegenerate (i, j)-bisimplices are precisely those of Z for which $i + j \le n$. Then observe that

$$Z = \bigcup_{n \ge -1} (Z^n \cup W)$$

and

$$K_\xi Z = \bigcup_{n \ge -1} K_\xi(Z^n \cup W).$$

Now, let $\Delta_n(Z, W) = \coprod \Delta[i, j]$, where the coproduct is taken over all $i + j = n$ and nondegenerate (i, j)-bisimplices of $Z^n \cup W$ which are not in $Z^{n-1} \cup W$. Using the same indexing, define $\partial \Delta_n(Z, W) = \coprod \partial \Delta[i, j]$. Since the inclusion $\partial \Delta_n(Z, W) \to \Delta_n(Z, W)$ is a disjoint union of maps of the form $\partial \Delta[i, j] \to \Delta[i, j]$, we can use Proposition 8.3.9 to see that

$$K_\xi \partial \Delta_n(Z, W) \to \Delta_n(Z, W)$$

is a Dwyer map of relative posets. So, if we apply the functor K_ξ to the pushout diagram

$$
\begin{array}{ccc}
\partial \Delta_n(Z, W) & \longrightarrow & Z^{n-1} \cup W \\
\downarrow & & \downarrow \\
\Delta_n(Z, W) & \longrightarrow & Z^n \cup W
\end{array}
$$

we can see from Proposition 8.1.13(1) that the square

$$
\begin{array}{ccc}
K_\xi \partial\Delta_n(Z, W) & \longrightarrow & K_\xi Z^{n-1} \cup W \\
\downarrow & & \downarrow \\
K_\xi \Delta_n(Z, W) & \longrightarrow & K_\xi Z^n \cup W
\end{array}
$$

is also a pushout diagram, and the right-hand vertical map is a Dwyer map. Then we can apply Proposition 8.1.12 to see that $K_\xi W \to K_\xi Z$ is a Dwyer map.

It remains only to show that $K_\xi W$ and $K_\xi Z$ are relative posets. Considering the special case when $W = \varnothing$, together with Proposition 8.1.13(3) and the fact that transfinite composition preserves relative inclusions of relative posets, we obtain that indeed $K_\xi W \to K_\xi Z$ is a map of relative posets. □

Lemma 8.3.11 [11, 10.2] *Suppose that*

$$
\begin{array}{ccc}
\mathcal{A} & \overset{f}{\longrightarrow} & C \\
i\downarrow & & \downarrow j \\
\mathcal{B} & \overset{g}{\longrightarrow} & \mathcal{D}
\end{array}
$$

is a pushout diagram of relative categories, in which i is a Dwyer map. Then in the Reedy model structure on $\mathcal{S}\mathcal{S}ets^{\Delta^{op}}$ and any of its localizations, we have:

1 *the induced map*

$$
N_\xi \mathcal{B} \amalg_{N_\xi \mathcal{A}} N_\xi C \to N_\xi \mathcal{D}
$$

is a weak equivalence, and

2 *if $N_\xi i$ is a weak equivalence, then so is $N_\xi j$, and if $N_\xi f$ is a weak equivalence, then so is $N_\xi g$.*

Proof Let us first look at the Dwyer map $i \colon \mathcal{A} \to \mathcal{B}$. Since $Z\mathcal{A}$ is a cosieve in $\mathcal{B}$, the image of a map

$$
[n]_{\min} \times [m]_{\max} \to \mathcal{B},
$$

for any $m, n \geq 0$, is only in $Z\mathcal{A}$ if and only if the image of the object $(0, 0)$ is only in $Z\mathcal{A}$. One can similarly determine whether the image of such a map is solely in $X\mathcal{A}$ or in the intersection $X\mathcal{A} \cap Z\mathcal{A}$ simply by determining where the image of $(0, 0)$ is. First applying the functor N, we can conclude that there is an isomorphism

$$
N\mathcal{B} \cong N(X\mathcal{A}) \amalg_{N(X\mathcal{A} \cap Z\mathcal{A})} N(Z\mathcal{A}).
$$

Since Dwyer maps are preserved under pushout, we can similarly conclude that

$$N\mathcal{D} \cong N(X C) \amalg_{N(X C \cap Z C)} N(Z C).$$

We use these decompositions to write the map

$$N\mathcal{B} \amalg_{N\mathcal{A}} N\mathcal{C} \to N\mathcal{D}$$

as a composite. First, the inclusions $\mathcal{A} \to Z\mathcal{A}$ and $C \to Z C$ induce a map

$$N\mathcal{B} \amalg_{N\mathcal{A}} N\mathcal{C} \to N\mathcal{B} \amalg_{N(Z\mathcal{A})} N(Z C).$$

Then applying the decompositions above, there are isomorphisms

$$N\mathcal{B} \amalg_{N(Z\mathcal{A})} N(Z C) \cong N(X\mathcal{A}) \amalg_{N(X\mathcal{A} \cap Z\mathcal{A})} N(Z\mathcal{A}) \amalg_{N(Z\mathcal{A})} N(Z C)$$
$$\cong N(X\mathcal{A}) \amalg_{N(X\mathcal{A} \cap Z\mathcal{A})} N(Z C).$$

Now if we apply Proposition 8.1.13(2), we obtain an isomorphism

$$N(X\mathcal{A}) \amalg_{N(X\mathcal{A} \cap Z\mathcal{A})} N(Z C) \cong N(X C) \amalg_{N(X C \cap Z C)} N(Z C).$$

But we have already determined that this space is precisely $N\mathcal{D}$.

Now, applying Lemma 8.3.4, we can replace the functor N by N_ξ. It remains to show that the first map in the composite,

$$N_\xi\mathcal{B} \amalg_{N_\xi\mathcal{A}} N_\xi\mathcal{C} \to N_\xi\mathcal{B} \amalg_{N_\xi(Z\mathcal{A})} N_\xi(Z C),$$

is a weak equivalence. Recall that the inclusions $\mathcal{A} \to Z\mathcal{A}$ and $C \to Z C$ are homotopy equivalences, so by Proposition 8.3.8 we have that the maps $N\mathcal{A} \to N(Z\mathcal{A})$ and $N\mathcal{C} \to N(Z C)$ are weak equivalences. It follows that the natural map

$$N_\xi\mathcal{B} \amalg_{N_\xi\mathcal{A}} N_\xi\mathcal{C} \to N_\xi\mathcal{D}$$

is a weak equivalence, establishing (1).

The first statement of (2) follows immediately from (1). The second statement follows since the Reedy model structure and its localizations are left proper. $\square$

Proposition 8.3.12   [11, 10.3] *The unit map η_ξ: id $\to N_\xi K_\xi$ of the adjoint pair (K_ξ, N_ξ) is a natural weak equivalence in the Reedy model structure on* $SSets^{\Delta^{op}}$, *and therefore also in any of its localizations.*

Proof   We need to show that, for any simplicial space W, the map η_ξ: $W \to N_\xi K_\xi W$ is a weak equivalence. Let us first prove the result for the special case where $W = \Delta[n, m]$.

Consider the diagram

$$\Delta[n,m] \xrightarrow{\eta_{\xi}} N_{\xi}K_{\xi} = N_{\xi}\xi([n]_{\min} \times [m]_{\max})$$

$$\eta \downarrow \qquad\qquad\qquad \downarrow \pi_*$$

$$NK\Delta[n,m] \xrightarrow{\pi^*} N_{\xi}K\Delta[n,m] = N_{\xi}([n]_{\min} \times [m]_{\max})$$

in which η is the unit of the adjoint pair (K,N) and $\pi = \pi_i \circ \pi_t$. One can check that there is a Reedy weak equivalence

$$NK\Delta[n,m] = N([n]_{\min} \times [m]_{\max}) \simeq \Delta[n,m].$$

The map π^* is a weak equivalence by Lemma 8.3.4. It suffices, then, to show that π_* is a weak equivalence. By Proposition 8.2.4(2), the map

$$\xi([n]_{\min} \times [m]_{\max}) \to [n]_{\min} \times [m]_{\max}$$

is a weak equivalence, and N_{ξ} preserves weak equivalences by Proposition 8.3.8, giving the desired result.

Now we consider the case of a more general simplicial space W. As in the proof of Proposition 8.3.10, we write $W = \bigcup_n W^n$ and $N_{\xi}K_{\xi}W^n$, and observe that it suffices to prove that the map

$$\eta_{\xi} \colon W^n \to N_{\xi}K_{\xi}W^n$$

is a weak equivalence for any $n \geq 0$. When $n = 0$, observe that we have such a weak equivalence. So, assume that this map η_{ξ} is a weak equivalence for any $k < n$. Again using notation from the proof of Proposition 8.3.10, consider the diagram

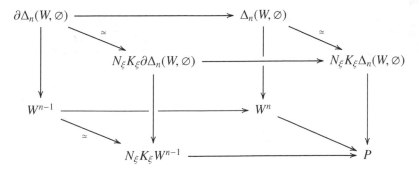

in which the front and back squares are pushouts. By the special case and the inductive hypothesis, the indicated maps are weak equivalences. We can thus conclude that the map $W^n \to P$ is a weak equivalence. Using Proposition 8.3.10 and Corollary 8.3.11, we can conclude that the natural map $P \to N_\xi K_\xi W^n$ is also a weak equivalence. □

Corollary 8.3.13 [11, 10.4] *A map of simplicial spaces $W \to Z$ is a Reedy weak equivalence if and only if the map $N_\xi K_\xi W \to N_\xi K_\xi Z$ is a Reedy weak equivalence.*

Theorem 8.3.14 [11, 6.1] *The adjunction*

$$K_\xi \colon SSets^{\Delta^{op}} \rightleftarrows RelCat \colon N_\xi$$

lifts every localization of the Reedy model structure to a Quillen equivalent cofibrantly generated left proper model structure on RelCat in which:

1 *a map is a weak equivalence if and only if its image under N_ξ is a weak equivalence in $SSets^{\Delta^{op}}$;*
2 *a map is a fibration if its image under N_ξ is a fibration in $SSets^{\Delta^{op}}$;*
3 *every cofibration is a Dwyer map; and*
4 *every cofibrant object is a relative poset.*

Proof We use Theorem 1.7.16 to prove that the Reedy model structure on $SSets^{\Delta^{op}}$ lifts to a model structure on $RelCat$ via the adjunction

$$K_\xi \colon SSets^{\Delta^{op}} \rightleftarrows RelCat \colon N_\xi.$$

Thus, we must prove that conditions (1) and (2) of that theorem hold. Condition (1) holds using the smallness of the objects of all the maps in the sets $K_\xi I$ and $K_\xi J$, where I and J denote the sets of generating cofibrations and generating acyclic cofibrations, respectively, for the Reedy model structure on $SSets^{\Delta^{op}}$.

To prove that condition (2) of Theorem 1.7.16 holds, we need to show that the functor N_ξ takes transfinite compositions of pushouts along maps in $K_\xi(J)$ to Reedy weak equivalences. Let us start by showing that the maps in $N_\xi K_\xi(J)$ are Reedy weak equivalences. First, recall that the maps in J are all monomorphisms, so by Proposition 8.3.10, the maps in $K_\xi(J)$ are Dwyer maps between relative posets. Furthermore, since the maps in J are Reedy weak equivalences, Proposition 8.3.10 tells us that the maps in $N_\xi K_\xi(J)$ are Reedy weak equivalences.

Now let us consider the case of a transfinite composition of pushouts along maps in $K_\xi(J)$. Combining Propositions 8.1.13 and 8.1.12, we see that transfinite compositions of pushouts along Dwyer maps are still Dwyer maps. By Corollary 8.3.11, the functor N_ξ preserves pushouts along Dwyer maps, and by Lemma 8.3.5 it preserves cofibrations. Therefore, the maps in $N_\xi K_\xi(J)$ are Reedy acyclic cofibrations, which are preserved under pushouts and transfinite compositions.

Thus, we have the result for the Reedy model structure on $SSets^{\Delta^{op}}$. The fact that it holds for any localization of the Reedy model structure is a result of applying Theorem 2.8.8.

Now let us verify the conditions (1)–(4) stated in Theorem 8.3.14. Properties (1) and (2) follow from the description of the lifted model structure in Theorem 1.7.16.

By Proposition 8.3.10, the generating cofibrations in $\mathcal{R}el\mathcal{C}at$ are Dwyer maps between relative posets. By Propositions 8.1.11, 8.1.12, and 8.1.13, Dwyer maps are closed under retracts, pushouts, and transfinite composition, from which we can conclude that all cofibrations in $\mathcal{R}el\mathcal{C}at$ are Dwyer maps, establishing (3) of Theorem 8.3.14. After observing that relative posets are preserved under pushout, and transfinite composition of monomorphisms preserves posets, we also obtain (4) of the theorem.

Now we prove that this adjunction defines not only a Quillen pair but also a Quillen equivalence of model categories. We need to prove that, for any simplicial space W, the unit map $W \rightarrow N_{\xi}(K_{\xi}W)^f$ is a Reedy weak equivalence. We know from Corollary 8.3.13 that $W \rightarrow N_{\xi}K_{\xi}W$ is a weak equivalence for any W. Consider the fibrant replacement $K_{\xi}W \rightarrow (K_{\xi}W)^f$, which is an acyclic cofibration. By (1) of the theorem, the map $N_{\xi}K_{\xi}W \rightarrow N_{\xi}(K_{\xi}W)^f$ is a weak equivalence, from which it follows that the composite $W \rightarrow N_{\xi}(K_{\xi}W)^f$ is a weak equivalence, as we wished to show.

Lastly, let us show that these model structures on $\mathcal{R}el\mathcal{C}at$ are all left proper. Consider any pushout diagram of simplicial spaces

where $U \rightarrow V$ is a cofibration and $U \rightarrow W$ is a weak equivalence. Using Corollary 8.3.13, consider the diagram

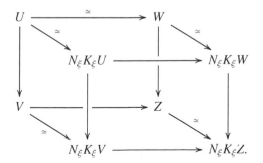

Now the map $N_\xi K_\xi U \to N_\xi K_\xi W$ is a weak equivalence by the two-out-of-three property. Using the fact that applying the left adjoint functor K_ξ to the original diagram results in a pushout diagram, we can apply Proposition 8.3.11 to see that $N_\xi K_\xi V \to N_\xi K_\xi Z$ is a weak equivalence. Then we can apply the two-out-of-three property again to see that $V \to Z$ is a weak equivalence, as we needed to show.   □

Since the model structure CSS is obtained as a localization of the Reedy model structure on simplicial spaces, the Quillen equivalence we wanted is simply a special case.

Corollary 8.3.15   *There is a model structure on the category of small relative categories which is Quillen equivalent to CSS.*

9

Comparing Functors to Complete Segal Spaces

In the various comparisons we have made, we have actually overdetermined the relationships between the model categories, in that we have more Quillen equivalences than we need to show that all the models for $(\infty, 1)$-categories are equivalent. We could ask, then, whether certain diagrams of Quillen equivalences commute even up to homotopy. For example, given a (fibrant) simplicial category, we can obtain a complete Segal space in two different ways: by applying the simplicial nerve to get a Segal category, then taking a fibrant replacement in CSS; or by applying the coherent nerve functor to obtain a quasi-category and then extending to get a complete Segal space.

In addition, we have another functor which is not given by one of the Quillen equivalences at all. If we take the simplicial nerve of a simplicial category, it is a Segal space (up to Reedy fibrant replacement), and hence we can apply the completion functor of Definition 5.5.7 to get a complete Segal space. We would like to compare the output of this functor to what we get from the chains of Quillen functors. The advantage of this procedure, while not part of a Quillen equivalence, is that it enables us to give an up-to-homotopy characterization of the resulting complete Segal space.

Lastly, if we begin with a model category, we have a number of ways to obtain a complete Segal space. We can take its simplicial localization to get a simplicial category and then apply one of the methods given above. However, we can also take the classification diagram, as described in Definition 8.3.1, to get a complete Segal space directly, at least up to Reedy fibrant replacement. Once again, we want to know that these methods result in weakly equivalent complete Segal spaces. Understanding these relationships is the subject of this chapter.

9.1 Classifying and Classification Diagrams

Recall the classifying diagram construction, which associates to any small category C a complete Segal space NC, from Definition 3.3.1. We saw that the classifying diagram of a category can be regarded as a more refined version of the nerve. Now that we have given a rigorous definition of complete Segal space, we can state the following result.

Proposition 9.1.1   [103, 3.8, 6.1] *If C is a small category, then its classifying diagram NC is a complete Segal space.*

Proof   First, we show that NC is Reedy fibrant. Using the definition of Reedy fibration, to check that NC is Reedy fibrant is equivalent to showing that for any $n \geq 0$ the map

$$(NC)_n = \mathrm{Map}(\Delta[n]^t, NC) \to \mathrm{Map}(\partial\Delta[n]^t, NC)$$

is a fibration of simplicial sets. If $n = 0$, then we need to check that the map

$$(NC)_0 \to \mathrm{Map}(\varnothing, NC) = \Delta[0]$$

is a fibration, or that $(NC)_0$ is a Kan complex. But $(NC)_0$ is defined to be the nerve of the groupoid $\mathrm{iso}(C)$ and hence a Kan complex.

If $n = 1$, then we need to show that the map

$$(NC)_1 \to (NC)_0 \times (NC)_0$$

induced by the source and target maps in C is a fibration. Let $(x, y) \in (NC)_{0,0} \times (NC)_{0,0}$, so that x and y are given by objects of C. Then the elements of the fiber over (x, y) in $(NC)_{1,0}$ correspond to functions $f \colon x \to y$ in C. If (x, y) and (x', y') are in the same connected component of $(NC)_0$, then there are isomorphisms $g \colon x \to x'$ and $h \colon y \to y'$ in C, from which we can show that there is a one-to-one correspondence between fibers over (x, y) and fibers over (x', y'). Since these fibers are discrete, being given by morphisms in a category, our map of interest is a covering space and hence a fibration.

In the case $n = 2$, we consider the map

$$\mathrm{Map}(\Delta[2]^t, NC) \to \mathrm{Map}(\partial\Delta[2]^t, NC).$$

Although there could be maps $\partial\Delta[2]^t \to NC$ which cannot be extended to a map $\Delta[2]^t \to NC$, our desired map is an inclusion of path components and therefore a fibration.

Finally, if $n \geq 3$, then associativity from C shows that the maps

$$\mathrm{Map}(\Delta[n]^t, NC) \to \mathrm{Map}(\partial\Delta[n]^t, NC)$$

are all isomorphisms. Thus we have shown that NC is Reedy fibrant.

Now observe from the definition of NC that the Segal maps

$$(NC)_n \to \underbrace{(NC)_1 \times_{(NC)_0} \cdots \times_{(NC)_0} (NC)_1}_{n}$$

are isomorphisms for all $n \geq 2$, which establishes that NC is a Segal space.

Finally, we need to show that NC is complete. If $I[1]$ denotes the groupoid with two objects and a single isomorphism between them, the equivalence of categories $I[1] \to [0]$ induces an equivalence of categories $\mathrm{iso}(C) \to \mathrm{iso}(C^{I[1]})$ and hence a weak equivalence on nerves. Since

$$(NC)_0 = \mathrm{nerve}(\mathrm{iso}\, C)$$

and

$$(NC)_{\mathrm{heq}} = \mathrm{nerve}(\mathrm{iso}\, C^{I[1]}),$$

we hence have that the map

$$(NC)_0 \to (NC)_{\mathrm{heq}}$$

is a weak equivalence, as we wished to show. □

But how might we assign a complete Segal space to a simplicial category? A first guess might be to take a simplicial space whose space in degree n is the diagonal of the simplicial nerve of the simplicial category obtained from $C^{[n]}$ by removing morphism components which do not consist of homotopy equivalences. Unfortunately, this construction is not homotopy invariant, as stated, so we need to find a more careful construction. Indeed, we can take the simplicial nerve, followed by the completion functor as defined in Proposition 5.5.10, to define a functor from the category of simplicial categories to the category of simplicial spaces with the desired properties. More precisely, from a simplicial category C we can take its simplicial nerve to obtain a simplicial space, followed by a fibrant replacement functor in the Segal space model structure, to obtain a Segal space W. From W we can then pass to a complete Segal space via the completion functor. We denote the resulting complete Segal space by $CS(C)$.

We first recall that this completion functor restricts to the classifying diagram in the case where C is a discrete category, a result which follows from Lemma 5.5.8.

Proposition 9.1.2 *If C is a discrete category, then $CS(C)$ is weakly equivalent to NC.*

Let us now recall the variant of the classifying diagram for a category with weak equivalences, as used in the previous chapter. In this case, rather than taking the subcategory iso(M) of isomorphisms of M, we take the subcategory of weak equivalences, denoted by we(M).

Definition 9.1.3 [103, 3.3] Let (M, W) be a category with weak equivalences. Its *classification diagram*, denoted by $N(M, W)$, is defined by

$$N(M, W)_n = \text{nerve}(\text{we}(M^{[n]})).$$

Unlike the classifying diagram, the classification diagram of a category with weak equivalences is not necessarily a complete Segal space, as it is not Reedy fibrant in general.

9.2 Some Results for Simplicial Categories

Before turning to the main results of this chapter, we first consider some results concerning simplicial categories arising as simplicial localizations of model categories.

First, we want to consider a general description of nerves of well-behaved subcategories of a model category. Since model categories are not small categories, their nerves are "large" simplicial sets, in that they are technically not simplicial objects in sets unless we move to a larger set-theoretic universe. We can make the following accommodation.

Definition 9.2.1 [56, 2.2] A simplicial set is *homotopically small* if each of its homotopy groups has a small underlying set.

The following result of Dwyer and Kan [56] shows that large simplicial categories can be replaced by suitably homotopically small ones. By "large" here we mean not only that the category need not have a set of objects, but also that the mapping spaces might also be large simplicial sets.

Proposition 9.2.2 [56, 2.3] *Let C be a large simplicial category, and let $\mathcal{E}$ be a small simplicial subcategory of C. There exists a small simplicial subcategory $\mathcal{D}$ of C which contains $\mathcal{E}$ and such that, for any objects x and y of $\mathcal{D}$, the inclusion $\text{Map}_{\mathcal{D}}(x, y) \to \text{Map}_C(x, y)$ is a weak equivalence of simplicial sets.*

We would like to find a homotopically small replacement for the nerve of a model category, and the following definition is one approach to doing so.

Definition 9.2.3 [54, 1.2] Let M be a model category. A *classification complex* of M is the nerve of any subcategory C of M such that:

1 every map in C is a weak equivalence,
2 if $f : X \to Y$ in M is a weak equivalence and either X or Y is in C, then f is in C, and
3 nerve(C) is homotopically small.

A *special classification complex* $sc(X)$ of an object X in M is a connected classification complex containing X.

Let M be a model category and x a fibrant and cofibrant object of M. Denote by $\text{Aut}^h_{L^H M}(x)$, or simply $\text{Aut}^h(x)$, the simplicial submonoid of $\text{Map}_{L^H M}(x, x)$ consisting of the components which are invertible in $\pi_0 \text{Map}_{L^H M}(x, x)$. We then consider its classifying complex $B \text{Aut}^h(x)$, which we now define.

Definition 9.2.4 Let M be a simplicial monoid. Its *classifying complex BM* is a simplicial set whose geometric realization is the classifying space of M.

Remark 9.2.5 The above definition is perhaps not very enlightening. If M is a discrete monoid, then its classifying complex is simply the nerve of M. Given a simplicial monoid M, we can take its nerve, a simplicial space with

$$\text{nerve}(M)_{n,m} = \text{Hom}([m], M_n).$$

Taking the diagonal of this simplicial space, we obtain a simplicial set, also often called the nerve of G.

From another perspective, a precise construction can be made for this classifying complex via the $\overline{W}G$ construction [62, V.4.4], [91]. However, we are not so concerned here with the precise construction as with the fact that such a classifying space exists, so for simplicity we will simply write BM for the classifying complex of M.

Lastly, we remark that the classifying space of a monoid does not serve a "classification" purpose as does the classifying space of a group. Nonetheless, as a construction it is completely analogous for monoids and for groups, so we continue this abuse of terminology.

The following result gives information about the mapping spaces in LM and their classifying complexes in the special case where W is all of M. As the localization in this case is a groupoid, one should compare these statements to the analogous ones for groupoids.

Proposition 9.2.6 [57, 5.5] *Suppose that* $W = M$ *and* nerve(M) *is connected.*

1 *The simplicial localization LM is a simplicial groupoid, so for all objects x and y, the simplicial sets $\text{Map}_{LM}(x, y)$ are all isomorphic. In particular, the simplicial sets $\text{Map}_{LM}(x, x)$ are all isomorphic simplicial groups.*

2 *The classifying complex* $B\,\mathrm{Map}_{LM}(x, x)$ *has the homotopy type of* $\mathrm{nerve}(M)$, *and thus each simplicial set* $\mathrm{Map}_{LM}(x, y)$ *has the homotopy type of the loop space* $\Omega(\mathrm{nerve}(M))$.

Observe that we can just as easily use the hammock localization of M, since the two constructions give Dwyer–Kan equivalent simplicial categories.

Now, we can introduce a second equivalent approach to the simplicial monoid of self-homotopy equivalences of an object in a model category.

Proposition 9.2.7 [56, 4.6] *Let M be a model category and W its subcategory of weak equivalences. The inclusion of* $\mathrm{Aut}^h_{L^H M}(x)$ *into* $\mathrm{Map}_{L^H W}(x, x)$ *induces a weak equivalence on their classifying complexes.*

Remark 9.2.8 Using the equivalence of the hammock localization and the simplicial localization, the previous result also gives us that $B\,\mathrm{Aut}^h_{L^H M}(x)$ is weakly equivalent to $B\,\mathrm{Map}_{LW}(x, x)$. Since $\mathrm{Map}_{LW}(x, x)$ is actually a simplicial group, rather than a simplicial monoid, we can often say more about its classifying complex.

The following proposition was proved by Dwyer and Kan [54, 2.3] in the case that M is a simplicial model category. However, the proof does not actually require the simplicial structure; in fact, their proof is essentially the one given below, with the extra step showing that the mapping spaces in the hammock localization are equivalent to those given by the simplicial structure of M by Proposition 3.5.10.

Proposition 9.2.9 [25, 6.3] *Let X be an object of a model category M. The classifying complex $B\,\mathrm{Aut}^h(x)$ is weakly equivalent to the special classification complex $sc(x)$, and the two can be connected by a finite zigzag of weak equivalences.*

Proof Let W be the subcategory of weak equivalences of M. Consider the connected component of $\mathrm{nerve}(W)$ containing x. For the rest of this proof, we assume that W is such that its nerve is connected. We further assume that $\mathrm{nerve}(W)$ is homotopically small, taking an appropriate subcategory if necessary, as described in Proposition 9.2.2.

In this case, by Proposition 9.2.6, the mapping spaces $\mathrm{Map}_{LW}(x, x)$ are all isomorphic. Furthermore, by the same result, $B\,\mathrm{Map}_{LW}(x, x)$ has the homotopy type of $\mathrm{nerve}(W)$. Thus, we can take $\mathrm{nerve}(W)$ as $sc(x)$.

Now, as in the statement of the proposition, we take $\mathrm{Aut}^h(x)$ to consist of the components of $\mathrm{Map}_{L^H M}(x, x)$ which are invertible in $\pi_0\,\mathrm{map}\,L^H M(x, x)$. By Proposition 9.2.7 and Remark 9.2.8, $B\,\mathrm{Map}_{LW}(x, x)$ and $B\,\mathrm{Aut}^h(x)$ are weakly equivalent simplicial sets. Since the former models $B\operatorname{Aut}^h(x)$, we can conclude that $sc(x)$ has the same homotopy type as $B\operatorname{Aut}^h(x)$.   □

9.3 Comparison of Functors

We have two functors $SC \to CSS$ given by the two different chains of Quillen equivalences,

$$SC \leftrightarrows SeCat_f \rightleftarrows SeCat_c \rightleftarrows CSS$$

and

$$SC \rightleftarrows QCat \leftrightarrows CSS.$$

It is a consequence of work of Joyal and Tierney [74, §§4 and 5] that the simplicial space obtained from a simplicial category via these functors is weakly equivalent to the one obtained from the previous composite functor. Observe that, for each composite, the resulting simplicial space is not Reedy fibrant in general, and so not a complete Segal space, but applying the fibrant replacement functor L_{CSS} results in a complete Segal space.

We now state the result that establishes the equivalence between the composite of the simplicial nerve with the completion functor $CS : SC \to CSS$ and the functor arising from the chain of Quillen equivalences factoring through the Segal category model structures. Let L_S denote a fibrant replacement functor in the Segal space model structure $SeSp$, and let L_{CSS} denote a fibrant replacement functor in the complete Segal space model structure CSS.

Theorem 9.3.1   [25, 6.1] *If C is a simplicial category, then the complete Segal spaces $CS(C)$ and $L_{CSS}(\mathrm{nerve}(C))$ are weakly equivalent in CSS.*

Proof   Consider the zigzag of maps of simplicial spaces

$$L_{CSS}(\mathrm{nerve}(C)) \leftarrow \mathrm{nerve}(C) \to L_S\,\mathrm{nerve}(C) \to CS(C).$$

The map on the left is the localization functor in CSS and so is a weak equivalence in CSS. The middle map is a weak equivalence in $SeSp$ and therefore also a weak equivalence in CSS, since the latter model category is a localization of the former. The map on the right is given by the completion, and it is a weak equivalence in CSS by construction. Therefore, the objects at the far left and right of this zigzag, both of which are complete Segal spaces, are weakly equivalent as objects of CSS.   □

Now, we look at the classification diagram construction for a model category M. We want to show that $N(M, W)$ is weakly equivalent to $CS(L^H M)$.

An initial problem here is that $N(M, W)$ is not necessarily Reedy fibrant, and therefore may not be a complete Segal space. We prove that a Reedy fibrant replacement of it, denoted by $N(M, W)^f$, is in fact a complete Segal space. We prove the following theorem very similarly to the way Rezk [103] proves it for simplicial model categories, using Proposition 9.2.9.

Theorem 9.3.2 [25, 6.2], [103, 8.3] *Let M be a model category, and let W denote its subcategory of weak equivalences. Then $N(M, W)^f$ is a complete Segal space. Furthermore, for any objects x, y of M, there is a weak equivalence of spaces $\mathrm{map}_{N(M, W)^f}(x, y) \simeq \mathrm{Map}_{L^H(M, W)}(x, y)$, and there is an equivalence of categories $\mathrm{Ho}(N(M, W)^f) \simeq \mathrm{Ho}(M)$.*

Proof For any $n \geq 0$, consider the category $M^{[n]}$ of functors $[n] \to M$, and consider it as a model category with the projective model structure. Observe that, given any functor $[m] \to [n]$, there is an induced functor $M^{[n]} \to M^{[m]}$.

Let $Y = (y_0 \to y_1 \cdots \to y_n)$ be a fibrant and cofibrant object of $M^{[n]}$, and consider its restriction $Y' = (y_0 \to y_1 \cdots \to y_{n-1})$ in $M^{[n-1]}$. Consider the map

$$M^{[n]} \to M \times M^{[n-1]}$$

which takes an object Y as above to the pair (y_n, Y'). It induces a map of simplicial sets

$$B\operatorname{Aut}^h_{L^H M^{[n]}}(Y) \to B\operatorname{Aut}^h_{L^H M}(y_n) \times B\operatorname{Aut}^h_{L^H M^{[n-1]}}(Y').$$

The homotopy fiber of this map is weakly equivalent to the union of those components of $\mathrm{Map}_{L^H M}(y_{n-1}, y_n)$ containing the conjugates of the map f_{n-1} : $y_{n-1} \to y_n$, or maps $j \circ f_{n-1} \circ i$, where i and j are self-homotopy equivalences.

Iterating this process, the homotopy fiber of the map

$$B\operatorname{Aut}^h_{L^H M^{[n]}}(Y) \to B\operatorname{Aut}^h_{L^H M}(y_n) \times \cdots \times B\operatorname{Aut}^h_{L^H M}(y_0)$$

is weakly equivalent to the union of the components of

$$\mathrm{Map}_{L^H M}(y_{n-1}, y_n) \times \cdots \times \mathrm{Map}_{L^H M}(y_0, y_1)$$

containing conjugates of the sequence of maps $f_i \colon y_i \to y_{i+1}$, for each $0 \leq i \leq n - 1$. However, applying Proposition 9.2.9 to the map in question shows that this simplicial set is also the homotopy fiber of the map

$$\mathrm{sc}(Y) \to \mathrm{sc}(y_n) \times \cdots \times \mathrm{sc}(y_0).$$

Let U denote the simplicial space $N(M, W)$, so that $U_n = \mathrm{nerve}(\mathrm{we}(M^{[n]}))$. Then, let V be a Reedy fibrant replacement of U, from which we get weak equivalences $U_n \to V_n$ for all $n \geq 0$.

For each $n \geq 0$, there exists a map $p_n : U_n \to U_0^{n+1} = \mathrm{cosk}_0(U)_n$. Then the homotopy fiber of p_n over any $(x_0, \ldots, x_n) \in U_0^{n+1}$, given by

$$\mathrm{map}_V(x_0, x_1) \times \cdots \times \mathrm{map}_V(x_{n-1}, x_n),$$

is weakly equivalent to

$$\mathrm{Map}_{L^H\mathcal{M}}(x_0^{cf}, x_1^{cf}) \times \cdots \times \mathrm{Map}_{L^H\mathcal{M}}(x_{n-1}^{cf}, x_n^{cf}),$$

where x_i^{cf} denotes a fibrant–cofibrant replacement of x_i in $\mathcal{M}$. It follows that, once we take the Reedy fibrant replacement V of U, it is a Segal space.

Now, consider the set $\pi_0 U_0$, which consists of the weak equivalence classes of objects in $\mathcal{M}$. Observe that we must have $\pi_0 V_0 \cong \pi_0 U_0$. Further, note that

$$\mathrm{Hom}_{\mathrm{Ho}(\mathcal{M})}(x, y) = \pi_0 \, \mathrm{Map}_{L^H\mathcal{M}}(x^{cf}, y^{cf}).$$

It follows that $\mathrm{Ho}(\mathcal{M})$ is equivalent to $\mathrm{Ho}(V)$.

It remains to show that V is a complete Segal space. Consider the space $V_{\mathrm{heq}} \subseteq V_1$, and define U_{heq} to be the preimage of V_{heq} under the fibrant replacement map $U \to V$. Since V is a Reedy fibrant replacement for U, it suffices to show that $U_0 \to U_{\mathrm{heq}}$ is a weak equivalence of simplicial sets. Notice that U_{heq} must consist precisely of the components of U_1 whose 0-simplices come from weak equivalences in $\mathcal{M}$. In other words, $U_{\mathrm{heq}} = \mathrm{nerve}(\mathrm{we}(\mathrm{we}(\mathcal{M}))^{[1]})$.

Consider the adjoint pair of functors

$$F : \mathcal{M}^{[1]} \rightleftarrows \mathcal{M} : G$$

where $F(x \to y) = x$ and $G(x) = \mathrm{id}_x$. This adjoint pair can be restricted to an adjoint pair

$$F : \mathrm{we}(\mathrm{we}(\mathcal{M})^{[1]}) \rightleftarrows \mathrm{we}(\mathcal{M}) : G.$$

Applying the nerve functor, we obtain a weak equivalence of simplicial sets which is precisely the weak equivalence $U_{\mathrm{heq}} \simeq U_0$ that we need. $\square$

Now that we have proved that the mapping spaces and homotopy categories agree for the Reedy fibrant replacement of $N(\mathcal{M}, \mathcal{W})$ and for $L^H\mathcal{M}$, it remains to show that they agree for $L^H\mathcal{M}$ and $CS(L^H\mathcal{M})$.

Theorem 9.3.3 [25, 6.4] *Let $\mathcal{M}$ be a model category. For any objects x and y of $L^H\mathcal{M}$, there is a weak equivalence of simplicial sets*

$$\mathrm{Map}_{L^H\mathcal{M}}(x, y) \simeq \mathrm{map}_{CS(L^H\mathcal{M})}(x, y),$$

and there is an equivalence of categories $\pi_0(L^H\mathcal{M}) \simeq \mathrm{Ho}(CS(L^H\mathcal{M}))$.

Proof   Given the hammock localization $L^H \mathcal{M}$ of the model category $\mathcal{M}$, we have the following composite map of simplicial spaces:

$$\text{nerve}(L^H \mathcal{M}) \to \text{nerve}(L^H \mathcal{M})^f \to CS(L^H \mathcal{M}).$$

For simplicity of notation, let $X = \text{nerve}(L^H \mathcal{M})$, so X^f is its Reedy fibrant replacement.

On the left-hand side, the mapping spaces of $X = \text{nerve}(L^H \mathcal{M})$ are precisely those of $L^H \mathcal{M}$, by the definition of the nerve functor. In the nerve, a mapping space $\text{map}_X(x, y)$, for some objects x and y of $\mathcal{M}$, is given by the fiber over (x, y) of the map $(d_1, d_0) \colon X_1 \to X_0 \times X_0$. Although these mapping spaces can be defined for X, there is no reason that they are homotopy invariant. When we take a Reedy fibrant replacement X^f of X, however, this map becomes a fibration by Proposition 2.6.11, and hence this fiber is actually a homotopy fiber and so homotopy invariant. For a general simplicial space, we cannot assume that the mapping spaces of the Reedy fibrant replacement are equivalent to the original ones. However, if the degree zero space of the simplicial space in question is discrete, then the map above is a fibration. Using an argument similar to the one used in the proof of Proposition 6.3.1, we can find a Reedy fibrant replacement functor which leaves the 0-space discrete. While the space in degree one might be changed in this process of passing to X^f, it is still weakly equivalent to X_1. In particular, the mapping spaces in X^f are weakly equivalent to those in X.

Since the objects of X^f are just the objects of $L^H \mathcal{M}$, which are simply the objects of $\mathcal{M}$, this equivalence of mapping spaces is sufficient to give an equivalence of homotopy categories $\text{Ho}(X) \simeq \text{Ho}(X')$.

The rightmost map of the above composite is given by the completion $i_{X^f} \colon X^f \to \widehat{X^f}$, which is a Dwyer–Kan equivalence. In other words, it induces weak equivalences on mapping spaces and an equivalence of homotopy categories. Thus, the composite map induces equivalences on mapping spaces and an equivalence on homotopy categories.   □

9.4 Complete Segal Spaces From Simplicial Categories

In this section, we give a thorough description of the weak equivalence type of complete Segal spaces which occur as images of Rezk's functor from the category of simplicial categories. We consider several different cases, beginning with ones for which we can use the classifying diagram construction, i.e., discrete categories, and then proceed to more general simplicial categories.

It should be noted that we characterize these complete Segal spaces up to weak equivalence, and so the resulting descriptions are of the homotopy type of the spaces in each simplicial degree. For example, in the case of a discrete category, we describe the corresponding complete Segal space in terms of the isomorphism classes of objects, rather than in terms of individual objects, in order to simplify the description. This characterization might not, then, be suitable for all purposes, but one need not reduce so far as we have here.

Furthermore, notice that determining the homotopy type of the spaces in degrees zero and one is sufficient to determine the homotopy type of all the spaces, since we are considering Segal spaces. Thus, we focus our attention on these spaces, adding in a few comments about how to continue the process with the higher-degree spaces.

For an object x of a simplicial category C, let $\langle x \rangle$ denote the weak equivalence class of x in C, and for a morphism $\alpha \colon x \to y$, let $\langle \alpha \rangle$ denote the weak equivalence class of α in the morphism category $C^{[1]}$. As before, let $\mathrm{Aut}^h(x)$ denote the space of self-maps of x which are invertible in $\pi_0 C$.

Theorem 9.4.1 [25, 7.3] *Let C be a simplicial category. The complete Segal space corresponding to C has the form*

$$\coprod_{\langle x \rangle} B\,\mathrm{Aut}^h(x) \Leftarrow \coprod_{\langle \alpha \colon x \to y \rangle} B\,\mathrm{Aut}^h(\alpha) \Lleftarrow \cdots .$$

Proof   We give the proof by working through successively more complicated cases. The beginning cases have been discussed previously in the discussion of the classifying diagram, but we repeat them here.

Case 1: The category C is a discrete groupoid. If $C = G$ is a group, then applying the classifying diagram construction results in a complete Segal space which equivalent to the constant simplicial space which has the simplicial set BG at each level. In particular, since all morphisms are invertible, we obtain essentially no new information at level 1 that we did not have already at level 0.

If C has more than one object but only one isomorphism class of objects, we get instead a simplicial space weakly equivalent to the constant simplicial space which is $B\,\mathrm{Aut}(x)$ at each level, for a representative object x of the single component of C. If C has more than one isomorphism class $\langle x \rangle$, then we instead get the constant simplicial space $\coprod_{\langle x \rangle} B\,\mathrm{Aut}(x)$.

Case 2: The simplicial category C is a discrete category. Since in the classifying diagram NC, the space $(NC)_0$ encodes the isomorphisms of C only, we still obtain $\coprod_{\langle x \rangle} B\,\mathrm{Aut}(x)$ at level 0. However, if C is not a groupoid, then there is new information at level 1. It is instead weakly equivalent to

$$\coprod_{\langle \alpha \rangle} B\,\mathrm{Aut}(\alpha)$$

where the α index the isomorphism classes of morphisms in C. If we wanted to retain information about the course and target of each α, we could decompose this space further as

$$\coprod_{\langle x \rangle, \langle y \rangle} \coprod_{\langle \alpha : \, x \to y \rangle} B\operatorname{Aut}(\alpha).$$

Let us denote by $(NC)_1(x, y)$ the fiber of the map $(d_1, d_0)\colon (NC)_1 \to (NC)_0 \times (NC)_0$ over (x, y). Then this space also fits into a fibration

$$\operatorname{Hom}(x, y) \to (NC)_1(x, y) \to B\operatorname{Aut}(x) \times B\operatorname{Aut}(y).$$

The space in dimension two is determined, then, by the spaces at levels 0 and 1. The subspace corresponding to isomorphism classes of objects $\langle x \rangle, \langle y \rangle, \langle z \rangle$, denoted $(NC)_2(x, y, z)$, fits into a fibration

$$\operatorname{Hom}(x, y) \times \operatorname{Hom}(y, z) \to (NC)_2(x, y, z) \to B\operatorname{Aut}(x) \times B\operatorname{Aut}(y) \times B\operatorname{Aut}(z).$$

The whole space $(NC)_2$, up to homotopy, looks like

$$\coprod_{\langle x \rangle, \langle y \rangle, \langle z \rangle} B\operatorname{Aut}\left(\coprod_{\langle \alpha : \, x \to y \rangle, \langle \beta : \, y \to z \rangle} \operatorname{Hom}(\alpha) \times \operatorname{Hom}(\beta) \right).$$

We could describe each $(NC)_n$ analogously.

Case 3: The simplicial category C is a simplicial groupoid. First, consider the case where we have a simplicial group G. We now need to use the functor CS rather than the classifying diagram, since we are in the simplicial setting. Let G_n denote the group of n-simplices of G. Then we can write the simplicial nerve of G as a homotopy colimit of its simplices

$$\operatorname{nerve}(G) = \operatorname{hocolim}_{\Delta^{op}}(\operatorname{nerve}(G_n)^t).$$

We claim that

$$CS(\operatorname{hocolim}_{\Delta^{op}}(\operatorname{nerve}(G_n)^t)) \simeq CS(\operatorname{hocolim}_{\Delta^{op}} CS(\operatorname{nerve}(G_n)^t)).$$

We prove the more general statement that, for any simplicial space

$$W = \operatorname{hocolim}_{\Delta^{op}} W_n,$$

there is a weak equivalence

$$CS(\operatorname{hocolim}_{\Delta^{op}} W_n) \simeq CS(\operatorname{hocolim}_{\Delta^{op}} CS\, W_n).$$

To prove this claim, first we know that the completion map

$$\operatorname{hocolim}_{\Delta^{op}} W_n \to CS(\operatorname{hocolim}_{\Delta^{op}} W_n)$$

is a weak equivalence. Furthermore, since in CSS any complete Segal space Z

is a local object and every object is cofibrant, we have a weak equivalence of spaces

$$\text{Map}(CS(\text{hocolim}_{\Delta^{op}} W_n), Z) \simeq \text{Map}(\text{hocolim}_{\Delta^{op}} W_n, Z).$$

So, for any complete Segal space Z, we have that

$$
\begin{aligned}
\text{Map}(CS \, \text{hocolim}_{\Delta^{op}} CS(W_n), Z) &\simeq \text{Map}(\text{hocolim}_{\Delta^{op}} CS(W_n), Z) \\
&\simeq \text{holim}_{\Delta} \text{Map}(CS(W_n), Z) \\
&\simeq \text{holim}_{\Delta} \text{Map}(W_n, Z) \\
&\simeq \text{Map}(\text{hocolim}_{\Delta^{op}} W_n, Z) \\
&\simeq \text{Map}(CS \, \text{hocolim}_{\Delta^{op}} W_n, Z).
\end{aligned}
$$

Note that the above calculation depends on the fact that

$$\text{Map}(\text{hocolim}_{\Delta^{op}} W_n, Z) \simeq \text{holim}_{\Delta} \text{Map}(W_n, Z),$$

which follows from working levelwise on simplicial sets.

Let us return to the special case of a simplicial group. Since G_n is a discrete group, completing its nerve is equivalent to taking the classifying diagram NG_n which, by Case 1, is weakly equivalent to the constant simplicial space BG_n. Thus we have

$$
\begin{aligned}
CS(\text{nerve}(G_n)) &\simeq CS[\text{hocolim}_{\Delta^{op}}(\text{nerve}(G_n))] \\
&\simeq CS[\text{hocolim}_{\Delta^{op}}(CS(\text{nerve}(G_n)))] \\
&\simeq CS[\text{hocolim}_{\Delta^{op}}(BG_n)] \\
&\simeq CS(BG) \\
&\simeq BG.
\end{aligned}
$$

So, we obtain a simplicial space weakly equivalent to the constant simplicial space with BG at each level. If we have a simplicial groupoid, rather than a simplicial group, we obtain the analogous result, replacing BG with

$$\coprod_{\langle x \rangle} B\,\text{Aut}(x).$$

Case 4: The simplicial category C has every morphism invertible up to homotopy. Alternatively stated, this case covers the situation in which $\pi_0(C)$ is a groupoid.

Recall from Theorem 4.2.4 the model structure SC_O on the category of categories with a fixed object set O, in which the cofibrant objects are retracts of free objects. Recall that taking a cofibrant replacement of C in this model structure SC_O gives a free replacement of C, denoted here by $F(C)$, which is weakly equivalent to C. Taking the localization of $F(C)$ with respect to all morphisms results in a simplicial groupoid. So, we have Dwyer–Kan equivalences

$$F(C)^{-1}F(C) \xleftarrow{\simeq} F(C) \xrightarrow{\simeq} C.$$

But $F(C)^{-1}F(C)$ is a simplicial groupoid weakly equivalent to C, so we have now reduced this situation to Case 3.

Note that, to write down a description of this complete Segal space in terms of the original category C, we need to take isomorphism classes of objects in $\pi_0(C)$, or weak equivalence classes, as well as self-maps which are invertible up to homotopy rather than strict automorphisms. While we still use $\langle x \rangle$ to denote the equivalence class of a given object, we use $\mathrm{Aut}^h(x)$ to signify homotopy automorphisms of x. Thus, the complete Segal space corresponding to C in this case is weakly equivalent to

$$\coprod_{\langle x \rangle} B\,\mathrm{Aut}^h(x)$$

at each level.

Case 5: The simplicial category C is arbitrary. First consider the subcategory of C containing all the objects of C and only the morphisms of C which are invertible up to homotopy. Apply Case 5 to get a complete Segal space, but take only the 0-space of it.

To find the space in degree one, first recall the definition of the completion functor as applied to a Segal space W:

$$\widehat{W} = L_{CSS}\widetilde{W}$$

where

$$\widetilde{W}_n = \mathrm{diag}([m] \mapsto (W^{E(m)^t})_n).$$

Recall further that $(W^{E(m)^t})_n = \mathrm{Map}(E(m)^t \times \Delta[n]^t, W)$. Thus, the Segal space we obtain (before applying the functor L_{CSS}) looks like

$$\mathrm{Map}(E(0)^t \times \Delta[0]^t, W) \Leftarrow \mathrm{Map}(E(1)^t \times \Delta[1]^t, W) \Leftleftarrows \mathrm{Map}(E(2)^t \times \Delta[2]^t, W) \cdots.$$

If the Segal space W is (a fibrant replacement of) the simplicial nerve snerve(C), then W_1 has n-simplices given by diagrams

$$
\begin{array}{ccc}
x & \longrightarrow & y \\
{\scriptstyle \simeq}\downarrow & & \downarrow{\scriptstyle \simeq} \\
x' & \longrightarrow & y'
\end{array}
$$

in the category C_n.

We can think of these diagrams as given by equivalences in the category $C_n^{[1]}$. Observe that, ranging over all $n \geq 0$, these categories do not assemble to a simplicial category, since their objects do not agree, but they do form a simplicial object in categories, and the definition of this simplicial nerve can be extended to this setting. Now we are back in the context in which every morphism is invertible up to homotopy. Verifying that the previous methods still work in this more general context of simplicial objects and categories, we can conclude that the space we obtain in degree one is

$$\coprod_{\langle x \rangle, \langle y \rangle} \coprod_{\langle \alpha : x \to y \rangle} B \operatorname{Aut}^h (\alpha).$$

The spaces in higher degrees can be described similarly to those in Case 2, using homotopy automorphisms. □

10

Variants on $(\infty, 1)$-Categories

In this final chapter, we look very briefly at where we can go from here. The notions we have considered can be extended in a number of ways, for example to operads or to higher categories. We can also impose additional structures on $(\infty, 1)$-categories, such as stability, or look at restrictions to simpler approximations. The topics we consider here are by no means comprehensive, either in their diversity or in detail.

10.1 Finite Approximations

In this section, we consider ways to look at simpler approximations to $(\infty, 1)$-categories. The narrative here is based on lecture notes by David Blanc and primarily considers work of his with a number of collaborators [31, §13]. Here, we work primarily in the setting of simplicial categories.

We have seen that $(\infty, 1)$-categories, in the form of different models, arise naturally from model categories, and that translating from the framework of model categories to that of $(\infty, 1)$-categories is often useful as it allows for more flexibility. When making computations in a model category, for practical purposes one might want to replace that model category with some smaller version, obtained either by limiting the objects or by simplifying the mapping spaces in some way. (In the last chapter, for example, we discussed taking homotopically small replacements.) One approach to simplifying mapping spaces is to take their nth Postnikov approximations. Since this construction defines a monoidal functor in $SSets$, it can be extended nicely on the level of mapping spaces in SC.

Let C be a small simplicial category with object set O. If $\mathcal{V}$ is a monoidal category and $F\colon SSets \to \mathcal{V}$ is a monoidal functor, we can obtain a category enriched in $\mathcal{V}$ with object set O. For example, let P^n be the nth Postnikov section $P^n \colon SSets \to P^n SSets$, where $P^n SSets$ denotes the category of simplicial sets with trivial homotopy groups above degree n. Then $P^n C$ is a category enriched in $P^n SSets$ with object set O; we refer to such categories as $(P^n SSets, O)$-*categories*.

We make the following observations.

- The category $P^0 C$ is equivalent to $\pi_0 C$, a category with object set O, but it is cofibrant as a simplicial category. In fact, it looks like $F_*(\pi_0 C)$.
- For $n = 1$, consider the adjoint functors

$$\widehat{\pi}_1 \colon SSets \xrightarrow{\longleftarrow} Gpds \colon \text{nerve}$$

  where $Gpds$ denotes the category of small groupoids and $\widehat{\pi}_1$ agrees with the fundamental groupoid functor on Kan complexes. Since these functors commute with products, they can be extended to an adjoint pair between the category of simplicial categories and the category of categories enriched in groupoids, or *track categories*. Then $P^1 C$ is equivalent to $\widehat{\pi}_1 C$ via applying the nerve of mapping spaces [20].
- The enriched category $P^2 C$ is equivalent to a *double track category* or category enriched in a certain type of double groupoid [35].
- Higher-order models are only conjectural.

In the context of Postnikov sections of simplicial sets, one considers k-*invariants*, which are characteristic classes classifying the various Postnikov sections. Likewise, we can consider the k-invariants of $(SSets, O)$-categories. As a result, $(P^n SSets, O)$-categories are important as they allow us to replace one model by (a hopefully simpler) one with the same weak homotopy type, since it is constructed inductively using essentially the same k-invariants. Furthermore, they possess homotopy invariant information themselves.

However, even $(P^n S, O)$-categories can still be unwieldy, so one additional remedy is to restrict the object sets as well. This restriction leads to the notion of a mapping algebra in the sense of [16, §9], which can be defined as follows. Let C be a simplicial model category and A an object of C. Define C_A to be the full subsimplicial category of C generated by A under suspensions and coproducts.

Definition 10.1.1 Let A be an object of a simplicial model category C. An A-*mapping algebra* is a simplicial functor $X \colon C_A \to SSets$. An A-mapping algebra is *realizable* if it takes an object A' of C_A to the mapping space $\mathrm{Map}_C(A', Y)$ for some fixed object Y of C.

We denote such a realizable mapping algebra by $\mathcal{M}_A Y$.

Definition 10.1.2   An A-mapping algebra is *realistic* if it preserves limits.

In particular, if X is a realistic A-mapping algebra, then $X(\Sigma A') = \Omega X(A')$ for any object A' of C_A, and $X(\vee A_i) = \prod_i X(A_i)$ where each A_i is an object of C_A. Furthermore, a realistic A-mapping algebra X is determined by the simplicial set $X(A)$.

Example 10.1.3   Let $C = \mathcal{T}op_*$, the category of pointed topological spaces, and $A = S^1$. Then $\mathcal{M}_{S^1} Y = \Omega Y$ for all Y, which has an A_∞-structure.

Any realistic S^1-mapping algebra X can be realized in this way, so that $X(A) = \Omega Y$ for some Y. When $A = S^k$, $\mathcal{M}_A Y$ is a k-fold loop space, equipped with an action of all mapping spaces between (wedges of) spheres on it and its iterated loops. A realistic A-mapping algebra is any space Z equipped with such an action. Since this action encodes the $E(k)$ structure, we see that any realistic A-mapping algebra is realizable.

For more general A, even $A = S^1 \vee S^2$, the situation is more complicated, as the purely algebraic approach of analyzing the group structure does not work. However, by enhancing the structure of an A-mapping algebra suitably, one can still recover Y from X up to A-equivalence [16, §10].

One can also map into A and its products and loop spaces to obtain a dual definition. The realizable version is denoted by $\mathcal{M}^A Y$.

We can think of these mapping algebras as simplicial categories as follows. If $O = \mathrm{ob}(C_A)$ and $O^+ = O \cup \{*\}$, then an A-mapping algebra X is just a $(SSets, O^+)$-category such that $X|_O = C_A$ and $\mathrm{Map}_X(*, -) = *$. If we apply the nth Postnikov section functor to an A-mapping algebra, we obtain an *n-A-mapping algebra*, which is defined to be a $(P^n SSets, O^+)$-category extending $P^n C_A$ as above.

Example 10.1.4   When $n = 0$, we obtain a $\pi_0 C_A$-algebra. If A is a homotopy cogroup object, then the result is an algebraic theory in the sense of Lawvere [82].

Example 10.1.5   If Y is a connected space, then $\pi_0 \Omega^k Y = \pi_k Y$, so for $A = S^1$, $P^0 \mathcal{M}_A Y$ encodes the homotopy groups of Y.

We would like to consider n-realistic A-mapping algebras. In particular, if X is realistic, we want to know if we can say anything about $P^n X$. In fact, in this case we recover an *n-stem* for Y [18, §1]. For example, $P^{n+k} X \langle k - 1 \rangle$ is the $(k - 1)$-connective cover of $P^{n+k} X$ whose homotopy groups are only in the range from k to $n + k$. However, since we have no way to deloop, there is no way to recover all of Y.

As an application, derived functors of realizable mapping algebras can be used to describe terms in the Adams spectral sequence [17, 18, 19].

10.2 Stable (∞, 1)-Categories

A natural question to ask is whether (∞, 1)-categories, in whatever model, can be equipped with additional structure. Here, we look at what it means for an (∞, 1)-category to be stable, but other possibilities include monoidal (∞, 1)-categories or (∞, 1)-categories enriched in some monoidal category. Many of these structures are addressed by Lurie [87] in the context of quasi-categories, and it is from there that we take the definition of stable quasi-categories below.

A starting point for this theory is the definition of a stable model category, which is taken to be a model category whose homotopy category naturally has the structure of a triangulated category [71, §7]. A central example in homotopy theory is that of spectra; like (∞, 1)-categories, there are many different models for spectra, all of which are given by a stable model category. Since each model for (∞, 1)-categories comes with an associated definition of homotopy category, we could take the same definition, that an (∞, 1)-category is stable if its homotopy category is triangulated. However, it is helpful to look in more detail at some of the structure which makes the homotopy category triangulated. Along the way, we get some glimpses of what category theory looks like in the context of quasi-categories. In particular, we refer to 0-simplices of a quasi-category as *objects* and to 1-simplices as *morphisms*.

Let us first review what it means for an ordinary category to be triangulated.

Definition 10.2.1   A category C is *additive* if

1 it is pointed, i.e., it has a zero object which is both initial and terminal,
2 it admits finite products, and
3 for any objects x and y of C, the set $\mathrm{Hom}_C(x, y)$ has the structure of an abelian group.

It is a consequence of these conditions that finite products and finite coproducts coincide, so in particular C admits finite coproducts.

Definition 10.2.2   [118, 10.2.1] An additive category $\mathcal{T}$ is *triangulated* if it is equipped with a shift functor $\Sigma \colon \mathcal{T} \to \mathcal{T}$ which is an equivalence of categories and a collection of *distinguished triangles*

$$x \xrightarrow{f} y \xrightarrow{g} z \xrightarrow{h} \Sigma x,$$

satisfying four axioms.

Now, we would like to describe quasi-categories whose homotopy categories have this structure. The first feature of a triangulated category is that it is pointed, so that it has an object which is both initial and terminal. We want to translate a notion of initial and terminal object into the homotopical setting of a quasi-category. In particular, an initial object need not have a unique map to every other object, but only a map which is unique up to homotopy.

Definition 10.2.3   Let K be a quasi-category. An object $x \in X_0$ is *initial* if the mapping space $\mathrm{Map}_K(x, y)$ is contractible for every object y of K. Dually, an object x is *terminal* if the mapping space $\mathrm{Map}_K(y, x)$ is contractible for every object y of K. A *zero object* of K is an object which is both initial and terminal. If X has a zero object, it is *pointed*.

In passing to the homotopy category, an initial object in a quasi-category becomes an initial object in the usual sense, and similarly for terminal objects and zero objects. Now that we have pointed quasi-categories, we need a shift functor and distinguished triangles. We first define what we mean by triangles.

Definition 10.2.4   Let K be a pointed quasi-category with zero object 0.

1 A *triangle* in K is given by a diagram

2 A *fiber sequence* in K is a triangle whose associated diagram is a pullback.
3 A *cofiber sequence* in K is a triangle whose associated diagram is a pushout.

Observe that this definition requires us to know what a pushout or a pullback is in the context of a quasi-category. The idea is that, like the example of initial and terminal objects, limits and colimits in quasi-categories satisfy appropriate universal properties up to homotopy. Precise definitions can be found in Lurie [88, 1.2.13].

Since we have a notion of a fiber or a cofiber sequence, we can define fibers or cofibers of a given morphism in a quasi-category.

Definition 10.2.5   Let $g \colon x \to y$ be a morphism in a quasi-category K.

1 A *fiber* of g is the pullback w in a fiber sequence

$$
\begin{array}{ccc}
w & \longrightarrow & x \\
\downarrow & & \downarrow{\scriptstyle g} \\
0 & \longrightarrow & y.
\end{array}
$$

2 A *cofiber* of g is the pushout z in a cofiber sequence

$$
\begin{array}{ccc}
x & \overset{g}{\longrightarrow} & y \\
\downarrow & & \downarrow \\
0 & \longrightarrow & z.
\end{array}
$$

Definition 10.2.6   A quasi-category K is *stable* if:

1 it is pointed,
2 every morphism in K admits a fiber and a cofiber, and
3 a triangle in K is a fiber sequence if and only if it is a cofiber sequence.

Now we are ready to describe the shift functor. Let x be an object of a stable quasi-category K. Take a cofiber of a map $x \to 0$ and denote it by Σx; this construction defines a functor $\Sigma\colon K \to K$. Dually, one could take a fiber of a map $0 \to x$ and denote it by Ω. The functors Σ and Ω define inverse equivalences on K, similarly to the suspension and loop functors on the category of spectra. More explicit details can be found in Lurie [87].

Lastly, we define distinguished triangles.

Definition 10.2.7   A *distinguished triangle* in a stable quasi-category is given by a diagram

$$
\begin{array}{ccccc}
x & \overset{f}{\longrightarrow} & y & \longrightarrow & 0 \\
\downarrow & & \downarrow{\scriptstyle g} & & \downarrow \\
0' & \longrightarrow & z & \overset{h}{\longrightarrow} & \Sigma x
\end{array}
$$

in which: the objects 0 and $0'$ are zero objects, and both squares are pushout diagrams. In particular, the outer rectangle is also a pushout diagram.

The following theorem is the first main result of Lurie [87].

Theorem 10.2.8   *Suppose K is a stable quasi-category. Then its homotopy category is a triangulated category, with shift functor induced by the functor Σ and distinguished triangles induced by those given in the above definition.*

Some important examples of stable quasi-categories are the quasi-category
of spectra, which can be obtained via a composite of the simplicial localization
and coherent nerve functors on any model category of spectra, and a quasi-
categorical version of the derived category of an abelian category, whose homotopy category recovers the classical derived category.

10.3 Dendroidal Objects

One generalization of $(\infty, 1)$-categories is to $(\infty, 1)$-operads. A colored operad,
or multi-category, is a generalization of a category in which morphisms are per-
mitted to have multiple (or no) input objects but still only one output object.
There are two primary approaches to $(\infty, 1)$-operads currently in the literature.
The first is the dendroidal approach, where the indexing category Δ is replaced
by a category Ω of trees, defined by Moerdijk and Weiss [93, 94]. There are
analogues of quasi-categories, called dendroidal sets, as well as analogues
of Segal categories and complete Segal spaces given by replacing Δ with Ω
and making appropriate changes to the Segal condition. Work of Cisinski and
Moerdijk [45, 46, 47] establishes that they all have Quillen equivalent model
structures which are in turn Quillen equivalent to the model structure on simplicial colored operads.

The second approach has been defined by Lurie [87]. Rather than modifying
the diagram Δ, this method consists in adding structure to a quasi-category
in the form of markings. However, the appropriate model category has been
shown to be equivalent to the dendroidal set model in recent work of Heuts,
Hinich, and Moerdijk [68], at least in the nonunital setting. Further work in
this direction is being done by Barwick [9] and by Chu, Haugseng, and Heuts
[43].

Here, we primarily focus on the dendroidal approach, as analogies can be
made fairly easily with the contents of this book. For a more extended survey,
we suggest the notes of Moerdijk [92].

Let us begin with a brief description of simplicial colored operads. A *simplicial (symmetric) colored operad* $\mathcal{P}$ is given by a set of colors C and, for each
$(n + 1)$-tuple $(c_1, \ldots, c_n; c)$ of elements of C, a simplicial set $\mathcal{P}(c_1, \ldots, c_n; c)$,
together with appropriate composition maps. The idea is that $\mathcal{P}(c_1, \ldots, c_n; c)$
consists of the operations whose inputs are the colors $c_1, \ldots, c_n$ and whose output is c. Such operations can be composed when colors match appropriately,
and there are unit maps $\Delta[0] \to \mathcal{P}(c, c)$ for all $c \in C$, as well as an action of
the symmetric group Σ_n on each $(c_1, \ldots, c_n; c)$. Composition is required to be
suitably equivariant, and associativity and unit axioms hold as usual.

Theorem 10.3.1 [46, 1.14], [108] *There is a model structure on the category of simplicial colored operads which is analogous to the model structure on simplicial categories.*

In particular, weak equivalences are defined via Dwyer–Kan equivalences, or morphisms which are suitably fully faithful and essentially surjective, and fibrations are defined by fibrations on spaces of operations but also satisfy an appropriate lifting condition.

But like simplicial categories, simplicial colored operads have limitations in practice due to their rigidity. We would like to have counterparts to quasi-categories, Segal categories, and complete Segal spaces which model operadic rather than categorical structures, and for which composition is only defined up to homotopy.

The idea behind dendroidal objects is to find a generalization of the category Δ suitable for working with colored operads rather than categories. The objects of this category are given by certain kinds of trees.

A *tree* is a (nonplanar) graph with no cycles. A tree must have at least one edge, specified as the *root*; edges meet at a vertex, but edges which only meet other edges on one end need not have a vertex on the other end. The root is always such an edge; any others are called *leaves*. The data of a tree consists of both the graph and the choice of root. As an example, for $n \geq 0$, the *corolla* C_n is the tree with one vertex and $n + 1$ edges, one of which is specified as the root.

Just as the ordered set $[n]$ can be thought of as a category freely generated by the ordering of the elements, from any tree T we can obtain a free colored operad. The set of colors is just the set of edges, so that each edge has a distinct color, and we have one generator for each vertex v. Choosing an order for the incoming edges $e_1, \ldots, e_n$, the generator v is in $(e_1, \ldots, e_n; e)$ where e is the outgoing edge of v. This colored operad is denoted by $\Omega(T)$.

The category Ω, defined to be the full subcategory of the category of colored operads with objects the free colored operads generated by trees, as described above, is the desired analogue of Δ for operads.

Linear trees with n vertices and $n + 1$ edges correspond to the objects of Δ by taking the edges of a linear tree to elements of a set with $n + 1$ elements. Thus, Δ is naturally a subcategory of Ω.

Definition 10.3.2 A *dendroidal set* is a functor $\Omega^{op} \to Sets$. More generally, if C is a category then a *dendroidal object in C* is a functor $\Omega^{op} \to C$.

We denote by *dSets* the category of dendroidal sets.

The following model structure is the analogue of the quasi-category model structure.

Theorem 10.3.3 [45, 46] *There is a model structure on the category of dendroidal sets in which the fibrant objects satisfy an inner Kan condition. Furthermore, there is a Quillen equivalence between this model category and the model structure on simplicial colored operads.*

There are also dendroidal analogues of Segal categories and complete Segal spaces, whose underlying objects are *dendroidal spaces*, or functors $\Omega^{op} \rightarrow SSets$. We denote by *dsSets* the category of dendroidal spaces. Notice that any simplicial set or dendroidal set can be regarded as a dendroidal space by taking it to be constant in the dendroidal or simplicial direction, respectively; in this way we consider $\Omega[S]$ as a discrete dendroidal space. A *Segal preoperad* is a dendroidal space X for which the image of the tree consisting only of the root is a discrete simplicial set.

Theorem 10.3.4 [47] *There is a model structure on the category of Segal preoperads in which the fibrant objects are the Segal operads. There is also a model structure on the category of all dendroidal spaces in which the fibrant objects are complete Segal. They are Quillen equivalent to one another and to the dendroidal set model structure.*

One might ask, given that dendroidal sets are developed as analogues of quasi-categories, whether there are conditions that make them more like spaces. Bašić and Nikolaus show that there is indeed a notion of Kan dendroidal sets, which model connective spectra [14].

10.4 Higher (∞, n)-Categories

In this section we look at some of the ways that the approaches to $(\infty, 1)$-categories can be generalized to models for more general (∞, n)-categories. A guiding principle in this work is that an (∞, n)-category should, in some sense, be a category enriched in $(\infty, n - 1)$-categories.

From that perspective, we could consider categories enriched in any of the models for $(\infty, 1)$-categories that we have presented. Thus, we obtain several different models for $(\infty, 2)$-categories. However, they are not all optimal from the perspective of homotopy theory. The model structure for simplicial categories, for example, is not cartesian closed, so we do not expect to be able to define a model structure on the category of small categories enriched in simplicial categories. Even if we took enriched categories in one of the models which is cartesian, such as complete Segal spaces, we would not expect a cartesian model structure and therefore further iteration would not produce enriched structures that could be equipped with a model structure.

Furthermore, part of the motivation for weaker models, even in the case of $(\infty, 1)$-categories, was that natural examples do not satisfy strict enrichment. Hence, even aside from homotopy-theoretic concerns there is reason to want weaker models. Therefore, we want to construct higher-order versions of the models which already have weak enrichment: quasi-categories, Segal categories, and complete Segal spaces.

We first consider Θ_n-spaces as higher-order complete Segal spaces. We begin by recalling the definition of the Θ-construction, as first described by Berger [23]. Let C be a small category, and define ΘC to be the category with objects $[m](c_1, \ldots, c_m)$ where $[m]$ is an object of Δ and each c_i is an object of C. A morphism

$$[m](c_1, \ldots, c_m) \to [q](d_1, \ldots, d_q)$$

is given by $(\delta, \{f_{ij}\})$ where $\delta \colon [m] \to [q]$ in Δ and $f_{ij} \colon c_i \to d_j$ are morphisms in C indexed by $1 \le i \le m$ and $1 \le j \le q$ where $\delta(i-1) < j \le \delta(i)$ [102, 3.2].

Inductively, let Θ_0 be the terminal category with a single object and no nonidentity morphisms, and then define $\Theta_n = \Theta\Theta_n$. Note that $\Theta_1 = \Delta$. The categories Θ_n have also been studied in unpublished work of Joyal, using a more direct definition.

Looking at the case of Θ_2, we can think of objects as objects of Δ whose arrows are labeled by other objects of Δ; for example, $[4]([2], [3], [0], [1])$ can be depicted as

$$0 \xrightarrow{[2]} 1 \xrightarrow{[3]} 2 \xrightarrow{[0]} 3 \xrightarrow{[1]} 4 \,,$$

but since these labels can also be interpreted as strings of arrows, we get a diagram such as

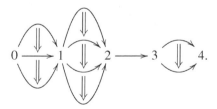

The elements of this diagram can be regarded as generating a strict 2-category by composing 1-cells and 2-cells whenever possible. In other words, the objects of Θ_2 can be seen as encoding all possible finite compositions that can

take place in a 2-category, much as the objects of Δ can be thought of as listing all the finite compositions that can occur in an ordinary category.

We can consider functors $X\colon \Theta_n^{op} \to \mathit{Sets}$ and ask that they satisfy higher-order analogues of the Segal and completeness conditions. Let us look, for example, at the object $[4]([2],[3],[0],[1])$ of Θ_2 that we considered above. One Segal condition on X would require that

$$X([4]([2],[3],[0],[1]))$$

$$\simeq X([1]([2])) \times_{X[0]} X([1]([3])) \times_{X[0]} X([1]([0])) \times_{X[0]} X([1]([1])).$$

In other words, we can break up the horizontal composition. However, another Segal condition gives us vertical composition, so that, for example,

$$X([1]([2])) \simeq X([1]([1])) \times_{X([1]([0]))} X([1]([1])).$$

The completeness conditions are also analogous to the one for complete Segal spaces. Indeed, the first one is essentially the same, saying that $X[0]$ should be weakly equivalent to the subspace of $X([1]([0]))$ consisting of homotopy equivalences. The second is similar, but up one dimension; it says that $X([1]([0]))$ is weakly equivalent to the subspace of $X([1]([1]))$ consisting of homotopy equivalences.

Of course, here we have only given an idea, and making these conditions precise is quite technical, especially for general values of n. Nonetheless, we give the following heuristic definition.

Definition 10.4.1 A functor $X\colon \Theta_n \to \mathit{SSets}$ is a Θ_n-*space* if it satsifies Segal and completeness conditions at all levels.

We can put Θ_n-spaces into the model category framework as follows.

Theorem 10.4.2 [102, 8.1] *There is a cartesian model structure, denoted by* $\Theta_n Sp$, *on the category of functors* $\Theta_n^{op} \to \mathit{SSets}$ *in which the fibrant objects are the* Θ_n-*spaces.*

As complete Segal spaces are known to be equivalent to simplicial categories, establishing them as models for $(\infty, 1)$-categories, $\Theta_{n+1} Sp$ should be Quillen equivalent to a model category whose objects are categories enriched in $\Theta_n Sp$, further strengthening the view that they are indeed models for $(\infty, n+1)$-categories.

The existence of the appropriate model structure for enriched categories can be regarded as a special case of a result of Lurie [88, A.3.2.4]. It is the natural generalization of the model structure SC for simplicial categories.

Theorem 10.4.3 [32, 3.11] *There is a cofibrantly generated model structure on the category* $\Theta_n Sp - Cat$ *of small categories enriched in* $\Theta_n Sp$ *in which the weak equivalences* $f\colon C \to \mathcal{D}$ *are given by the appropriate analogues of Dwyer–Kan equivalences.*

We claim that there is a chain of Quillen equivalences connecting $\Theta_n S p\text{–}Cat$ to $\Theta_{n+1} S p$:

$$\Theta_n S p - Cat \leftrightarrows SeCat_f(\Theta_n S p) \rightleftarrows SeCat_c(\Theta_n S p) \rightleftarrows CSS(\Theta_n S p) \leftrightarrows \Theta_{n+1} S p.$$

Definition 10.4.4 A *Segal precategory* in $\Theta_n S p$ is a functor $X \colon \Delta^{op} \to \Theta_n S p$ such that X_0 is a discrete object in $\Theta_n S p$, i.e., a constant Θ_n-diagram of sets. It is a $\Theta_n S p$-*Segal category* if, additionally, the Segal maps

$$\varphi_k \colon X_k \to \underbrace{X_1 \times_{X_0} \cdots \times_{X_0} X_1}_{k}$$

are weak equivalences in $\Theta_n S p$ for all $k \geq 2$.

Theorem 10.4.5 [32, 6.9, 6.12] *There are two cofibrantly generated model structures* $SeCat_c(\Theta_n S p)$ *and* $SeCat_f(\Theta_n S p)$ *on the category of Segal precategories in* $\Theta_n S p$ *whose fibrant objects are Segal categories in* $\Theta_n S p$. *The former has all monomorphisms as cofibrations, whereas the latter has fewer cofibrations. The identity functor induces a Quillen equivalence between them.*

Theorem 10.4.6 [32, 7.6] *A higher-order version of the simplicial nerve functor induces a Quillen equivalence between* $(\Theta_n S p)$*–Cat and* $SeCat_f(\Theta_n S p)$.

To complete the chain of Quillen equivalences, we need to understand complete Segal objects in $\Theta_n S p$. The rough idea is to impose Segal and completeness conditions on functors $X \colon \Delta^{op} \to \Theta_n S p$. This structure is perhaps the most subtle to define precisely, which we do not do here.

Theorem 10.4.7 [33] *There is a simplicial model structure* $CSS(\Theta_n S p)$ *on the category of functors* $\Delta^{op} \times \Theta_n^{op} \to SSets$ *in which the fibrant objects are the complete Segal space objects in* $\Theta_n S p$.

We can continue our chain of Quillen equivalences as follows.

Theorem 10.4.8 [33] *There is a Quillen equivalence between* $SeCat_c(\Theta_n S p)$ *and* $CSS(\Theta_n S p)$.

Finally, we need to establish a Quillen equivalence between $CSS(\Theta_n S p)$ and $\Theta_{n+1} S p$. To do so, let us first consider the functor $d \colon \Delta \times \Theta_n \to \Theta_{n+1}$ defined by $([n], \theta) \mapsto [n](\theta, \ldots, \theta)$. We get an induced functor $d^* \colon SSets^{\Theta_{n+1}^{op}} \to SSets^{\Delta^{op} \times \Theta_n^{op}}$ and its right Kan extension d_*.

Corollary 10.4.9 [33] *The adjoint pair* (d^*, d_*) *induces a Quillen equivalence of localized model categories*

$$d_* \colon CSS(\Theta_{n-1} S p) \leftrightarrows \Theta_n S p \colon d^*.$$

However, this Quillen equivalence can be extended to a chain, by iterating the application of the adjoints (d_*, d^*).

Observe that the functor d can be iterated to obtain a chain of functors connecting Δ^n and Θ_n:

$$\Delta^n \to \Delta^{n-1} \times \Delta \to \Delta^{n-2} \times \Theta_2 \to \cdots \to \Delta \times \Theta_{n-1} \to \Theta_n.$$

This chain induces a string of Quillen pairs

$$SSets^{(\Delta^{op})^n} \leftrightarrows SSets^{(\Delta^{op})^{n-1} \times \Delta^{op}} \leftrightarrows \cdots \leftrightarrows SSets^{\Theta_n^{op}}$$

on the level of injective model structures.

Imposing appropriate Segal and completeness conditions, let us write the corresponding chain of localized model structures

$$CSS^n(SSets) \leftrightarrows CSS^{n-1}(\Theta_1 S p) \leftrightarrows \cdots \leftrightarrows CSS(\Theta_{n-1} S p) \leftrightarrows CSS^0(\Theta_n S p).$$

Observe that, in the case where $i = 0$, we have $CSS^0(\Theta_{n-1} S p) = \Theta_{n-1} S p$. At the other extreme, if $i = n$, this description of the fibrant objects coincides with the Barwick–Lurie definition of n-fold complete Segal spaces [89].

Corollary 10.4.10   [33] *There is a chain of Quillen equivalences*

$$CSS^n(SSets) \leftrightarrows CSS^{n-1}(\Theta_1 S p) \leftrightarrows \cdots \leftrightarrows CSS(\Theta_{n-1} S p) \leftrightarrows \Theta_n S p.$$

One might ask if there is a corresponding analogue of quasi-categories. This question was asked by Joyal and led to the first definition of the categories Θ_n, but trying to describe the correct horn-filling-type conditions proved to be difficult. The problem was resolved by Ara, who induced a model structure for Θ_n-sets from the model structure for Θ_n-spaces, in such a way that the two are Quillen equivalent [1].

Here we have only given a route through the various approaches. Other approaches include the Segal n-categories of Hirschowitz and Simpson [70] and Pellissier [97], the axiomatic approach of Barwick and Schommer-Pries [13], and the n-relative categories of Barwick and Kan [12].

References

[1] Ara, D., Higher quasi-categories vs higher Rezk spaces, *J. K-Theory* 14 (2014), no. 3, 701–749.

[2] Ara, D., On the homotopy theory of Grothendieck ∞-groupoids, *J. Pure Appl. Algebra* 217 (2013), 1237–1278.

[3] Ara, D., and Métayer, F., The Brown–Golasiński model structure on strict ∞-groupoids revisited, *Homology, Homotopy Appl.* 13 (2011), no. 1, 121–142.

[4] Awodey, S., *Category Theory, Oxford Logic Guides 49*, Clarendon Press, 2006.

[5] Ayala, D., Francis, J., and Rozenblyum, N., Factorization homology I: higher categories (2015), available at arXiv:1504.04007.

[6] Ayala, D., Francis, J., and Rozenblyum, N., A stratified homotopy hypothesis (2015), available at arXiv:1502.01713.

[7] Badzioch, B., Algebraic theories in homotopy theory, *Ann. of Math. (2)* 155 (2002), no. 3, 895–913.

[8] Baez, J.C., and Dolan, J., Higher-dimensional algebra. III, *n*-categories and the algebra of opetopes, *Adv. Math.* 135 (1998), no. 2, 145–206.

[9] Barwick, C., From operator categories to topological operads (2013), available at arXiv:1302.5756.

[10] Barwick, C., On left and right model categories and left and right Bousfield localizations, *Homology, Homotopy Appl.* 12 (2010), no. 2, 245–320.

[11] Barwick, C., and Kan, D.M., Relative categories: another model for the homotopy theory of homotopy theories, *Indag. Math. (N.S.)* 23 (2012), no. 1–2, 42–68.

[12] Barwick, C. and Kan, D.M., *n*-relative categories: a model for the homotopy theory of *n*-fold homotopy theories, *Homology, Homotopy Appl.* 15 (2013), no. 2, 281–300.

[13] Barwick, C., and Schommer-Pries, C., On the unicity of the homotopy theory of higher categories (2011), available at arXiv:1112.0040.

[14] Bašić, M., and Nikolaus, T., Dendroidal sets as models for connective spectra, *J. K-Theory* 14 (2014), no. 3, 387–421.

[15] Batanin, M.A., Monoidal globular categories as a natural environment for the theory of weak *n* categories, *Adv. Math.* 136 (1998), 39–103.

261

[16] Baues, H.-J., and Blanc, D., Comparing cohomology obstructions, *J. Pure Appl. Algebra* 215 (2011), 1420–1439.

[17] Baues, H.-J., and Blanc, D., Higher order derived functors and the Adams spectral sequence, *J. Pure Appl. Algebra* 219 (2015), no. 2, 199–239.

[18] Baues, H.-J., and Blanc, D., Stems and spectral sequences, *Algebr. Geom. Topol.* 10 (2010), 2061–2078.

[19] Baues, H.-J., and Jibladze, M., Secondary derived functors and the Adams spectral sequence, *Topology* 45 (2006), 295–324.

[20] Baues, H.-J., and Wirsching, G., Cohomology of small categories, *J. Pure Appl. Algebra* 38 (1985), 187–211.

[21] Beke, T., Sheafifiable homotopy model categories, *Math. Proc. Cambridge Philos. Soc.* 129 (2000), 447–475.

[22] Bénabou, J., Introduction to bicategories, *Reports of the Midwest Category Seminar*, Springer, 1967, pp. 1–77.

[23] Berger, C., Iterated wreath product of the simplex category and iterated loop spaces, *Adv. Math.* 213 (2007), 230–270.

[24] Bergner, J.E., A characterization of fibrant Segal categories, *Proc. Amer. Math. Soc.* 135 (2007), 4031–4037.

[25] Bergner, J.E., Complete Segal spaces arising from simplicial categories, *Trans. Amer. Math. Soc.* 361 (2009), 525–546.

[26] Bergner, J.E., Equivalence of models for equivariant $(\infty, 1)$-categories, *Glasg. Math. J.* 59 (2017), no. 1, 237–253.

[27] Bergner, J.E., A model category structure on the category of simplicial categories, *Trans. Amer. Math. Soc.* 359 (2007), 2043–2058.

[28] Bergner, J.E., Rigidification of algebras over multi-sorted theories, *Algebr. Geom. Topol.* 6 (2006), 1925–1955.

[29] Bergner, J.E., Simplicial monoids and Segal categories, *Contemp. Math.* 431 (2007), 59–83.

[30] Bergner, J.E., Three models for the homotopy theory of homotopy theories, *Topology* 46 (2007), 397–436.

[31] Bergner, J.E., Workshop on the homotopy theory of homotopy theories (2011), available at arXiv:1108.2001.

[32] Bergner, J.E., and Rezk, C., Comparison of models for (∞, n)-categories, I, *Geom. Topol.* 17 (2013), 2163–2202.

[33] Bergner, J.E., and Rezk, C., Comparison of models for (∞, n)-categories, II (2014), available at arXiv:1406.4182.

[34] Blanc, D., and Paoli, S., Segal-type algebraic models of n-types, *Algebr. Geom. Topol.* 14 (2014), no. 6, 3419–3491.

[35] Blanc, D., and Paoli, S., Two-track categories, *J. K-Theory* 8 (2011), no. 1, 59–106.

[36] Boardman, J.M., and Vogt, R.M., *Homotopy Invariant Algebraic Structures on Topological Spaces, Lecture Notes in Mathematics 347*, Springer, 1973.

[37] Bousfield, A.K., and Kan, D.M., *Homotopy Limits, Completions, and Localizations, Lecture Notes in Mathematics 304*, Springer, 1972.

[38] Brown, R., Higgins, P.J., and Sivera, R., *Nonabelian Algebraic Topology*, with contributions by C.D. Wensley and S.V. Soloviev, *EMS Tracts in Mathematics 15*, European Mathematical Society, Zürich, 2011.

[39] Bullejos, M., Cegarra, A.M., and Duskin, J., On cat^n-groups and homotopy types, *J. Pure Appl. Algebra* 86 (1993), no. 2, 135–154.

[40] Cabello, J.G., and Garzón, A.R., Closed model structures for algebraic models of n-types, *J. Pure Appl. Algebra* 103 (1995), no. 3, 287–302.

[41] Cheng, E., Weak n-categories: comparing opetopic foundations, *J. Pure Appl. Algebra* 186 (2004), no. 3, 219–231.

[42] Cheng, E., Weak n-categories: opetopic and multitopic foundations, *J. Pure Appl. Algebra* 186 (2004), no. 2, 109–137.

[43] Chu, H., Haugseng, R., and Heuts, G., Two models for the homotopy theory of ∞-operads (2016), available at arXiv:1606.03826.

[44] Cisinski, D.-C., Batanin higher groupoids and homotopy types, *Categories in Algebra, Geometry and Mathematical Physics, Contemporary Mathematics, 431*, American Mathematical Society, 2007, pp. 171–186.

[45] Cisinski, D.-C., and Moerdijk, I., Dendroidal Segal spaces and ∞-operads, *J. Topol.* 6 (2013), no. 3, 675–704.

[46] Cisinski, D.-C., and Moerdijk, I., Dendroidal sets and simplicial operads, *J. Topol.* 6 (2013), no. 3, 705–756.

[47] Cisinski, D.-C., and Moerdijk, I., Dendroidal sets as models for homotopy operads, *J. Topol.* 4 (2011), no. 2, 257–299.

[48] Cordier, J.M., and Porter, T., Vogt's theorem on categories of homotopy coherent diagrams, *Math. Proc. Cambridge Philos. Soc.* 100 (1986), 65–90.

[49] Dugger, D., Combinatorial model categories have presentations, *Adv. Math.* 164 (2001), no. 1, 177–201.

[50] Dugger, D., *A Primer on Homotopy Colimits*, available at pages.uoregon.edu/ddugger/hocolim.pdf.

[51] Dugger, D., and Spivak, D.I., Mapping spaces in quasicategories, *Algebr. Geom. Topol.* 11 (2011), 263–325.

[52] Dugger, D., and Spivak, D.I., Rigidification of quasicategories, *Algebr. Geom. Topol.* 11 (2011), 225–261.

[53] Dwyer, W.G., and Kan, D.M., Calculating simplicial localizations, *J. Pure Appl. Algebra* 18 (1980), 17–35.

[54] Dwyer, W.G., and Kan, D.M., A classification theorem for diagrams of simplicial sets, *Topology* 23 (1984), 139–155.

[55] Dwyer. W.G., and Kan, D.M., Equivalences between homotopy theories of diagrams, *Algebraic Topology and Algebraic K-Theory, Annals of Mathematics Studies 113*, Princeton University Press, 1987, pp. 180–205.

[56] Dwyer, W.G., and Kan, D.M., Function complexes in homotopical algebra, *Topology* 19 (1980), 427–440.

[57] Dwyer, W.G., and Kan, D.M., Simplicial localizations of categories, *J. Pure Appl. Algebra* 17 (1980), no. 3, 267–284.

[58] Dwyer, W.G., Kan, D.M., and Smith, J.H., Homotopy commutative diagrams and their realizations, *J. Pure Appl. Algebra* 57 (1989), 5–24.

[59] Dwyer, W.G., and Spalinski, J., Homotopy theories and model categories, *Handbook of Algebraic Topology*, Elsevier, 1995.

[60] Friedman, G., An elementary illustrated introduction to simplicial sets, *Rocky Mountain J. Math.* 42 (2012), no. 2, 353–423.

[61] Gabriel, P., and Zisman, M., *Calculus of Fractions and Homotopy Theory, Ergebnisse der Mathematik und ihrer Grenzgebiete, 35*, Springer, 1967.

[62] Goerss, P.G., and Jardine, J.F., *Simplicial Homotopy Theory, Progress in Mathematics, 174*, Birkhäuser, 1999.

[63] Gordon, R., Power, A.J., and Street, R., Coherence for tricategories, *Mem. Amer. Math. Soc.* 117 (1995), no. 558.

[64] Gurski, M.N., An algebraic theory of tricategories, Ph.D. Thesis, University of Chicago, 2006.

[65] Gurski, N., Nerves of bicategories as stratified simplicial sets, *J. Pure Appl. Algebra* 213 (2009), no. 6, 927–946.

[66] Hatcher, A., *Algebraic Topology*, Cambridge University Press, 2002.

[67] Hermida, C., Makkai, M., and Power, J., On weak higher dimensional categories. I, *J. Pure Appl. Algebra* 154 (2000), no. 1–3, 221–246.

[68] Heuts, G., Hinich, V., and Moerdijk, I., On the equivalence between Lurie's model and the dendroidal model for infinity-operads, *Adv. Math.* 302 (2016), 869–1043.

[69] Hirschhorn, P. S., *Model Categories and Their Localizations, Mathematical Surveys and Monographs 99*, American Mathematical Society, 2003.

[70] Hirschowitz, A., and Simpson, C., Descente pour les *n*-champs (1998), available at arXiv:9807049.

[71] Hovey, M., *Model Categories, Mathematical Surveys and Monographs 63*, American Mathematical Society, 1999.

[72] Ilias, A., Model structure on the category of small topological categories, *J. Homotopy Relat. Struct.* 10 (2015), no. 1, 63–70.

[73] Joyal, A., Quasi-categories and Kan complexes, *J. Pure Appl. Algebra* 175 (2002), 207–222.

[74] Joyal, A., and Tierney, M., Quasi-categories vs Segal spaces, *Contemp. Math.* 431 (2007), 277–326.

[75] Kachour, C., Operadic definition of non-strict cells, *Cah. Topol. Géom. Différ. Catég.* 52 (2011), no. 4, 269–316.

[76] Kazhdan, D., and Varshavskiĭ, Ya., The Yoneda lemma for complete Segal spaces, *Funktsional. Anal. i Prilozhen.* 48 (2014), no. 2, 3–38; translation in *Funct. Anal. Appl.* 48 (2014), no. 2, 81–106.

[77] Kock, J., Joyal, A., Batanin, M., and Mascari, J.-F., Polynomial functors and opetopes, *Adv. Math.* 224 (2010), 2690–2737.

[78] Lack, S., A Quillen model structure for 2-categories, *K-Theory* 26 (2002), 171–205.

[79] Lack, S., A Quillen model structure for bicategories, *K-Theory* 33 (2004), no. 3, 185–197.

[80] Lack, S., A Quillen model structure for Gray-categories, *J. K-Theory* 8 (2011), 183–221.

[81] Lafont, Y., Métayer, F., and Worytkiewicz, K., A folk model structure on omega-cat, *Adv. Math.* 224 (2010), 1183–1231.

[82] Lawvere, F.W., Functorial semantics of algebraic theories, *Proc. Natl. Acad. Sci. USA* 50 (1963), 869–872.

[83] Leinster, T., *Higher Operads, Higher Categories, London Mathematical Society Lecture Note Series 298*, Cambridge University Press, 2004.

[84] Leinster, T., Operads in higher-dimensional category theory, *Theory Appl. Categ.* 12 (2004), no. 3, 73–194.

[85] Leinster, T., A survey of definitions of *n*-category, *Theory Appl. Categ.* 10 (2002), 1–70.

[86] Loday, J.-L., Spaces with finitely many nontrivial homotopy groups, *J. Pure Appl. Algebra* 24 (1982), no. 2, 179–202.

[87] Lurie, J., *Higher Algebra*, available at www.math.harvard.edu/~1lurie/papers/HA.pdf.

[88] Lurie, J., *Higher Topos Theory, Annals of Mathematics Studies 170*, Princeton University Press, 2009.

[89] Lurie, J., On the classification of topological field theories, *Current Developments in Mathematics, 2008*, International Press, 2009, pp. 129–280.

[90] Mac Lane, S., *Categories for the Working Mathematician, Second Edition, Graduate Texts in Mathematics 5*, Springer, 1997.

[91] May, J.P., *Simplicial Objects in Algebraic Topology*, University of Chicago Press, 1967.

[92] Moerdijk, I., Lectures on dendroidal sets, *Simplicial Methods for Operads and Algebraic Geometry, Advanced Courses in Mathematics – CRM, Barcelona, Birkhäuser*, 2010, pp. 1–118.

[93] Moerdijk, I., and Weiss, I., Dendroidal sets, *Algebr. Geom. Topol.* 7 (2007), 1441–1470.

[94] Moerdijk, I., and Weiss, I., On inner Kan complexes in the category of dendroidal sets, *Adv. Math.* 221 (2009), no. 2, 343–389.

[95] Morrison, S., and Walker, K., Higher categories, colimits, and the blob complex, *Proc. Natl. Acad. Sci. USA* 108 (2011), no. 20, 8139–8145.

[96] Paoli, S., Weakly globular cat^n-groups and Tamsamani's model, *Adv. Math.* 222 (2009), no. 2, 621–727.

[97] Pellissier, R., Catégories enrichies faibles (2003), available at arXiv:0308246.

[98] Porter, T., *n*-types of simplicial groups and crossed *n*-cubes, *Topology* 32 (1993), no. 1, 5–24.

[99] Quillen, D., Higher algebraic K-theory. I, *Algebraic K-Theory, I: Higher K-Theories (Proc. Conf., Battelle Memorial Inst., Seattle, Wash., 1972), Lecture Notes in Mathematics 341*, Springer, 1973.

[100] Quillen, D., *Homotopical Algebra, Lecture Notes in Mathematics 43*, Springer, 1967.

[101] Reedy, C.L., *Homotopy Theory of Model Categories*, available at www-math.mit.edu/~psh.

[102] Rezk, C., A cartesian presentation of weak *n*-categories, *Geom. Topol.* 14 (2010), 521–571.

[103] Rezk, C., A model for the homotopy theory of homotopy theory, *Trans. Amer. Math. Soc.* 353 (2001), no. 3, 973–1007.

[104] Riehl, E., *Categorical Homotopy Theory, New Mathematical Monographs 24*, Cambridge University Press, 2014.

[105] Riehl, E., and Verity, D., The 2-category theory of quasi-categories, *Adv. Math.* 280 (2015), 549–642.

[106] Riehl, E., and Verity, D., Fibrations and Yoneda's lemma in an ∞-cosmos, *J. Pure Appl. Algebra* 221 (2017), no. 3, 499–564.

[107] Riehl, E., and Verity, D., Kan extensions and the calculus of modules for ∞-categories, *Algebr. Geom. Topol.* 17 (2017), no. 1, 189–271.

[108] Robertson, M., The homotopy theory of simplicially enriched multicategories (2011), available at arXiv:1111.4146.

[109] Segal, G., Categories and cohomology theories, *Topology* 13 (1974), 293–312.

[110] Simpson, C., *Homotopy Theory of Higher Categories, New Mathematical Monographs 19*, Cambridge University Press, 2012.

[111] Street, R., The algebra of oriented simplexes, *J. Pure Appl. Algebra* 49 (1987), no. 3, 283–335.

[112] Strøm, A., The homotopy category is a homotopy category, *Arch. Math. (Basel)* 23 (1972), 435–441.

[113] Tamsamani, Z., Sur les notions de *n*-categorie et *n*-groupoíde non-stricte via des ensembles multi-simpliciaux, *K-Theory* 16 (1999), no. 1, 51–99.

[114] Thomason, R.W., Cat as a closed model category, *Cah. Topol. Géom. Différ. Catég.* 21 (1980), no. 3, 305–324.

[115] Toën, B., Homotopical and higher categorical structures in algebraic geometry (a view towards homotopical algebraic geometry) (2003), available at arXiv:0312262.

[116] Toën, B., Vers une axiomatisation de la théorie des catégories supérieures, *K-Theory* 34 (2005), no. 3, 233–263.

[117] Verity, D.R.B., Weak complicial sets I: basic homotopy theory, *Adv. Math.* 219 (2008), no. 4, 1081–1149.

[118] Weibel, C. A., *An Introduction to Homological Algebra, Cambridge Studies in Advanced Mathematics 38*, Cambridge University Press, 1994.

Index

$(\infty, 1)$-category, 77
(∞, n)-category, 77, 256
I-cofibration, 30
I-fibration, 30
I-injective, 30
Θ-construction, 257
Θ_n-space, 258
∞-groupoid, 76
n-arrow, 97
n-category, 75
n-coskeleton, 47
n-simplex, 35
  in a simplicial set, 35
  topological, 37
n-skeleton, 47
nth Postnikov approximation, 57
2-category, 75

accessible category, 55
accessible class of maps, 55
accessible functor, 55
acyclic cofibration, 18
  closed under pushout, 21
  cofibrantly generated, 31
  generating, 30
acyclic fibration, 18
  closed under pullback, 22
  Reedy, 51
additive category, 251
adjoint functors, 7, 12, 25, 38
  forgetful-free, 60, 79
  skeleton–coskeleton, 48
associativity, 4, 11
atomic morphism, 97

bead, 174
bicategory, 76

bisimplicial set, 37, 45
boundary of an n-simplex, 35
Bousfield localization, 57

cartesian closed category, 63
  relative categories, 214
cartesian model category, 63
  Θ_n-spaces, 258
  complete Segal spaces, 111
  recognition of, 64
  Segal categories, 145
  Segal spaces, 104
  simplicial sets, 63
  simplicial spaces, 64
categorical equivalence, 114
categorical homotopy, 113
category, 4
  accessible, 55
  additive, 251
  cartesian closed, 63
  closed monoidal, 12
  closed symmetric monoidal, 12
  enriched, 13
  equivalence of, 7, 10
  filtered, 9
  free, 79
  locally presentable, 54
  monoidal, 11
  of chain complexes, 14
  of components, 80
  of dendroidal sets, 255
  of gadgets, 189
  of groups, 5
  of relative categories, 214
  of sets, 5
  of simplices, 39
  of simplicial sets, 36

of small categories, 6
opposite, 6
small, 5
symmetric monoidal, 12
triangulated, 251
with all small colimits, 9
with all small limits, 8
category Θ_n, 257
category with weak equivalences, 78, 213
chain complexes, 15
groups, 15
topological spaces, 15
cellular model category, 32
localization, 57
topological spaces, 32
chain complexes, 14
characteristic relative functor, 216
classification complex, 236
classification diagram, 221, 236
classifying complex, 237
classifying diagram, 72, 104, 234
classifying space of a group, 67
closed monoidal category, 12, 32
closed symmetric monoidal category, 12
codegeneracy map, 35
coequalizer, 9
coface map, 35
cofiber, 252
cofiber sequence, 252
cofibrant object, 20
complete Segal spaces, 111
fixed-object Segal categories, 135
relative categories, 230
Segal categories, 139
Segal spaces, 104
simplicial categories, 86, 97
simplicial sets, 38
topological spaces, 21
cofibrant replacement, 20
cofibrantly generated model category, 30
from an adjoint pair, 31
recognition of, 31, 56
cofibration, 18
closed under pushout, 21
cofibrantly generated, 31
complete Segal spaces, 111
fixed-object Segal categories, 134, 135
fixed-object Segal precategories, 134
fixed-object simplicial categories, 85
generating, 30
in projective model structure, 46

localized model structure, 57
quasi-categories, 160
Reedy, 49
relative categories, 230
Segal categories, 139, 148
Segal spaces, 104
simplicial sets, 38
topological spaces, 18
coherent nerve, 100, 171
colimit, 9
directed, 9
filtered, 9
of fixed-object Segal precategories, 131
of fixed-object simplicial categories, 84
combinatorial model category, 55
localization, 57
simplicial sets, 55
compact object, 32
complete Segal space, 110
complete Segal space model structure, 111
completion
of a category, 120
of a Segal space, 120, 121
composition, 4
in a Segal space, 108
coproduct, 9
cosieve, 215
cosimplicial identities, 35
cosimplicial object, 37
cover, 104
CW approximation theorem, 17
cylinder object, 23

degeneracy map, 35
degenerate simplex, 35, 36
dendroidal category Ω, 255
dendroidal complete Segal space, 256
dendroidal object, 255
dendroidal set, 255
dendroidal space, 256
derived category, 15
diagram, 7
diagram category, 10
directed colimit, 9
directed poset, 9
discrete nerve, 114
discretization, 145
distinguished triangle, 251
in a stable quasi-category, 253
Dwyer inclusion, 216
Dwyer map, 216
Dwyer–Kan equivalence

of Segal precategories, 139
of Segal spaces, 116
of simplicial categories, 81
of simplicial sets, 198
effective monomorphism, 32
enriched category, 13, 75, 256
in simplicial sets, 42
equalizer, 8
equivalence
in a quasi-category, 162, 173
Joyal, 159, 160
of categories, 7, 10
essentially surjective functor, 7
face map, 35
faithful functor, 7
fiber, 252
fiber sequence, 252
fibrant object, 20
complete Segal spaces, 111
fixed-object Segal categories, 136, 137
localized model structure, 57, 58
quasi-categories, 160
Segal categories, 142, 149
Segal spaces, 104
simplicial categories, 96
simplicial sets, 38
topological spaces, 21
fibrant replacement, 20
fibration, 18
closed under pullback, 22
fixed-object Segal precategories, 131
fixed-object simplicial categories, 85
Hurewicz, 18
Joyal, 160
Kan, 160
Reedy, 49
relative categories, 230
Serre, 18
simplicial categories, 87
simplicial sets, 38
topological spaces, 18
filtered colimit, 9
filtered colimits, 55
flagged necklace, 181
flanked flagged necklace, 182
forgetful functor, 7
free category, 79, 85
free functor, 7
free map of simplicial categories, 85
free product of categories, 85
free resolution of a category, 171

full functor, 7
full subcategory, 6
fully faithful functor, 7
functor, 6
accessible, 55
adjoint, 7
essentially surjective, 7
faithful, 7
forgetful, 7
free, 7
full, 7
fully faithful, 7
homotopy initial, 42
reflecting weak equivalences, 26
representable, 11
functor category, 10
functorial factorization, 19
from small object argument, 30
generating acyclic cofibration, 30
fixed-object Segal precategories, 132, 135
projective model structure, 47
Reedy, 50
Segal categories, 140, 148
simplicial categories, 87
simplicial sets, 38
topological spaces, 30
generating cofibration, 30
fixed-object Segal precategories, 133, 135
projective model structure, 47
Reedy, 51
Segal categories, 127, 128, 139, 148
simplicial categories, 87
simplicial sets, 38
topological spaces, 30
geometric realization, 38
groupoid, 5
nerve of, 68
hammock localization, 80
homotopically small simplicial set, 236
homotopy
for quasi-categories, 159
for topological spaces, 23
in a model category, 24
in a Segal space, 108
of relative functors, 215
of simplicial spaces, 224
homotopy category, 25
as a localization, 25
of a Segal space, 108
of topological spaces, 17
homotopy colimit, 40

homotopy equivalence, 14
  in a Segal space, 109
  in a simplicial category, 86
  of quasi-categories, 160
  of relative categories, 215
  of simplicial spaces, 224
homotopy initial functor, 42
homotopy limit, 40
homotopy mapping space, 44
homotopy monomorphism, 111
homotopy theory, 27
  of homotopy theories, 82
horn of an n-simplex, 35
Hurewicz fibration, 18

initial object, 5, 10
  in a quasi-category, 252
initial subdivision, 219
injective model structure, 20, 45
inner anodyne map, 161
inner fibrant simplicial set, 161
inner fibration, 161
inner horn, 70
inner Kan complex, 71
internal hom object, 12

join of simplicial sets, 194
joint, 174
Joyal equivalence, 159, 160
Joyal fibration, 160

Kan complex, 38, 69
Kan fibration, 160
Kan weak equivalence, 160

leaf, 255
left adjoint functor, 7
  preserves colimits, 10
left Bousfield localization, 57
left derived functor, 26
left homotopy, 23
left lifting property, 19
  preserved by transfinite composition, 29
left proper model category, 28
  complete Segal spaces, 111
  localization, 57
  quasi-categories, 160
  relative categories, 230
  Segal categories, 144, 149
  Segal spaces, 104
  simplicial categories, 99
limit, 8
  of fixed-object Segal precategories, 130
  of fixed-object simplicial categories, 84

local equivalence, 56
local object, 56
localization, 14
  homotopy category as, 25
  of a model category, 57
  of a Segal precategory, 138
locally constant, 210
locally presentable category, 54

mapping algebra, 249
mapping space
  in a quasi-category, 194, 198
  in a Segal space, 107
  in a simplicial category, 42
  in a simplicial set, 195
model category, 18
  Θ_n-spaces, 258
  cartesian, 63
  categories enriched in Θ_n-spaces, 258
  cellular, 32
  chain complexes, 27
  cofibrantly generated, 30
  combinatorial, 55
  complete Segal objects in Θ_n-spaces, 259
  complete Segal spaces, 111
  dendroidal sets, 256
  determined by cofibrations and fibrant objects, 21
  fixed-object Segal categories, 134, 135
  fixed-object Segal precategories, 131, 134
  fixed-object simplicial categories, 85
  for chain complexes, 20
  for topological spaces, 20
  homotopy category of, 25
  left proper, 28
  localization, 57
  monoidal, 32
  proper, 28
  quasi-categories, 160
  Quillen equivalent, 26
  relative categories, 230
  right proper, 28
  Segal categories, 139, 148
  Segal categories in Θ_n-spaces, 259
  Segal spaces, 104
  simplicial, 43
  simplicial categories, 86
  simplicial colored operads, 255
  simplicial sets, 38
  simplicial spaces, 45, 48
  simplicial spaces, projective, 46
  symmetric monoidal, 33

monoidal category, 11
monoidal model category, 32
morphism, 4
    atomic, 97
    in a quasi-category, 251
multiplicative system, 15

natural isomorphism, 10
natural transformation, 10
necklace, 174
    flagged, 181
    flanked, 182
nerve
    of a category, 36, 66
    of a simplicial category, 99

object, 4
    initial, 5, 10
    of a quasi-category, 251
    of a Segal space, 107
    small, 29
    terminal, 5, 10
    zero, 5
operad, 254
opposite category, 6, 35
ordered simplicial set, 176
Ore's condition, 16
outer horn, 70
overcategory, 6
    model structure, 21

path object, 24
product, 8
projective model structure, 20, 46
proper model category, 28
    fixed-object simplicial categories, 85
    simplicial spaces, 46
    topological spaces, 28
pullback, 8
pullback-corner map, 43
pushout, 9
pushout-product map, 43, 63

quasi-category, 71, 157
    pointed, 252
    stable, 253
Quillen equivalence, 26
    categories enriched in and Segal categories
        in Θ_n-spaces, 259
    complete Segal spaces and relative
        categories, 232
    dendroidal sets and simplicial operads, 256
    fixed-object Segal category model
        structures, 135

fixed-object simplicial categories and Segal
    categories, 138
    localization of, 59
    quasi-categories and complete Segal spaces,
        211
    quasi-categories and simplicial categories,
        206
    Segal categories and complete Segal spaces,
        147
    Segal categories and simplicial categories,
        156
    Segal category model structures, 150
    simplicial sets and topological spaces, 39
Quillen pair, 25
    localization of, 59
    Segal categories and complete Segal spaces,
        147
    simplicial categories and Segal categories,
        151
Quillen's theorem A, 42

realistic mapping algebra, 250
realizable mapping algebra, 249
reduced hammock, 80
reduction functor, 126
Reedy category, 47
Reedy model structure, 48
    same as injective for simplicial spaces, 53
relative I-cell complex, 29
relative category, 213
    maximal, 214
    minimal, 214
relative functor, 213
relative inclusion, 213
relative poset, 214
representable functor, 11, 35, 45
right adjoint functor, 7
    preserves limits, 10
right derived functor, 26
right homotopy, 24
right lifting property, 19
right proper model category, 28
    simplicial categories, 98
rigidification functor, 171, 172
root, 255

Segal category, 125
    in Θ_n-spaces, 259
Segal condition, 103
Segal map, 36, 54, 103
Segal operad, 256
Segal precategory, 125
    in Θ_n-spaces, 259

Segal preoperad, 256
Segal space, 103
Segal space model structure, 104
Serre fibration, 18
sieve, 215
simplex category, 37
simplicial category, 42, 83
  as a simplicial localization, 82
simplicial computad, 97
simplicial functor, 83
simplicial group, 37
simplicial indexing category Δ, 34
simplicial localization, 79
simplicial model category, 43, 82
  complete Segal spaces, 111
  fixed-object simplicial categories, 85
  Segal categories, 144
  Segal spaces, 104
  simplicial sets, 43
  simplicial spaces, 45, 46
simplicial monoid, 86
simplicial nerve, 99
simplicial object, 37
  simplicial category as, 42
simplicial resolution, 79
simplicial set, 35
  as a colimit, 38
  geometric realization of, 38
  model structure, 38
simplicial space, 45
  as a homotopy colimit, 47
singular functor, 38
small category, 5
small object, 29
small object argument, 30
source, 4
space of homotopy equivalences, 110
special anodyne map, 162
special classification complex, 236
special horn, 162
special inner fibration, 164
special left horn, 162
special right horn, 162
spine, 174
stable quasi-category, 253
strict n-category, 75
strict local equivalence, 59
strictly local object, 59, 151
strong deformation retraction of relative
  categories, 215
strong retract of simplicial categories, 85

subcategory, 6
  full, 6
subdivision, 220
symmetric colored operad, 254
symmetric monoidal category, 12
symmetric monoidal model category, 33
  simplicial sets, 39
  topological spaces, 33

target, 4
terminal object, 5, 10
  in a quasi-category, 252
terminal subdivision, 218
Thomason model structure, 71
topological space
  as an ∞-groupoid, 76
  localization of, 57
total left derived functor, 27
total right derived functor, 27
totally nondegenerate necklace, 182
track category, 249
transfinite composition, 29
tree, 255
triangle, 252
triangulated category, 251
tricategory, 76
two-fold subdivision, 220
two-out-of-three property, 19
  for Dwyer–Kan equivalences of Segal
    spaces, 116
  for Joyal equivalences, 169

undercategory, 6
  model structure, 21
underlying category of a relative category, 213
unit, 5, 11
universe axiom, 15

Warsaw circle, 17
weak ∞-groupoid, 76
weak n-category, 76
weak equivalence, 14, 18
  between fibrant objects in a localized model
    category, 57
  fixed-object Segal categories, 134
  fixed-object Segal precategories, 131, 134
  fixed-object simplicial categories, 85
  in a relative category, 213
  Kan, 160
  localized model structure, 57
  of complete Segal spaces, 111
  of quasi-categories, 160
  of relative categories, 230

of Segal categories, 148
of simplicial categories, 86
of simplicial sets, 38
of simplicial spaces, 45
Reedy, 49
Segal categories, 139
Segal spaces, 104
weak homotopy equivalence, 14
   but not a homotopy equivalence, 17

weakly contractible, 39
Whitehead's theorem, 17

Yoneda embedding, 11
Yoneda lemma, 11, 36

zero object, 5
   in a quasi-category, 252

Printed in the United States
by Baker & Taylor Publisher Services